IEE ELECTROMAGNETIC WAVES SERIES 6

SERIES EDITORS: J. R. WAIT, G. MILLINGTON AND E. D. R. SHEARMAN

Waveguide tapers transitions and couplers

Previous volumes in this series

Volume 1	Geometrical theory of diffraction for electromagnetic waves Graeme L. James
Volume 2	Electromagnetic waves and curved structures Leonard Lewin, David C. Chang and Edward F. Kuester
Volume 3	Microwave homodyne systems Ray J. King
Volume 4	Radio direction-finding P.J.D. Gething
Volume 5	ELF communications antennas Michael L. Burrows

Waveguide tapers transitions and couplers

F. SPORLEDER
German Federal Post Office
Darmstadt
West Germany

and

H.-G. UNGER
Professor of Electrical Engineering
University of Braunschweig
West Germany

PETER PEREGRINUS LTD.
On behalf of the
Institution of Electrical Engineers

Published by: The Institution of Electrical Engineers, London
and New York
Peter Peregrinus Ltd., Stevenage, UK, and New York

ISBN: 0 906048 16 8

Typeset at the Alden Press Oxford London and Northampton
Printed in England by A. Wheaton & Co., Ltd., Exeter

Contents

		Page
1	**Introduction**	**1**
	References	8
2	**Coupled wave equations**	**10**
	2.1 Field equations and sources	10
	2.2 Wave propagation and excitation in cylindrical waveguides	12
	2.3 Coupled wave equations for nonuniform wave-guides	20
	2.4 Characteristics of coupled wave equations	27
	2.5 Coupled wave equations for curved waveguides	33
	2.6 Waveguides filled with homogeneous, isotropic and lossless matter	36
	2.7 Multihole couplers	47
	2.8 Coupled dielectric waveguides	55
	2.9 Dielectric waveguide transitions	70
	References	91
3	**Taper analysis with horn modes**	**92**
	3.1 Nonuniform sector horn	94
	3.2 Nonuniform round horn	110
	References	123
4	**Abrupt waveguide transitions**	**124**
	4.1 Abrupt transition between cylindrical waveguides	124
	4.2 Abrupt waveguide to horn transition	133
	4.3 Convergence properties of coupled wave equations	140
	References	150
5	**Transitions and couplers with one parasitic wave**	**151**
	5.1 TEM matching sections	151
	5.2 Optimum coupling distribution for short coupling sections	163
	References	176

6 Gradual waveguide transitions **177**
6.1 Highly overmoded waveguide transitions with low mode conversion 177
6.2 Circular and rectangular waveguide tapers 182
6.3 Quasidiagonalisation of coupled wave equations 189
6.4 Waveguide taper design with coupled horn waves 203
6.5 Coupling of modes below cutoff 209
6.6 Horn transitions 220
References 228

7 Waveguide tapers with spurious mode degeneracies **229**
7.1 Sector transducer 231
7.2 Rectangular-to-sector transducer 249
References 254

8 Wave coupling with complete and nearly complete power transfer **255**
8.1 Mode-selective directional couplers 256
8.2 Adiabatic coupling 268
References 281

9 Optical waveguide tapers **282**
9.1 Monomode waveguide tapers 284
9.2 Monomode to multimode taper 291
9.3 Multimode to multimode taper 295
Reference 299

Index **300**

Chapter 1

Introduction

A tapered waveguide is an elongated guiding structure for electromagnetic waves that gradually becomes narrower towards one end. Such a waveguide taper may not only undergo a change in the size of its cross-section in the direction of propagation in a gradual fashion, but also in its cross-sectional form. More general still, the guiding structure may be tapered in the sense that the whole distribution of matter by which it is formed changes gradually in the direction of wave propagation.

In nature we encounter such gradual changes in matter, as for example, when electromagnetic energy in the form of light or radio waves propagates through the atmosphere, which changes gradually owing to changes in temperature and pressure or due to its state of ionisation being possibly different from point to point. More often, however, we deal with waveguide tapers that we specifically construct to suit particular purposes.

As one class of simple examples we have ordinary two-wire transmission lines for TEM-waves which, on occasion, we taper in form or size, or even with respect to both, to connect transmission lines of different sizes and forms or lines of different characteristic impedances. The TEM-wave which is incident on a transmission line taper reflects at least partially, and ordinarily the line should taper in such a fashion that the reflection stays below a tolerable level, throughout the band of operating frequencies.

Transmission lines of different characteristic impedances are also connected to each other via uniform line sections with intermediate values of their characteristic impedance, and abrupt transitions between them. By suitably choosing the lengths and characteristic impedances of these intermediate line sections the different lines are matched to each other. The quarter-wave transmission-line transformer represents the simplest example of such stepped impedance transformers.

Tapers and transitions occur also in other types of waveguides. Notable examples are hollow-tube metallic waveguides, mainly for microwaves, and dielectric waveguides, which prove very effective for optical waves. Hollow-tube metallic waveguides are usually to transmit only one mode of propagation which is not in every application the fundamental mode. A typical example is the circular electric (H_{01}-) mode of the round waveguide. The H_{01}-mode has not the lowest cutoff. H_{01}-waveguide tapers must therefore be designed with regard not only to low reflection but also to low power loss to other propagating, but unwanted modes.

When we taper the cross-section of a metallic waveguide from one form to another we have, in principle, the possibility to transform a mode of one waveguide cross-section to some other mode in the other cross-section. The tapered transition should accomplish this transformation with little loss to reflection and with little power conversion to unwanted modes.

Another type of waveguide transition is the waveguide bend. In it the waveguide cross-section usually does not change except for its axial direction. By bending the waveguide gently the transition from one direction to another is accomplished in a gradual fashion, thus avoiding too much power loss to, and interference with, reflections and other unwanted modes. Reflection and power conversion to unwanted modes in waveguide bends may be even more effectively reduced by gradually tapering the curvature from its zero values at both ends to a suitable maximum curvature in a centre section of the bend. For these reasons curved waveguides and waveguide bends belong in the context of waveguide tapers and transitions.

Reflection and mode conversion in tapered transmission lines and in waveguide transitions is kept low by destructive interference of parasitic wave components. To interfere destructively with one another the parasitic wave components must differ sufficiently in phase. Destructive interference requires the transition to be sufficiently long. Cost and weight limit the transition length. We therefore need tapers which fulfill design requirements in the shortest possible length.

The design aspects are quite different for optical waveguide tapers. Their microscopic dimensions allow reduction of mode conversion by destructive interference in tapers which extend over many light wavelengths and yet are relatively short in absolute terms. Taper length is therefore not quite so critical here.

The open structures of dielectric optical waveguide tapers, in contrast to metallic waveguide tapers, not only reflect part of the incident wave and convert it to unwanted modes, they also radiate incident wave energy into the surrounding space. The analysis and design of optical waveguide transitions must therefore take this radiation into account, in addition to reflection and mode conversion. Transitions between monomode optical waveguides have radiation even as the main detrimental effect that limits their efficiency.

All waveguide tapers and transitions have in common the interaction of their modes of propagation, including in the case of optical waveguides, the radiation modes. These modes propagate independently one from another in the adjoining uniform waveguides; along the non-uniform regions of the transition, however, they interact with each other.

Such mode interaction by coupling and the ensuing power transfer occur also between modes in different waveguides when these waveguides run parallel and overlap or otherwise influence each other electromagnetically.

The analysis of coupled waveguides and the design of waveguide couplers belong therefore also in this context. Coupling between parallel metallic waveguides is usually accomplished through apertures in a common partition wall. Open dielectric waveguides couple by way of the transversely evanescent fields of their guided modes.

In parallel transmission lines such as parallel microstrips on a common substrate or several wire pairs in a common cable, as well as between individual optical fibre guides in a fibre cable, the coupling is usually unintentional and the interaction and conversion must be kept below specified crosstalk levels.

Actual waveguide couplers, however, are required to convert a specified fraction of the power incident in a particular mode of one of the coupled waveguides to a certain mode of the other waveguide without too much power loss to other modes or to radiation. The analysis and design problems of such coupling correspond, not only in principle but even in many details, to those for waveguide tapers and transitions. In both situations is the mode interaction to be controlled in similar ways.

Waveguide tapers and transitions as well as waveguide couplers perform physically by means of coupling and interaction of their modes of propagation. For their analysis and design we therefore resort also to a representation of the electromagnetic field in terms of these modes of propagation. With a series expansion of the electromagnetic fields in terms of waveguide modes, Maxwell's equations, under the boundary conditions of the particular structure, may be converted to coupled wave equations or generalised transmission line equations. Whereas Maxwell's equations represent a limited system of second-order partial-differential equations for the components of the electric and magnetic field vectors, the coupled wave equations or the equivalent generalised transmission-line equations consist, in general, of infinitely many first-order ordinary coupled differential equations, two for each mode, with its forward and backward travelling components.

The field expansion in terms of waveguide modes and the subsequent conversion of Maxwell's equations to coupled wave equations should be done in such a fashion that the boundary conditions of the waveguide transition with all its inhomogeneities are fully taken into account. There are several methods to accomplish this conversion to coupled wave equations. Which one to use depends on the type of waveguide and the nonuniformity of the transition. One method which serves well under quite general circumstances uses effective sources. In applying this method, an infinitesimally short section of the waveguide transition is considered, including its inhomogeneity. Any one of the waveguide modes is assumed to be incident on to this short section. Fictitious sources of electric and magnetic currents are then impressed such that the incident mode propagates through this section by itself without interacting with other modes. If these sources alone were impressed then the incident mode would propagate through this short section and exit from it without exciting any other modes, despite the nonuniformity of this section. We subsequently return from this arrangement with fictitious sources to the actual source-free arrangement by impressing additional source distribution of equal magnitude but opposite direction to compensate the fictitious sources. These additional sources excite other modes in the forward and backward direction including, in general, some reflection in the mode of incidence.

By way of these effective sources, and their relation to the incident mode, we thus obtain the coupling between this incident mode and all other modes, in the forward and backward direction, that interact with it owing to the nonuniformity. The method of effective sources turns out to apply quite generally without having any restrictions as to the degree of nonuniformity. Extensive use of this method is therefore made throughout this book.

Another method which has previously been used[2] to convert Maxwell's equations for nonuniform waveguides to generalised transmission line equations, starts with series expansions of the electromagnetic field in terms of normal modes of a suitable waveguide as reference structure. Then, by partial integration of the finite system of second-order partial-differential equations, which Maxwell's equations represent, an infinite system of ordinary-differential equations of first-order is obtained for the electric and magnetic field amplitudes of the normal modes or for their forward and backward travelling wave amplitudes. This conversion to generalised transmission-line equations by partial integration is not in all cases applicable, and when feasible, has in some cases certain restrictions as to the degree of nonuniformity. It will, nevertheless, also be used in this text, not only as an alternative to the method of effective sources, but also in cases where it is more suitable than this method.

When we represent the electromagnetic field of a waveguide transition or coupler in terms of the fields of normal modes of a reference structure, we have still to decide what type of waveguide to choose locally along the transition or coupler as a reference structure. For practical reasons we should always work with normal modes of such reference structures that, at both ends of the transition or coupler, become identical with the input and output waveguides of the arrangement. With this choice of reference system the normal modes for our field representation will, at the input and output parts, be the modes of propagation of the connecting waveguides. Under working conditions the transition or coupler is usually excited by one or at most a few of such modes, and we are interested in the power that exits in these modes for a given excitation.

If at all feasible we will therefore work with a reference system that is defined at both ends by the connecting waveguides. In between, however, along the transition or coupler, we are not bound to any particular reference waveguide yet. As a general guideline, however, we should choose a local reference structure that corresponds in as many details as possible to the local cross-section of the transition and, perhaps, in addition even its rate of change in propagation direction. The closer the local reference system resembles the waveguide arrangement the less interaction between the local normal modes of the reference system will be caused by its deviation from the waveguide arrangement. Coupled wave equations or generalised transmission-line equations with only weak coupling between modes are more readily solved and with a higher degree of accuracy than those which exhibit strong coupling.

An obvious choice of local reference waveguide for a gradual waveguide transition is a cylindrical, and hence uniform, waveguide that is identical in cross-section and in the distribution of matter over it, to the local cross-section and distribution of matter in the waveguide transition. The normal modes of this reference waveguide, while they propagate independently of each other in the cylindrical and uniform reference structure, couple to each other in the actual transition due to the reference structure changing its cross-section and distribution of matter from point to point along the transition.

A more natural choice yet, than the locally cylindrical reference guide, is a reference structure that not only has the same cross-section as the waveguide transition at the respective location, but also the same rate of change.

For the example of a circular waveguide taper this more natural reference structure would be a right circular cone whose surface is tangent to the circular waveguide taper in the reference cross-section. The normal modes of propagation of the right circular cone have spherical phase fronts. In the circular waveguide taper they couple to each other wherever the taper changes its flare angle.

It appears from this example that such a more natural reference structure exists not just for any gradual waveguide transition, but only for waveguide tapers of specific geometries. Formally, however, it will always be possible to transform from a given system of local normal modes, such as those of the locally cylindrical system, to a more natural system, by way of partial integration of the coupled wave equations. The new modes in the system of coupled wave equations that results from partial integration, will couple to each other only due to the rate of change of the coupling in the given reference system. The fields of these new modes conform better to the fields in the waveguide transition and interact less with each other.

Which local reference system to choose along the waveguide transition and coupler, or to which more natural system to transform to by partial integration, depends very much on the type of waveguide, the form of transition or coupling, and what the objectives are in the analysis or design.

Waveguide tapers, transitions and couplers perform important functions in microwave circuits and in guiding structures and arrangements for optical waves. An extensive literature is therefore available that treats the analysis and the design of such tapers, transitions and couplers. This holds in particular for the large variety of forms of transitions and couplers. So many special methods have been developed for the analysis and design of special types of this large variety of structures, that it is quite impossible to include them all in this text. We rather attempt here to introduce the more fundamental and general aspects of the theory for transitions and couplers and to this end select very carefully typical examples.

The list of references with which we conclude the Introduction includes therefore only those contributions which represent major advances towards the present state of the art in waveguide tapers, transitions and couplers.

A powerful method to analyse wave propagation and coupling in these structures is based on the concept of generalised transmission line or coupled wave equations. This concept was first proposed and studied in detail by Schellkunoff.[1] Reiter[2] and Katzenelenbaum[3] have also made significant contributions to advance this method, while Solymar[4] derived simplified but still quite general expressions for wave coupling in these equations.

Utilising this method of coupled wave equations many authors have analysed various inhomogeneous waveguide structures. As far as it concerns curved waveguiding structures a complete bibliography may be found in Reference 5.

Improved and optimum designs for TEM-transmission line tapers with minimum reflection were first presented by Klopfenstein[6] and Collin.[7] Tapered transitions with low mode conversion between overmoded waveguides of round and rectangular waveguides were analysed and designed[8,9] by taking only the most strongly coupled unwanted mode into account, and assuming both modes to be far above cutoff. The design has been further improved by Hecken[10,11] who arrived at practically optimum waveguide tapers.

Wave propagation and reflection near cutoff in a nonuniform waveguide was first studied by Schnetzler[12] and later on in more detail by Tang.[13] Tang also applied this phenomenon to design waveguide highpass filters by suitably tapering the waveguide in the cutoff region.[14] Hinken improved on this design by developing an adjustable waveguide highpass filter.[15]

Wave propagation through tapered waveguide transitions was analysed in natural co-ordinates[16] and by expanding the fields in these co-ordinates in terms of horn waves.[17] Bahar[18] investigated the coupling between these horn waves by matching fields on the boundary between two horns of different flare angles. With sufficient accuracy this coupling was also determined[19] by using the method of effective sources and alternatively by converting Maxwell's equations to generalised transmission-line equations in terms of horn waves.

Field expansions in terms of horn waves was also used[20,21] to synthesise improved waveguide transitions. In the same Reference 21 it is shown furthermore, that analysing tapered waveguide transitions in terms of cylindrical modes of the local cross-section yields eventually the same overall scattering matrix, as when analysing it in terms of horn waves.

The design methods which have been developed[7,8] for tapered waveguide transitions with round and rectangular cross-sections may be inadequate for transitions between overmoded waveguides of substantially different cross-sectional form. A typical example is the transition from the fundamental mode of the rectangular waveguide to the H_{01}-mode of the round waveguide. Design rules for this particular transition are given in Reference 22. Such transitions and still more complicated transition structures may be analysed and designed by utilising computers on a large scale.[23]

Wave propagation in coupled waveguides and its application to the design of waveguide couplers has been treated comprehensively in Reference 24. Adiabatic coupling constitutes an extension of this coupled wave theory. It was first described in Reference 25. More recently the adiabatic coupler proves very useful for optical waveguides and a very elegant analysis is given in Reference 26.

Coupling between metallic waveguides is usually accomplished by apertures in a common partition wall. For small apertures Bethe developed a general coupling theory.[27] Of the many different applications for this small-hole coupling, only Reference 28 is quoted here because the results of Reference 27 are extended in Reference 28 to take into account specific waveguide and hole configurations.

Coupling between modes in parallel optical waveguides as well as in optical waveguide tapers has been analysed in a number of publications using various methods. We make reference here in particular to Marcuse,[29] who first utilised radiation modes in a systematic fashion to take into account the radiated power in optical-waveguide tapers, and to Snyder,[30] who formulated the coupled mode theory in particular for nonuniform optical fibres.

A number of significant contributions to mode coupling between parallel optical waveguides are surveyed in Reference 31.

References

1 SCHELKUNOFF, S. A.: 'Conversion of Maxwell's equations into generalized telegraphist's equations', *Bell Syst. Tech. J.*, 1955, **34**, pp. 995–1043

2 REITER, G.: 'Generalized telegraphist's equation for waveguides of varying cross-section', *Proc. IEE.*, 1959, **106B**, (13), pp. 54–57

3 KATZENELENBAUM, B. S.: 'On the theory of nonuniform waveguides with slowly changing parameters'. Congres Internatl. Circuits et Antennes Hyperfrequences, Paris, France, October 1957

4 SOLYMAR, L.: 'Spurious mode generation in non-uniform waveguide', *IEE Trans.*, 1959, **MTT-7**, pp. 379–838

5 LEWIN, L., CHANG, D. C., and KNESTER, E. F.: 'Electromagnetic waves and curved structures' (Peter Peregrinus Ltd., 1977)

6 KLOPFENSTEIN, R. W.: 'A transmission line taper of improved design', *Proc. IRE*, 1956, **44**, pp. 31–35

7 COLLIN, R. E.: 'The optimum tapered transmission line matching section', *ibid.*, 1956, **44**, pp. 539–548

8 UNGER, H.-G.: 'Circular waveguide taper of improved design', *Bell Syst. Tech. J.*, 1958, **37**, pp. 899–912
9 TANG, C. C. H.: 'Optimization of waveguide tapers capable of multimode propagation', *IEEE Trans.*, 1961, **MTT-9**, pp. 442–452
10 HECKEN, R. P.: 'A near-optimum matching section without discontinuities', *ibid.*, 1972, **MTT-20**, pp. 734–739
11 HECKEN, R. P., and ANUFF, A.: 'On the optimum design of tapered waveguide transitions', *ibid.*, 1973, **MTT-21**, pp. 374–380
12 SCHNETZLER, K.: 'Die Reflexion der Grundwelle an den Knickstellen eines Hohlleiters, insbesondere bei einem stetigen Übergang von einem rechteckigen auf einen runden Hohlleiter', *AEU,* 1960, **14,** pp. 177–182
13 TANG, C. C. H.: Mode conversion in tapered waveguides at and near cutoff', *IEEE Trans.*, 1966, **MTT-14**, pp. 233–239
14 TANG, C. C. H.: 'Nonuniform waveguide high-pass filters with extremely steep cutoff', *ibid.*, 1964, **MTT-12**, pp. 300–309
15 HINKEN, J. H.: 'An adjustable waveguide high-pass filter', *AEU* 1975, **29**, pp. 284–286
16 POKROVSKII, V. L., ULINICH, F. R., and SAVVINHYKH, S. K.: 'The theory of tapered waveguides', *Radiotekh & Elektron.*, 1959, **4**, pp. 161–171
17 UNGER, H.-G.: 'Wave propagation in horns and through horn junctions', *AEU*, 1965, **19**, pp. 459–468
18 BAHAR, E.: 'Wave scattering in nonuniform waveguides with large flare angles and near cutoff', *IEEE Trans.*, 1968, **MTT-16**, pp. 503–510
19 UNGER, H.-G., SAAD, L., and SPORLEDER, F.: 'Electromagnetic waveguide taper and horn analysis in natural coordinates'. Symposia Mathematica, Volume XVIII, Publicazione dell' Instituto Nazionale di Alta Matematica Roma
20 BAHAR, E., and CRAIN, G.: 'Synthesis of multimode waveguide transition sections', *Proc. IEE*, 1968, **115**, pp. 1395–1397
21 SPORLEDER, F.: 'Waveguide transition design of improved accuracy', *AEU*, 1976, **30**, pp. 289–296
22 SPORLEDER, F., and SAAD, L.: 'Broadband sector transducer', *ibid.*, 1976, **30**, pp. 253–259
23 SAAD, S. S., DAVIES, J. B., and DAVIES, O. J.: 'Computer analysis of gradually tapered waveguide with arbitrary cross-sections', *IEEE Trans.*, 1977, **MTT-26**, pp. 437–440
24 MILLER, S. E.: 'Coupled wave theory and waveguide applications', *Bell Syst. Tech. J.*, 1954, **33**, pp. 661–719
25 LOUISELL, W. H.: 'Analysis of the single tapered mode coupler', *ibid.*, 1955, **34**, pp. 853–870
26 SMITH, R. B.: 'Coupling efficiency of tapered coupler', *Electron. Lett.*, 1975, **11**, pp. 204–206
27 BETHE, H. A.: 'Theory of diffraction by small holes', *Phys. Rev.*, 1944, **66**, pp. 163–182
28 SPORLEDER, F.: 'Broadband modeselective directional couplers' *in* 'Trunk telecommunications by guided waves', IEE Conf. Publ. 71, 1971
29 MARCUSE, D.: 'Theory of dielectric optical waveguides' (Academic Press, New York, 1974)
30 SNYDER, A. W.: 'Coupled mode theory for optical fibers', *J. Opt. Soc. Am.*, 1972, **62**, pp. 1267–1277
31 KUESTER, E. F., and CHANG, D. C.: 'Nondegenerate surface-wave mode coupling between dielectric waveguides', *IEEE Trans.*, 1975, **MTT-23**, pp. 877–882

Chapter 2

Coupled wave equations

2.1 Field equations and sources

The wave-guiding, wave-coupling, and wave-transforming structures which we consider here are usually made of materials and operated at low-enough field intensities so that all their characteristics appear to be linear. In addition, even if these materials are not all isotropic and not everywhere homogeneous they have at least symmetrical tensors $\boldsymbol{\varepsilon}(\boldsymbol{r})$ of permittivity and $\boldsymbol{\mu}(\boldsymbol{r})$ of permeability, so that in general the law of reciprocity applies. Furthermore, under these conditions, only the steady state with harmonic time variation of frequency f or circular frequency $\omega = 2\pi f$ for fields and sources needs to be considered.

Thus the complex electric-field vector $\boldsymbol{E}$ and the complex magnetic field vector $\boldsymbol{H}$ denote the complex amplitude vectors of the fields, the time dependence of which is given by $\exp(j\omega t)$. Similarly the complex vectors $\boldsymbol{J}$ and $\boldsymbol{M}$ denote the electric and magnetic current densities, respectively, which excite these fields.

With this notation Maxwell's equations appear as

$$\nabla \times \boldsymbol{H} = j\omega\boldsymbol{\varepsilon}\cdot\boldsymbol{E} + \boldsymbol{J} \tag{2.1.1a}$$

$$-\nabla \times \boldsymbol{E} = j\omega\boldsymbol{\mu}\cdot\boldsymbol{H} + \boldsymbol{M} \tag{2.1.1b}$$

where $\boldsymbol{J}$ and $\boldsymbol{M}$ represent impressed sources. Any currents induced by the fields are taken into account by imaginary contributions to the complex components of $\boldsymbol{\varepsilon}$ and $\boldsymbol{\mu}$.

Actually time harmonic fields can only persist if generated by real sources. The connection between the sources and the generated fields is uniquely determined by Maxwell's equations if the distribution of media is specified in the region where the field is nonzero.

In the present context we have no need to follow the transmission paths from the real sources to the field distribution in the area of our direct concern, namely the nonuniform sections of waveguides.

This connection between the real sources and the excited fields is in general much too complex for practical microwave networks. To find simple models for microwave transmission lines we will rather introduce fictitious sources, and it is necessary to know appropriate principles for calculating the fields from these fictitious sources.

A very simple statement can be made if, on a surface S, a surface distribution of currents $\boldsymbol{J}_s$ and $\boldsymbol{M}_s$ is present, as shown in Fig. 2.1. $\boldsymbol{n}$ is

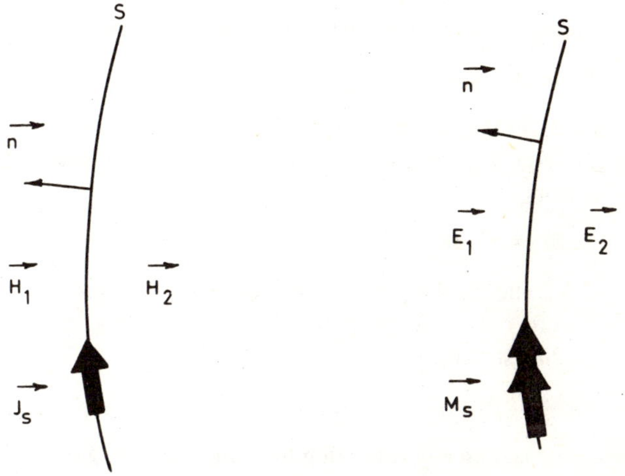

Fig. 2.1. *Surface currents causing field discontinuities*

the unit vector normal to the surface pointing into region 1. At this surface S Maxwell's equations require field discontinuities according to (e.g. Reference 1)

$$\boldsymbol{n} \times (\boldsymbol{H}_1 - \boldsymbol{H}_2) = \boldsymbol{J}_s \tag{2.1.2a}$$

$$(\boldsymbol{E}_1 - \boldsymbol{E}_2) \times \boldsymbol{n} = \boldsymbol{M}_s \tag{2.1.2b}$$

where the subscripts 1 and 2 denote the side of S on which $\boldsymbol{E}$ and $\boldsymbol{H}$ are evaluated.

Another relation between sources and fields, which is frequently needed, is given by the reciprocity theorem. Consider two sets of source distributions $\boldsymbol{J}_a$, $\boldsymbol{M}_a$ and $\boldsymbol{J}_b$, $\boldsymbol{M}_b$ existing alternatively in the same distribution of media. Denote the field produced by the sources $\boldsymbol{J}_a$, $\boldsymbol{M}_a$ by $\boldsymbol{E}_a$, $\boldsymbol{H}_a$, respectively and the field produced by $\boldsymbol{J}_b$, $\boldsymbol{M}_b$ by $\boldsymbol{E}_b$, $\boldsymbol{H}_b$. Then for any point for which the fields are source-free the the condition

$$\nabla(\boldsymbol{E}_a \times \boldsymbol{H}_b - \boldsymbol{E}_b \times \boldsymbol{H}_a) = 0 \tag{2.1.3a}$$

follows from Maxwell's equations (Reference 1). If eqn. 2.1.3*a* is integrated throughout a source-free volume with surface S we have

$$\oiint_S (\boldsymbol{E}_a \times \boldsymbol{H}_b - \boldsymbol{E}_b \times \boldsymbol{H}_a)\cdot d\boldsymbol{S} = 0 \tag{2.1.3b}$$

where $d\boldsymbol{S}$ is the surface element vector.

By definition, the reaction of the field a on source b is

$$\langle a, b\rangle = \iiint (\boldsymbol{E}_a \cdot \boldsymbol{J}_b - \boldsymbol{H}_a \cdot \boldsymbol{M}_b)dV \tag{2.1.4a}$$

where the integration extends over all regions of space with

$$\boldsymbol{J}_b \neq 0 \quad \text{and} \quad \boldsymbol{M}_b \neq 0$$

In this notation the reaction of field a on source b is equal to the reaction of field b on source a, only if $\boldsymbol{\varepsilon}$ and $\boldsymbol{\mu}$ in all space are symmetrical tensors.

$$\langle a, b\rangle = \langle b, a\rangle \tag{2.1.4b}$$

Eqns. 2.1.3 and 2.1.4 represent two forms of the reciprocity theorem which will turn out to be quite useful for investigating the field excitation in guiding structures.

2.2 Wave propagation and excitation in cylindrical waveguides

To describe cylindrical waveguides and the field in them let us use a Cartesian co-ordinate system x, y, z. Cylindrical waveguides are here understood to be guiding structures which have a straight axis (the z-axis) and for which the distribution of matter is a z-invariant function of x and y. All matter is again assumed linear with $\boldsymbol{\varepsilon}(x,y)$ and $\boldsymbol{\mu}(x,y)$ being symmetrical tensors.

When we refer to practical cylindrical waveguides we can always assume, without any loss of generality, that the regions under consideration are surrounded by a perfectly-conducting screen, the inner surface of which is likewise cylindrical, and hence z-invariant. In the case of an open guiding structure such as a dielectric cylinder the distance between the screen and the guiding area can be made large enough for the screen not to interfere with the guided waves. In this case the screen is only introduced to allow general and simple formulations of certain wave characteristics.

Note that the above definition of a cylindrical guide includes most of the transmission lines such as coaxial lines, hollow-tube waveguides, striplines, dielectric-lined circular waveguides and dielectric waveguides such as planar optical waveguides and fibres. Even periodic structures, for example helical circular transmission lines or corrugated waveguides, can be regarded as uniform cylindrical waveguides if the period length of the nonuniform elements is short compared to the wavelength and if the nonuniform elements are replaced by equivalent uniform distributions of matter with the respective symmetric $\boldsymbol{\varepsilon}$ - and $\boldsymbol{\mu}$ -tensors.

In general, Cartesian co-ordinates are appropriate only in describing fields in cylindrical guides of rectangular cross-sectional geometry. We will nevertheless use these co-ordinates at present for all cylindrical guides to simplify the notation and evaluation. The results will mostly be independent of the particular co-ordinate system.

We will not undertake here to derive in detail the field propagation in cylindrical guides from Maxwell's equations. We rather remind the reader, that any field in a uniform waveguide can be described in terms of eigenfields of normal modes. In the uniform cylindrical guide these modes propagate independently of one another in both axial directions. Any particular normal mode is characterised by its propagation constant and its modal field, which in turn is a solution of the source-free Maxwell's equations for the given distribution of matter and the proper boundary conditions at the surrounding conducting wall.

The field of the ith normal mode may be written in the following form:

$$\boldsymbol{E}^{(\pm)}(x,y,z) = A_i^{(\pm)}(z_0) \exp\left[\mp \gamma_i (z-z_0)\right] \boldsymbol{E}_i^{(\pm)}(x,y) \tag{2.2.1a}$$

$$\boldsymbol{H}^{(\pm)}(x,y,z) = A_i^{(\pm)}(z_0) \exp\left[\mp \gamma_i (z-z_0)\right] \boldsymbol{H}_i^{(\pm)}(x,y) \tag{2.2.1b}$$

$A_i^{(\pm)}(z_0)$ is the complex amplitude of the mode at $z = z_0$.

γ_i is its propagation constant. $\boldsymbol{E}_i^{(\pm)}$ and $\boldsymbol{H}_i^{(\pm)}$ denote the normal-mode fields as a function of the cross-sectional co-ordinates x and y. They are independent of z in the formulation of eqns. 2.2.1. The plus and minus signs in the superscripts of A, $\boldsymbol{E}$ and $\boldsymbol{H}$ indicate waves travelling in the positive and negative z-direction, respectively. To generalise some of the following representations we will on occasion replace the superscripts (+) and (−) by (p) or (q). p or q will then be equal to +1 or −1, if they appear as factors. In this notation eqn. 2.2.1a for example appears as

$$\boldsymbol{E}^{(p)}(x,y,z) - A_i^{(p)}(z_0) \exp\left[-p \gamma_i (z-z_0)\right] \boldsymbol{E}_i^{(p)}(x,y) \tag{2.2.1c}$$

In any waveguide for which, according to the assumptions of the first section, the law of reciprocity applies, the fields of different normal modes obey certain orthogonality relations. These relations follow directly from the Lorentz reciprocity theorem[2] and in their most general form appear as

$$\iint_{x,y} (\boldsymbol{E}_i^{(+)} \times \boldsymbol{H}_k^{(-)} - \boldsymbol{E}_k^{(-)} \times \boldsymbol{H}_i^{(+)}] \cdot \boldsymbol{n}_z \, dS = -2R_i\delta_{ik} \tag{2.2.2a}$$

$$\iint_{x,y} [\boldsymbol{E}_i^{(+)} \times \boldsymbol{H}_k^{(+)} - \boldsymbol{E}_k^{(+)} \times \boldsymbol{H}_i^{(+)}] \cdot \boldsymbol{n}_z \, dS = 0 \tag{2.2.2b}$$

$$\iint_{x,y} [\boldsymbol{E}_i^{(-)} \times \boldsymbol{H}_k^{(-)} - \boldsymbol{E}_k^{(-)} \times \boldsymbol{H}_i^{(-)}] \cdot \boldsymbol{n}_z \, dS = 0 \tag{2.2.2c}$$

The surface integrals in these equations extend over all of the cross-section of the particular waveguide. δ_{ik} denotes the Kronecker symbol, while $-2R_i$ may be regarded as a normalisation factor, which we are still free to choose.

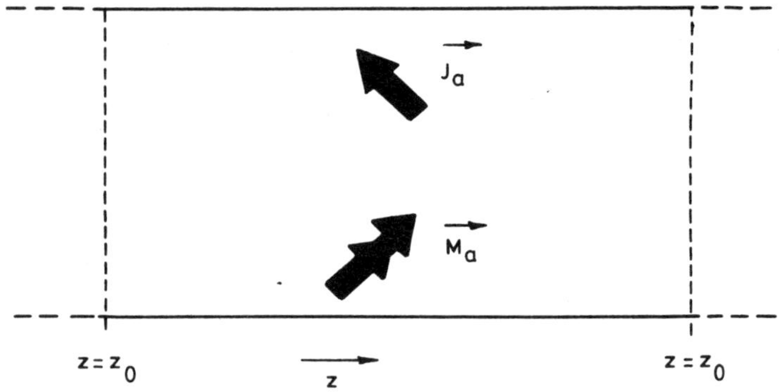

Fig. 2.2. *Uniform cylindrical waveguide containing sources*

Let us now consider, the excitation of normal modes by a given distribution of sources inside the waveguide. According to Fig. 2.2 we assume that all impressed sources of electric current density $\boldsymbol{J}_a = \boldsymbol{J}_a(x, y, z)$ and magnetic current density $\boldsymbol{M}_a = \boldsymbol{M}_a(x, y, z)$ are distributed between the two cross-sections of the wave-guide at $z = z_0'$ and at $z = z_0 > z_0'$. We furthermore let the waveguide extend to infinity on both sides, or be terminated in matched loads for all modes, which these sources excite. Under these conditions all normal modes excited by these sources at $z \geqslant z_0$ travel in the positive z-direction and their amplitudes are given by

$$A_i^{(+)}(z_0) = -\frac{1}{2R_i} \iiint_{x,y,z<z_0} \exp\,[\gamma_i(z-z_0)] \times$$

$$(E_i^{(-)} \cdot J_a - H_i^{(-)} \cdot M_a) dV \tag{2.2.3a}$$

For $z < z_0'$ we have only waves from these sources which travel in the negative z-direction. Their amplitudes in turn are given by

$$A_i^{(-)}(z_0') = -\frac{1}{2R_i} \iiint_{x,y,z>z_0'} \exp\,[-\gamma_i(z-z_0')] \times$$

$$(E_i^{(+)} \cdot J_a - H_i^{(+)} \cdot M_a) dV \tag{2.2.3b}$$

These relations for the excitation of normal modes in uniform waveguides follow likewise from the law of reciprocity.[2] They hence imply the normal mode fields to obey the orthogonality according to eqn. 2.1.2.

In view of subsequent applications let us also list here the special case of current distributions which excite only waves in one particular direction, either forward or backward. If the sources consist only of surface currents in the waveguide cross-section at $z = z_0$ with distributions according to

$$i_S^{(p)}(x,y) = A_i^{(p)}(z_0) J_{Si}^{(p)} = p A_i^{(p)}(z_0)(n_z \times H_i^{(p)}) \tag{2.2.4a}$$

$$m_S^{(p)}(x,y) = A_i^{(p)}(z_0) M_{Si}^{(p)} = p A_i^{(p)}(z_0)(E_i^{(p)} \times n_z) \tag{2.2.4b}$$

where n_z denotes the unit vector in the positive z-direction, then in the case of $p = +1$ no fields are excited in the region $z < z_0$, and in the region with $z > z_0$ only the forward travelling wave of mode index i is excited. Its amplitude is given by

$$A_i^{(+)}(z_0) \exp\,[-\gamma_i(z-z_0)]$$

In the case of $p = -1$ in eqn. 2.2.4, these surface currents excite only the backward travelling wave of mode index i in the region with $z < z_0$ and with an amplitude according to

$$A_i^{(-)}(z_0) \exp\,[\gamma_i(z-z_0)]$$

Let us now demonstrate, by way of an example, how these results will be applied in the following sections to analyse wave propagation in nonuniform waveguides. Fig. 2.3 shows a cylindrical waveguide which is uniform except for a small inhomogeneity at $z = z_1$. Let the nth normal mode be incident with amplitude A_n. Without the inhomogeneity the undisturbed field E_0, H_0 in the waveguide at $z = z_1$ is given by

$$E_0 = A_n(z_1)E_n \tag{2.2.5a}$$

$$H_0 = A_n(z_1)H_n \tag{2.2.5b}$$

As an example let the inhomogeneity be a small obstacle. To calculate its effect on the field propagation it can be replaced by effective sources J_a, M_a the magnitude of which are proportional to the actual field E_0', H_0' in the vicinity of the obstacle.

$$J_a = \mathbf{p}_e E_0' \tag{2.2.6a}$$

$$M_a = \mathbf{p}_m H_0' \tag{2.2.6b}$$

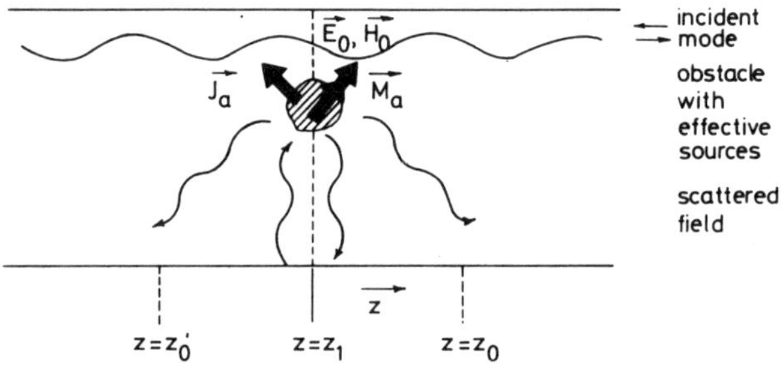

Fig. 2.3. *Scattering of a small obstacle*

Under general circumstances $\mathbf{p}_e$ and $\mathbf{p}_m$ are tensors which are non-zero only in the region of the obstacle. We assume their values to be known. Thus, if E_0', H_0' is known, the problem of calculating the scattering of the obstacle is solved by way of eqns. 2.2.3. The total field E_0', H_0' is, however, not identical with the incident field E_0, H_0 because, according to Fig. 2.3, the field excited by J_a, M_a and scattered back from the waveguide walls contributes to the total field. Even for small obstacles this contribution cannot, in general, be neglected. We therefore have to use a slightly different approach to calculate the scattering at the inhomogeneity. For this approach we introduce on the right-hand side of the cross-section at $z = z_0$ of Fig. 2.3 additional sources, which force the total scattered field on the left-hand side of $z = z_1$ to be zero. The amplitudes of the normal modes excited by these additional sources are chosen such that they compensate the field which is scattered by the obstacle in the negative z-direction. With these additional sources as shown in Fig. 2.4 the obstacle scatters fields only into the region $z > z_1$. These scattered fields contain, however, modes of amplitudes $\Delta A^{(+)}$ and $\Delta A^{(-)}$ propagating in the positive and negative z-direction, respectively.

According to eqns. 2.2.3 these amplitudes are, at $z = z_0$, given by

$$\Delta A_i^{(\pm)}(z_0) = \mp \frac{1}{2R_i} \iiint_{x,y,z<z_0} \exp\left[\pm \gamma_i(z - z_0)\right]$$

$$(E_i^{(\mp)} \cdot J_a - H_i^{(\mp)} \cdot M_a)\,dV \tag{2.2.7}$$

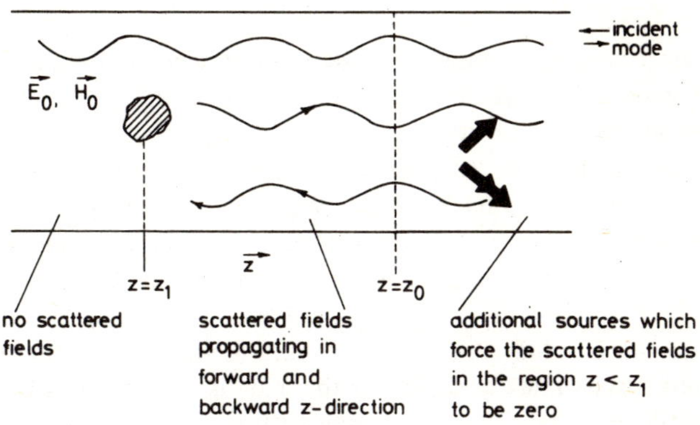

Fig. 2.4. *Model for calculating the transmission matrix*

Since on the left-hand side of the obstacle the total field E_0', H_0' is identical with the eigenfield of the incident nth normal mode E_0, H_0, and since this total field must be continuous, we can now assume

$$E_0' = E_0, \quad H_0' = H_0$$

if only the obstacle is sufficiently small. Thus

$$J_a = \mathbf{p}_e E_n A_n(z_1) \tag{2.2.8a}$$

$$M_a = \mathbf{p}_m H_n A_n(z_1) \tag{2.2.8b}$$

is valid, and with eqns. 2.2.7 and 2.2.8 the complete transmission matrix of the waveguide with the obstacle in Fig. 2.3 is determined. During the following investigations we shall primarily be interested in this transmission matrix.

One remark has to be added concerning the convergence properties of the field excitation model of Fig. 2.4 which results in eqns. 2.2.7 and 2.2.8. If all the propagating and nonpropagating modes in the region $z < z_1$ are to be compensated, additional sources of infinite magnitude may be required at $z > z_0$ if the inhomogeneity is of finite size. In this case only a limited number of normal modes can be considered in the transmission matrix, because for an infinite number of modes the series representations might diverge.

In the case of lossless waveguides the modes which are far below cutoff with their evanescent fields decreasing rapidly need not be compensated on the left-hand side of the obstacle. They can be assumed to be terminated on either side of the obstacle. The eigenfields of these modes represent the near fields very close to the sources, which will not suffer any significant backscattering from waveguide boundaries.

In the following investigations we will consider the field scattering by inhomogeneities which extend only over an infinitesimally short distance in the axial direction. In this case a large number of evanescent modes may contribute to the coupling of the modes near or above cutoff. Nevertheless, the required additional sources at $z > z_0$ will remain finite since the distance between their location and the inhomogeneity under consideration can be assumed to be infinitesimally small.

Before we conclude this Section a quite useful specialisation of the preceding results must be added. If the material in a guide is isotropic or exhibits only transverse anisotropy with $\mu_{xz} = \mu_{zx} = \mu_{yz} = \mu_{zy} = 0$ and $\epsilon_{xz} = \epsilon_{zx} = \epsilon_{yz} = \epsilon_{zy} = 0$ then the guide has reflection symmetry with respect to the longitudinal or z-direction.[2] This means that, if at a perfectly-conducting short-circuit plane, which extends over any entire cross-section z = constant, a single mode travelling in the positive z-direction is reflected, the reflected field consists only of the field of the respective incident mode travelling in the negative z-direction and vice versa. In this case the following relations between the transverse fields $\boldsymbol{E}_{ti}^{(\pm)}$ and $\boldsymbol{H}_{ti}^{(\pm)}$ of the mode i may be introduced

$$\boldsymbol{E}_{ti}^{(+)} = \boldsymbol{E}_{ti}^{(-)} = \boldsymbol{E}_{ti} \tag{2.2.9a}$$

$$\boldsymbol{H}_{ti}^{(+)} = -\boldsymbol{H}_{ti}^{(-)} = \boldsymbol{H}_{ti} \tag{2.2.9b}$$

Under these conditions the orthogonality relation 2.2.2 can be written as

$$\iint_{x,y} (\boldsymbol{E}_{ti} \times \boldsymbol{H}_{tk}) \cdot \boldsymbol{n}_z \, dS = R_i \delta_{ik} \tag{2.2.10}$$

If, in addition, the materials making up or filling the guide are all lossless, the following orthogonality relation is also valid:

$$\iint_{x,y} (\boldsymbol{E}_{ti} \times \boldsymbol{H}_{tk}^*) \cdot \boldsymbol{n}_z \, dS = P_i \delta_{ik} \tag{2.2.11}$$

where $\boldsymbol{H}_{tk}^*$ is the complex conjugate value of $\boldsymbol{H}_{tk}$.

Eqn. 2.2.11 can be verified by calculating the real or active part of the transmitted complex power

$$\mathrm{Re}\,(P) = \mathrm{Re}\left[\iint_{x,y} (\boldsymbol{E} \times \boldsymbol{H}^*)\cdot \boldsymbol{n}_z\, dS\right] \tag{2.2.12a}$$

of any field $\boldsymbol{E}$, $\boldsymbol{H}$, which is given in terms of the normal mode fields, and by realising that

$$\frac{d}{dz}\mathrm{Re}\,(P) = 0 \tag{2.2.12b}$$

must be valid for any value of z and any combination of normal modes.

This consideration leads also to a number of power relations for lossless uniform waveguides. The total complex power transmitted by any mode with subscript i is given by

$$\begin{aligned} P &= \iint_{x,y} [(A_i^{(+)} + A_i^{(-)})\boldsymbol{E}_{ti}] \times [(A_i^{(+)} - A_i^{(-)})^*\boldsymbol{H}_{ti}^*]\cdot \boldsymbol{n}_z\, dS \\ &= P_i(A_i^{(+)} + A_i^{(-)})(A_i^{(+)} - A_i^{(-)})^* \end{aligned} \tag{2.2.13}$$

When we introduce

$$A_i^{(\pm)} = A_i^{(\pm)}(z) = A_i^{(\pm)}(z_0)\exp\,[\mp \gamma_i(z - z_0)]$$

according to eqn. 2.2.1*a* and take into consideration, that eqn. 2.2.12*b* must be satisfied for any combination of $A_i^{(+)}$ and $A_i^{(-)}$, we obtain

$$\mathrm{Re}\,[(\gamma_i + \gamma_i^*)P_i] = 0 \quad \mathrm{Re}\,[(\gamma_i - \gamma_i^*)P_i] = 0 \tag{2.2.14}$$

Thus for a mode with $\mathrm{Re}\,(P_i) \neq 0$ the value of γ_i must be purely imaginary and P_i must be purely real. Such a mode is propagating or said to be above cutoff.

For a mode with $\mathrm{Re}\,(P_i) = 0$ the value of γ_i must be purely real; this mode is evanescent and said to be below cutoff.

Thus for any mode above cutoff

$$\mathrm{Im}\,(P_i) = 0 \quad \mathrm{Re}\,(\gamma_i) = 0 \tag{2.2.15a}$$

is valid, while for a mode below cutoff

$$\mathrm{Re}\,(P_i) = 0 \quad \mathrm{Im}\,(\gamma_i) = 0 \tag{2.2.15b}$$

holds.

For lossless waveguides which have reflection symmetry with respect to the z-co-ordinate a simple relation between R_i of eqn. 2.2.10 and P_i of eqn. 2.2.11 can be given

$$R_i = |P_i| \tag{2.2.16}$$

The normal mode fields of these waveguides are usually normalised such that

$$R_i = 1 \qquad (2.2.17)$$

and eqns. 2.2.9 are satisfied. Note that if a mode goes through cutoff (e.g. by way of variation of frequency), a discontinuity of its fields must occur according to eqns. 2.2.11 and 2.2.15.

2.3 Coupled wave equations for nonuniform waveguides

Fig. 2.5*a* shows the type of nonuniform waveguide which we consider in this Section. We assume its axis to be straight and let it coincide with the z-axis of suitable cylindrical co-ordinates. The cross-section is defined as the area within the ideally conducting boundaries of a plane surface to which the z-axis is perpendicular.

We distinguish between two different characteristics of the cross-sections which may vary as a function of z. The variation of the distribution of matter within the cross-section will be characterised by a z-dependent distribution of $\boldsymbol{\epsilon}$ or $\boldsymbol{\mu}$. The rate at which the cross-section at $z = z_1$ changes its form and size is given by the finite angle ϕ which the plane tangential to the wall makes with the z-direction. Fig. 2.5*a* shows such a longitudinal section and the angle ϕ in a particular cross-section at $z = z_1$.

The field within the nonuniform waveguide will be presented in terms of the local normal modes of a cylindrical waveguide corresponding in cross-section to the particular local cross-section of the nonuniform waveguide. Due to the change of cross-sectional properties these modes change along the waveguide and in addition interact with each other. It is the subject of this Section to determine the general coupling characteristics.

For this purpose we select a short section of the given nonuniform waveguide between $z_1 - \Delta z \leqslant z \leqslant z_1$ and join uniform cylindrical waveguides to it on both sides with the respective local characteristics as shown in Fig. 2.5*b*.

To calculate the transmission matrix of the short nonuniform waveguide we let a normal mode of order n be incident from one of the cylindrical waveguides. The field of this mode is not the correct solution of Maxwell's equations in the nonuniform section. We can, however, introduce fictitious sources which are chosen such that Maxwell's equations are satisfied by the normal mode fields together with these fictitious sources. To return subsequently to the actual

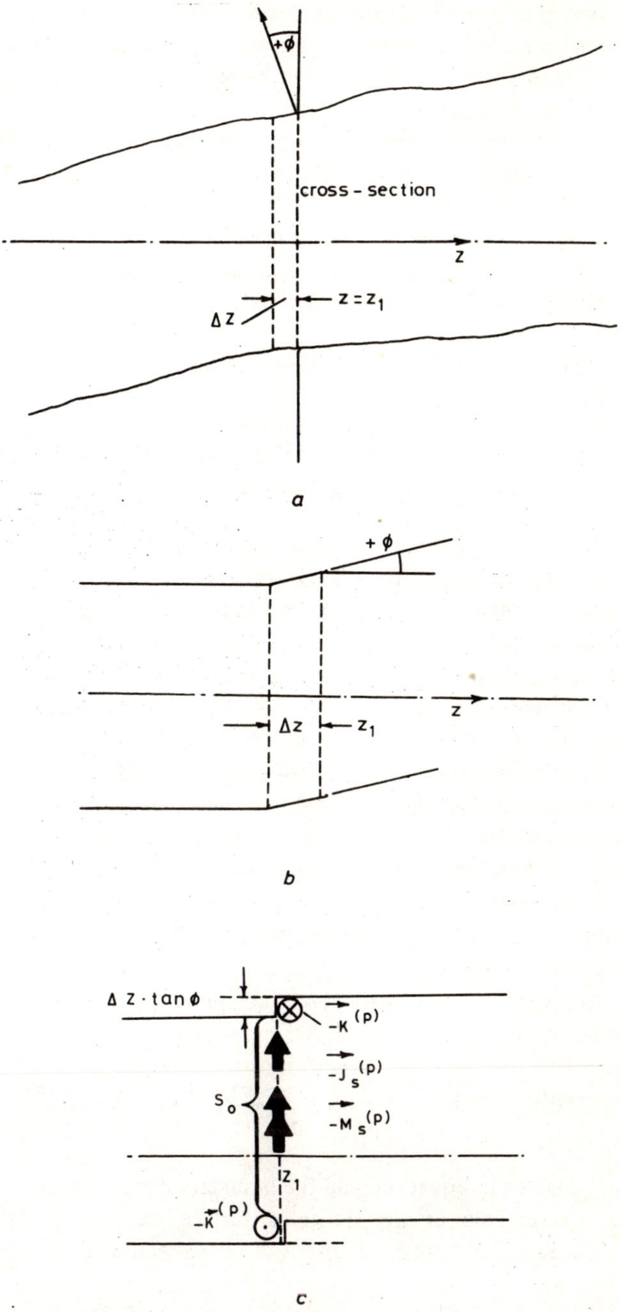

Fig. 2.5. *Inhomogeneous section with straight axis*

source-free problem we introduce another set of source distributions equal in magnitude but of opposite sign which compensate the fictitious sources. These latter sources can then be regarded as the sources which cause spurious mode generation.

To simplify the calculation process we shall use the structure in Fig. 2.5*c* instead of that in Fig. 2.5*b*. In this arrangement the continuous transition which connects the cylindrical waveguides in Fig. 2.5*b* is replaced by a step between these waveguides at $z = z_1$. This simplification will not lead to any error because for gradually changing characteristics we shall be interested only in mode-coupling terms of first-order proportional to Δz.

With the notation of the preceding Section 2.2 the field of the nth incident normal mode is given by

$$\boldsymbol{E} = A_n^{(p)}(z_1) \exp\left[-p\gamma_n(z - z_1)\right] \boldsymbol{E}_n^{(p)}(x, y) \qquad (2.3.1a)$$

$$\boldsymbol{H} = A_n^{(p)}(z_1) \exp\left[-p\gamma_n(z - z_1)\right] \boldsymbol{H}_n^{(p)}(x, y) \qquad (2.3.1b)$$

where $A_n^{(p)}(z_1)$ is again the amplitude at $z = z_1$. $\boldsymbol{E}_n, \boldsymbol{H}_n$ represents the normal mode field which for a given cross-section depends only on the transverse co-ordinates x and y. γ_n is the propagation constant of the mode n. For (p) being $(+)$ and $p = +1$ eqns. 2.3.1 describe the field of a wave propagating in the positive z-direction. For (p) being $(-)$ and $p = -1$ a wave in the negative z-direction is indicated.

Note that in Fig. 2.5*c* at $z = z_1$ the same transverse co-ordinates x, y apply for the cross-sections on both sides of the step discontinuity. When represented in terms of these transverse co-ordinates the modal field $\boldsymbol{E}_n$, $\boldsymbol{H}_n$ of the nth mode in the waveguide at $z < z_1$ differs slightly from the field of the corresponding mode in the waveguide at $z > z_1$. Eqns. 2.3.1 therefore contain a field discontinuity at $z = z_1$ which, according to eqn. 2.1.2, requires additional surface currents of magnitude

$$-\boldsymbol{J}_S^{(p)} = \boldsymbol{n}_z \times \left[\boldsymbol{H}_n^{(p)}(z > z_1) - \boldsymbol{H}_n^{(p)}(z < z_1)\right] A_n^{(p)}(z_1) \qquad (2.3.2a)$$

$$-\boldsymbol{M}_S^{(p)} = \left[\boldsymbol{E}_n^{(p)}(z > z_1) - \boldsymbol{E}_n^{(p)}(z < z_1)\right] \times \boldsymbol{n}_z A_n^{(p)}(z_1) \qquad (2.3.2b)$$

to satisfy Maxwell's equations and the boundary conditions.

The components of the fictitious surface currents $-\boldsymbol{J}_S^{(p)}$ and $-\boldsymbol{M}_S^{(p)}$ of eqns. 2.3.2 are functions of the cross-sectional co-ordinates x and y.

$\boldsymbol{E}_n^{(p)}(z > z_1)$, $\boldsymbol{H}_n^{(p)}(z > z_1)$ and $\boldsymbol{E}_n^{(p)}(z < z_1)$, $\boldsymbol{H}_n^{(p)}(z < z_1)$ denote the eigenfields of the nth mode in the cross-sections $z > z_1$ and $z < z_1$, respectively. They vanish outside the conducting boundaries.

Within the part S_0 of the surface which is common to the cross-sections on both sides of the connection the normal mode fields are nearly identical. The fictitious sources according to eqns. 2.3.2 are therefore rather small. At the steps in the boundary, however, the eigenfield changes from a finite magnitude within the cross-section to zero in the conducting wall and outside. Thus at these steps rather strong currents are required which need to be investigated in more detail. The electrical current densities parallel to and on the conducting walls are of no significance because they do not excite any fields; for these currents the conducting wall acts as a short circuit. To consider the respective magnetic current densities $-\boldsymbol{M}_S^{(p)}$ at the step we note that the electric normal mode field $\boldsymbol{E}_{tn}^{(p)}$ at the conducting boundary of the cross-section consists only of a component normal to the boundary of the cross-section. Thus this magnetic current density $-\boldsymbol{M}_S^{(p)}$ is directed parallel to the boundary.

For small values of the quantity $\Delta z \tan \phi$ at the step the magnetic current density can be lumped into a magnetic line current of magnitude $-\boldsymbol{K}^{(p)}$ in the cross-section $z = z_1$ parallel to the boundary:

$$-\boldsymbol{K}^{(p)} = \boldsymbol{E}_n^{(p)} \times \boldsymbol{n}_z \tan \phi \Delta z A_n^{(p)}(z_1) \qquad (2.3.3a)$$

If this additional line current is introduced then the surface currents of eqns. 2.3.2a have only to be taken in the common part S_0 of the cross-section. In terms of the normal mode fields in Fig. 2.5a or 2.5b they may be written in the following form:

$$-\boldsymbol{J}_S^{(p)} = \boldsymbol{n}_z \times \frac{d\boldsymbol{H}_n^{(p)}}{dz} \Delta z A_n^{(p)}(z_1) \qquad (2.3.3b)$$

$$-\boldsymbol{M}_S^{(p)} = \frac{d\boldsymbol{E}_n^{(p)}}{dz} \times \boldsymbol{n}_z \Delta z A_n^{(p)}(z_1) \qquad (2.3.3c)$$

To return to the original source-free problem we introduce additional currents of magnitude $+\boldsymbol{K}^{(p)}$, $+\boldsymbol{J}_S^{(p)}$, $+\boldsymbol{M}_S^{(p)}$. They can be regarded as sources of spurious modes. To calculate the amplitudes $\Delta A_m^{(q)}(z_1)$ of the spurious modes which they excite we use eqn. 2.2.7.

$$\Delta A_m^{(q)}(z_1) = -\frac{q}{2R_m}\left[\iint\limits_{x,y} (E_m^{(-q)}\cdot J_S^{(p)} - H_m^{(-q)}\cdot M_S^{(p)})dS \right.$$

$$\left. -\oint_s K^{(p)}\cdot H_m^{(-q)}ds \quad + O(\Delta z^2)\right] \tag{2.3.4}$$

The line integral $\oint_s ds$ extends along the boundary of a cross-section at $z = z_1$. (q) is again $(\pm)$ and $(-q)$ is $(\mp)$ if it appears as superscript, while q as a factor has the values $q = \pm 1$.

$O(\Delta z^2)$ is a term proportional to Δz^2. The physical meaning of this term may be explained as follows.

Eqn. 2.2.7 has been derived for sources in a uniform waveguide. In the present case, however, the waveguide is not quite uniform. As shown in Fig. 2.5*c* it has the small step at $z = z_1$. We therefore need to examine the error which is contained in eqn. 2.3.4 due to this step.

To excite in the waveguide, with its discontinuity at $z = z_1$ of Fig. 2.5*c*, only the field of the ith mode travelling e.g. in the backward z-direction we must introduce sources according to eqns. 2.3.2 at $z = z_1$ in addition to the sources $J_{Si}^{(-)}$ and $M_{Si}^{(-)}$ of Fig. 2.2 and of eqns. 2.2.4. Since the magnitude of these additional sources is proportional to Δz the fields E_c, H_c which are excited by the corresponding compensating sources are likewise proportional to Δz. Thus the sources $J_{Si}^{(-)}$ and $M_{Si}^{(-)}$ by themselves excite the eigenfield of the ith mode and the scattered fields E_c, H_c. If, according to eqn. 2.2.7, the reaction of this field on the sources of eqns. 2.3.3 is calculated, the contribution of the scattered field E_c, H_c is found to be proportional to Δz.[2]

This contribution is indicated in eqns. 2.3.4 by the term $O(\Delta z^2)$. A similar additional term occurs if eqn 2.3.4 together with eqns. 2.3.3, which have been derived for the structure in Fig. 2.5*c*, are applied to calculate the coupling of the structure in Fig. 2.5*b*.

Eqn. 2.3.4 together with eqn. 2.3.3 describe the field distortion which occurs at $z = z_1$ in Fig. 2.5*b* if the nth mode is incident. In the limiting case $\Delta z \rightarrow 0$ we can set up a complete system of equations which describe the field propagation from $z = z_1 - \Delta z$ to $z = z_1$. In these equations the normal propagation of the incident mode between $z_1 - \Delta z < z < z_1$ has to be considered as well, so that we obtain

$$\frac{d}{dz}\begin{bmatrix} A_1^{(+)} \\ A_2^{(+)} \\ \cdot \\ \cdot \\ A_1^{(-)} \\ A_2^{(-)} \\ \cdot \\ \cdot \end{bmatrix} =$$

$$\begin{bmatrix} -\gamma_1 + \kappa_{11}^{(++)} & \kappa_{12}^{(++)} & \dots & \kappa_{11}^{(+-)} & \kappa_{12}^{(+-)} & \dots \\ \kappa_{21}^{(++)} & -\gamma_2 + \kappa_{22}^{(++)} & \dots & \kappa_{21}^{(+-)} & \kappa_{22}^{(+-)} & \dots \\ \cdot & \cdot & & \cdot & \cdot & \\ \cdot & \cdot & & \cdot & \cdot & \\ \kappa_{11}^{(-+)} & \kappa_{12}^{(-+)} & \dots & \gamma_1 + \kappa_{11}^{(--)} & \kappa_{12}^{(--)} & \dots \\ \kappa_{21}^{(-+)} & \kappa_{22}^{(-+)} & \dots & \kappa_{21}^{(--)} & \gamma_2 + \kappa_{22}^{(--)} & \dots \\ \cdot & \cdot & & \cdot & \cdot & \\ \cdot & \cdot & & \cdot & \cdot & \end{bmatrix} \cdot \begin{bmatrix} A_1^{(+)} \\ A_2^{(+)} \\ \cdot \\ \cdot \\ A_1^{(-)} \\ A_2^{(-)} \\ \cdot \\ \cdot \end{bmatrix} \tag{2.3.5}$$

where we assume that every subscript 1, 2, . . . denotes one particular normal mode.

The coupling coefficients κ of eqn. 2.3.5 are calculated by introducing eqns. 2.3.3 into eqn. 2.3.4 dividing by Δz and going to the limit of $\Delta z \to 0$:

$$\lim \frac{\Delta A_m^{(q)}(z_1)}{\Delta z} = \kappa_{mn}^{(q,p)}(z_1) A_m^{(p)}(z_1) \tag{2.3.6}$$

The coupling coefficients are thus obtained as

$$2R_m \kappa_{mn}^{(q,p)} = q \iint_{x,y} \left(\frac{d\boldsymbol{H}_n^{(p)}}{dz} \times \boldsymbol{E}_m^{(-q)} - \boldsymbol{H}_m^{(-q)} \times \frac{d\boldsymbol{E}_n^{(p)}}{dz} \right) \cdot \boldsymbol{n}_z \, dS$$

$$- q \oint_s (\boldsymbol{H}_m^{(-q)} \times \boldsymbol{E}_n^{(p)}) \cdot \boldsymbol{n}_z \tan \phi \, ds \tag{2.3.7}$$

In these expressions the surface integral has to be extended over the entire cross-section while the line integral $\oint ds$ denotes integration along the boundary of the respective cross-section.

The result in eqn. 2.3.7 can be further simplified for waveguides which exhibit reflection symmetry with respect to the z-co-ordinate.

With eqn. 2.2.10 we obtain

$$\frac{d}{dz}\iint (\boldsymbol{H}_{tm} \times \boldsymbol{E}_{tn})\cdot\boldsymbol{n}_z dS = -\delta_{mn}\frac{dR_m}{dz}$$

$$= \oint (\boldsymbol{H}_{tm} \times \boldsymbol{E}_{tn})\cdot\boldsymbol{n}_z \tan\phi ds + \iint \left(\frac{d\boldsymbol{H}_{tm}}{dz} \times \boldsymbol{E}_{tn}\right)\cdot\boldsymbol{n}_z dS$$

$$+ \iint \left(\boldsymbol{H}_{tm} \times \frac{d\boldsymbol{E}_{tn}}{dz}\right)\cdot\boldsymbol{n}_z dS \qquad (2.3.8)$$

Introducing this result and eqns. 2.2.9 into eqn. 2.3.7 yields

$$\kappa_{mn}^{(q,p)} = \frac{1}{2R_m}\iint_{x,y}\left(\boldsymbol{E}_{tn} \times \frac{d\boldsymbol{H}_{tm}}{dz} - qp\boldsymbol{E}_{tm} \times \frac{d\boldsymbol{H}_{tn}}{dz}\right)\cdot\boldsymbol{n}_z dS$$

$$-\delta_{mn}\frac{dR_m}{dz}\frac{1}{2R_m} \qquad (2.3.9)$$

Note that for the evaluation of eqn. 2.3.9 or 2.3.7 any change of the transverse co-ordinate system along the waveguide must be considered in the expressions

$$\frac{d\boldsymbol{H}_{ti}}{dz} = \frac{d\boldsymbol{H}_{ti}(x,y)}{dz} \quad \text{and} \quad \frac{d\boldsymbol{E}_{ti}}{dz} = \frac{d\boldsymbol{E}_{ti}(x,y)}{dz}$$

Eqns. 2.3.5 are the so-called coupled wave equations. They represent an infinite set of coupled transmission lines, one for each of the local normal modes. Usually only approximate solutions of this system can be calculated. In the following chapters we will discuss procedures to find such solutions.

2.4 Characteristics of coupled wave equations

Following the general method of analysis as it was outlined in the previous Section the coupled wave equations can be set up for any non-uniform waveguide with its cross-sectional parameters varying continuously in the axial direction. We need only require that any point inside the guide can be uniquely assigned to a surface S_0, for which a complete set of normal modes with the corresponding fields is known. In the previous Section this surface S_0 was assumed to be a plane local cross-section, but for some nonuniformities in waveguides it is more suitable to base the analysis on surfaces S_0 with curvature. The surfaces S_0 are hence not necessarily plane cross-sections. In Chapter 3 of this book circular cylinder surfaces as well as spherical surfaces will be used. In the following calculations of the present Section $\iint dS_0$ is meant to denote integration over these entire surfaces S_0, while $\boldsymbol{n}_z$, the unit vector perpendicular to this surface, is a vector pointing in the direction of propagation of the field. In the previous Section the waveguide axis was assumed to be straight and to coincide with the z-axis of a suitable system of cylindrical co-ordinates. A more general form of coupled wave equations can account also for curvature of the waveguide axis: in that case the z-axis is likewise curved.

By modifying the coupled wave equations 2.3.5 according to

$$\frac{d}{dz}\begin{bmatrix} A_1^{(+)} \\ A_2^{(+)} \\ \cdot \\ \cdot \\ A_1^{(-)} \\ A_2^{(-)} \\ \vdots \end{bmatrix} =$$

$$\begin{bmatrix} -\gamma_1^{(+)}+\kappa_{11}^{(++)} & \kappa_{12}^{(++)} & \cdots & \kappa_{11}^{(+-)} & \kappa_{12}^{(+-)} & \cdots \\ \kappa_{21}^{(++)} & -\gamma_2^{(+)}+\kappa_{22}^{(++)} & \cdots & \kappa_{21}^{(+-)} & \kappa_{22}^{(+-)} & \cdots \\ \cdot & \cdot & & \cdot & \cdot & \\ \cdot & \cdot & & \cdot & \cdot & \\ \kappa_{11}^{(-+)} & \kappa_{12}^{(-+)} & \cdots & -\gamma_1^{(-)}+\kappa_{11}^{(--)} & \kappa_{12}^{(--)} & \cdots \\ \kappa_{21}^{(-+)} & \kappa_{22}^{(-+)} & \cdots & \kappa_{21}^{(--)} & -\gamma_2^{(-)}+\kappa_{22}^{(--)} & \cdots \\ \vdots & \vdots & & \vdots & \vdots & \end{bmatrix} \begin{bmatrix} A_1^{(+)} \\ A_2^{(+)} \\ \cdot \\ \cdot \\ A_1^{(-)} \\ A_2^{(-)} \\ \vdots \end{bmatrix}$$

(2.4.1a)

we take into account that the propagation constant of a mode travelling in the forward direction may be different from that of the corresponding mode travelling in the backward direction.

In the following chapters it will on occasion be useful to refer to a more schematic notation of the coupled wave equations. For this reason we will use positive subscripts for the modes travelling in the positive z-direction and negative subscripts for the backward travelling modes according to

$$A_n = A_n^{(+)}, \; A_{-n} = A_n^{(-)}$$

The coupled wave equations can furthermore be written in matrix notation

$$\frac{d}{dz}A = (-\gamma + \kappa)A \tag{2.4.1b}$$

where A is a column vector containing the individual wave amplitudes A_n as its components. γ is a diagonal matrix with $\gamma_{nn} = \gamma_n = \gamma_{|n|}^{(\pm)}$ as its diagonal and only elements. The element κ_{nm} of the coupling matrix κ describes the coupling from the mth mode to the nth mode.

In the present Section, however, we will only use the notation of eqn. 2.4.1a for the coupled wave equations.

A number of basic characteristics of the coupled wave equations are obtained by applying the reciprocity theorem according to eqn. 2.1.3b. The two different source-free field distributions in eqn. 2.1.3b will be denoted by the subscripts m and n and by the superscripts p and q. The closed surface is chosen to consist of two surfaces S_0 representing waveguide cross-sections with the z-values z_0 and $z_0 + \Delta z$. The surface integral of eqn. 2.1.3b is closed along the conducting wall between the two surfaces S_0. Since the tangential components of the E-fields $\boldsymbol{E}_n$ and $\boldsymbol{E}_m$ vanish at the perfectly-conducting wall, eqn. 2.1.3b appears as

$$-\iint_{z=z_0} (\boldsymbol{E}_n \times \boldsymbol{H}_m - \boldsymbol{E}_m \times \boldsymbol{H}_n)\cdot \boldsymbol{n}_z dS_0$$

$$= \iint_{z=z_0+\Delta z} (\boldsymbol{E}_n \times \boldsymbol{H}_m - \boldsymbol{E}_m \times \boldsymbol{H}_n)\cdot \boldsymbol{n}_z dS_0 \tag{2.4.2}$$

Let us now at $z = z_0$ choose $\boldsymbol{E}_m, \boldsymbol{H}_m$ $(\boldsymbol{E}_n, \boldsymbol{H}_n)$ to be identical to the field $\boldsymbol{E}_m^{(q)}, \boldsymbol{H}_m^{(q)} (\boldsymbol{E}_n^{(p)}, \boldsymbol{H}_n^{(p)})$ of the mth (nth) normal mode with amplitude $A_m^{(q)}(z_0) = A_m^{(q)}, (A_n^{(p)}(z_0) = A_n^{(p)})$ travelling either in the positive or negative z-direction:

$$\underline{z = z_0} \quad E_m = A_m^{(q)} E_m^{(q)}, \quad H_m = A_m^{(q)} H_m^{(q)} \tag{2.4.3a}$$

$$E_n = A_n^{(p)} E_n^{(p)}, \quad H_n = A_n^{(p)} H_n^{(p)} \tag{2.4.3b}$$

For $\Delta z \to 0$ the corresponding fields at $z = z_0 + \Delta z$ can then directly be calculated using the coupled wave eqns. 2.4.1 with

$$\Delta A_k = \frac{dA_k}{dz} \Delta z$$

At $z = z_0 + \Delta z$ we obtain

$$E_k = A_k^{(r)} E_k^{(r)}$$
$$+ \Delta z\, A_k^{(r)} [-\gamma_k^{(r)} E_k^{(r)} + \sum_i \kappa_{ik}^{(+,r)} E_i^{(+)} + \sum_i \kappa_{ik}^{(-,r)} E_i^{(-)}] \tag{2.4.4a}$$

$$H_k = A_k^{(r)} H_k^{(r)}$$
$$+ \Delta z A_k^{(r)} [-\gamma_k^{(r)} H_k^{(r)} + \sum_i \kappa_{ik}^{(+,r)} H_i^{(+)} + \sum_i \kappa_{ik}^{(-,r)} H_i^{(-)}] \tag{2.4.4b}$$

where k stands for m when (r) is meant to denote (q), but k stands for n when (r) is meant to be (p).

Note, that the fields $E_i^{(\pm)}, H_i^{(\pm)}$ of the normal modes at $z = z_0$ differ from those at $z = z_0 + \Delta z$. When we introduce eqns. 2.4.3 and 2.4.4 into eqn. 2.4.2 for $n \neq m$ and make use of the orthogonality relations 2.2.2 we arrive at

$\underline{m \neq n}$:

$$A_n^{(p)} A_m^{(q)} \Delta z \iint_{z = z_0 + \Delta z} \{E_n^{(p)} \times H_n^{(+)} \kappa_{nm}^{(+,q)} + E_n^{(p)} \times H_n^{(-)} \kappa_{nm}^{(-,q)}$$
$$+ E_m^{(+)} \times H_m^{(q)} \kappa_{mn}^{(+,p)} + E_m^{(-)} \times H_m^{(q)} \kappa_{mn}^{(-,p)}$$
$$- E_m^{(q)} \times H_m^{(+)} \kappa_{mn}^{(+,p)} - E_m^{(q)} \times H_m^{(-)} \kappa_{mn}^{(-,p)}$$
$$- E_n^{(+)} \times H_n^{(p)} \kappa_{nm}^{(+,q)} - E_n^{(-)} \times H_n^{(p)} \kappa_{nm}^{(-,q)}\} \cdot n_z dS_0$$
$$+ O(\Delta z^2) = 0 \tag{2.4.5}$$

Eqn. 2.4.5 must be satisfied for any values of $A_n^{(+)} A_m^{(+)}, A_n^{(+)} A_m^{(-)}, A_n^{(-)} A_m^{(+)}$ and $A_n^{(-)} A_m^{(-)}$. In the limiting case $\Delta z \to 0$ eqn. 2.4.5 yields for

$$n \neq m: \quad A_n^{(+)} A_m^{(+)} \neq 0 \quad R_n \kappa_{nm}^{(-+)} = R_m \kappa_{mn}^{(-+)} \tag{2.4.6a}$$

$$A_n^{(-)} A_m^{(-)} \neq 0 \quad R_n \kappa_{nm}^{(+-)} = R_m \kappa_{mn}^{(+-)} \tag{2.4.6b}$$

$$A_m^{(-)} A_n^{(+)} \neq 0 \quad R_n \kappa_{nm}^{(--)} = -R_m \kappa_{mn}^{(++)} \tag{2.4.6c}$$

with

$$-R_n = \tfrac{1}{2} \iint (\boldsymbol{E}_n^{(+)} \times \boldsymbol{H}_n^{(-)} - \boldsymbol{E}_n^{(-)} \times \boldsymbol{H}_n^{(+)}) \cdot \boldsymbol{n}_z \, dS_0$$

according to eqn. 2.2.2.

An additional relation for the coupling coefficients of the coupled wave equations can be obtained by specifying $A_n^{(p)} = A_n^{(+)}$ and $A_m^{(q)} = A_n^{(-)}$. In this case eqn. 2.4.2 with

$$\iint\limits_{z=z_0+\Delta z} (\boldsymbol{E}_n^{(+)} \times \boldsymbol{H}_n^{(-)} - \boldsymbol{E}_n^{(-)} \times \boldsymbol{H}_n^{(+)}) \cdot \boldsymbol{n}_z \, dS_0$$

$$-\iint\limits_{z=z_0} (\boldsymbol{E}_n^{(+)} \times \boldsymbol{H}_n^{(-)} - \boldsymbol{E}_n^{(-)} \times \boldsymbol{H}_n^{(+)}) \cdot \boldsymbol{n}_z \, dS_0 = -2 \frac{dR_n}{dz} \Delta z$$

leads for $m = n$ and in the case of $A_n^{(+)} A_n^{(-)} \neq 0$ to the relation

$$-\gamma_n^{(-)} - \gamma_n^{(+)} + \kappa_{nn}^{(--)} + \kappa_{nn}^{(++)} = \frac{1}{R_n} \frac{dR_n}{dz} \tag{2.4.6d}$$

Eqns. 2.4.6 represent relations which are all a consequence of reciprocity.

Another set of relations follows for lossless nonuniform waveguides when all its material exhibits longitudinal isotropy. In this case the transmitted real power Re (P) of any field distribution is determined by the amplitudes $A_i^{(+)}$ and $A_i^{(-)}$ of normal modes and according to eqn. 2.2.12*a* can be calculated from

$$\begin{aligned} \text{Re}\,(P) &= \sum_i \text{Re}\,[P_i(A_i^{(+)} + A_i^{(-)})(A_i^{(+)} - A_i^{(-)})^*] \\ &= \sum_i \{\text{Re}\,(P_i)[A_i^{(+)}(A_i^{(+)})^* - A_i^{(-)}(A_i^{(-)})^*] \\ &\qquad - 2\,\text{Im}\,(P_i)\,\text{Im}\,[A_i^{(-)}(A_i^{(+)})^*]\} \end{aligned} \tag{2.4.7}$$

Re (x) and Im (x) denote the real part and the imaginary part of the quantity x, respectively.

Since the power must be conserved along the lossless waveguide the following relation holds:

$$\frac{d}{dz}\{\mathrm{Re}\,(P)\} = 0$$

$$= \sum_i \Bigg\{ \frac{d}{dz}\{\mathrm{Re}\,(P_i)\}[A_i^{(+)}(A_i^{(+)})^* - A_i^{(-)}(A_i^{(-)})^*]$$

$$+ 2\mathrm{Re}\,(P_i)\,\mathrm{Re}\left[\frac{dA_i^{(+)}}{dz}(A_i^{(+)})^* - \frac{dA_i^{(-)}}{dz}(A_i^{(-)})^*\right]$$

$$-2\frac{d\,\mathrm{Im}\,(P_i)}{dz}\,\mathrm{Im}\,[A_i^{(-)}(A_i^{(+)})^*]$$

$$-2\mathrm{Im}\,(P_i)\,\mathrm{Im}\left[\frac{dA_i^{(-)}}{dz}(A_i^{(+)})^* - \frac{dA_i^{(+)}}{dz}(A_i^{(-)})^*\right]\Bigg\} \quad (2.4.8a)$$

with

$$\frac{dA_i^{(p)}}{dz} = -\gamma_i^{(p)}A_i^{(p)} + \sum_k (\kappa_{ik}^{(p,+)}A_k^{(+)} + \kappa_{ik}^{(p,-)}A_k^{(-)}) \quad (2.4.8b)$$

$$p = \pm 1$$

from eqn. 2.4.1. Eqns. 2.4.8 must be satisfied for any combination of $A_i^{(+)}$ and $A_i^{(-)}$. The following results will be evaluated realising that for lossless waveguides according to eqn. 2.2.15 either $\mathrm{Re}\,(P_i) = 0$ or $\mathrm{Im}\,(P_i) = 0$ is valid.

If we assume that in eqns. 2.4.8 only $A_n^{(p)}$ differs from zero for the respective value of z, we obtain for $\mathrm{Im}\,(P_n) = 0$

$$\frac{1}{2\,\mathrm{Re}\,(P_n)}\frac{d\,\mathrm{Re}\,(P_n)}{dz} + \mathrm{Re}\,(-\gamma_n^{(p)} + \kappa_{nn}^{(p,p)}) = 0$$

with

$$p = \pm 1 \quad (2.4.9a)$$

while for $\mathrm{Re}\,(P_n) = 0$, we obtain $\mathrm{Im}\,(\kappa_{nn}^{(-+)}) = 0$ (2.4.10)

If with $A_n^{(p)}A_m^{(p)} \neq 0$ both $A_n^{(p)}$ and $A_m^{(p)}$ are different from zero for $n \neq m$, we obtain

$$\mathrm{Re}\,(P_n)\kappa_{nm}^{(p,p)} = -\mathrm{Re}\,(P_m)(\kappa_{mn}^{(p,p)})^* \quad (2.4.9b)$$

Additional relations for coupling coefficients are obtained if a pair of amplitudes out of $A_n^{(\pm)}$ and $A_m^{(\pm)}$ and only this pair is different from zero. They are evaluated here only for $\mathrm{Im}\,(P_n) = \mathrm{Im}\,(P_m) = 0$:

$$A_n^{(+)}A_m^{(-)} \neq 0, \quad n = m \quad \text{or} \quad n \neq m$$

$$\text{Re}\,(P_n)\kappa_{nm}^{(+-)} = \text{Re}\,(P_m)(\kappa_{mn}^{(-+)})^* \tag{2.4.9c}$$

Before we conclude this discussion of general characteristics of coupled wave equations let us examine these results for a lossless waveguide which exhibits reflection symmetry. In this case the fields of all normal modes can always be so normalised that

$$R_n = \text{constant} = 1 \tag{2.4.11a}$$

For this normalisation we obtain as a consequence of the law of reciprocity the following relations:

$n = m$ or $n \neq m$:

$$\kappa_{nm}^{(+-)} = \kappa_{mn}^{(+-)}, \quad \kappa_{nm}^{(-+)} = \kappa_{mn}^{(-+)} \tag{2.4.11b}$$

$n \neq m$:

$$\kappa_{nm}^{(++)} = -\kappa_{mn}^{(--)} \tag{2.4.11c}$$

$n = m$:

$$-\gamma_n^{(-)} + \kappa_{nn}^{(--)} = -(-\gamma_n^{(+)} + \kappa_{nn}^{(++)}) \tag{2.4.11d}$$

If in a lossless cylindrical waveguide the normal modes i of a set are propagating and if they are normalised according to eqn. 2.4.11*a* then the forward and backward travelling components of each mode together transmit the real power

$$\text{Re}\,(P_i) = A_i^{(+)}(A_i^{(+)})^* - A_i^{(-)}(A_i^{(-)})^* \tag{2.4.12a}$$

Under these conditions conservation of power requires

$$\frac{d}{dz}\left[\sum_i \text{Re}\,(P_i)\right] = \frac{d}{dz}\,\text{Re}\,(P) = 0 \tag{2.4.12b}$$

and the following relations derive from eqns. 2.4.8

$n = m$ or $n \neq m$:

$$\kappa_{mn}^{(+-)} = (\kappa_{nm}^{(-+)})^* \tag{2.4.12c}$$

$n \neq m$:

$$\kappa_{nm}^{(++)} + (\kappa_{mn}^{(++)})^* = \kappa_{nm}^{(--)} + (\kappa_{mn}^{(--)})^* = 0 \tag{2.4.12d}$$

$n = m$:

$$\text{Re}\,(-\gamma_n^{(+)} + \kappa_{nn}^{(++)}) = \text{Re}\,(-\gamma_n^{(-)} + \kappa_{nn}^{(--)}) = 0 \tag{2.4.12e}$$

Note that these relations follow from eqns. 2.4.8 only if the condition 2.4.12*a* is satisfied. All these relations in eqns. 2.4.6, 2.4.9–2.4.12 are quite useful in evaluating coupling between modes or check solutions for nonuniform waveguides.

Let us now compare these relations with the results of the previous Section, in particular with eqn. 2.3.9. Eqns. 2.4.6 can be verified directly. If the nonuniform section is lossless and the modes under consideration are above cutoff eqns. 2.4.9*a* and 2.4.9*b* can also be verified keeping in mind that for propagating modes

$$\mathrm{Re}\,(\gamma_n^{(\pm)}) = 0 \quad \text{and} \quad P_n = |R_n|$$

is valid according to eqns. 2.2.16.

2.5 Coupled wave equations for curved waveguides

Fig. 2.6*a* shows a waveguide section with its axis curved. We assume that between the cross-sections which are spaced by Δz the waveguide axis is curved with curvature c. The radius of curvature r of the axis must be assumed to be larger than the critical distance x_0 between the axis and the conducting wall (Fig. 2.6*a*)

$$r > x_0 \tag{2.5.1}$$

Any cross-section is again identified by the corresponding z-value of its location along the axis. For the section under consideration near $z = z_1$ the transverse co-ordinates x and y, which describe any point of a given cross-section, are chosen such that the curved axis is contained in the xz-plane. The curvature c will be taken positive if the axis curves towards the negative x-direction. To derive coupled wave equations for the curved waveguide near $z = z_1$ we consider only the section between $z_1 - \Delta z/2 < z < z_1 + \Delta z/2$ and join at both of its ends straight cylindrical waveguides with the same cross-section as the curved section according to Fig. 2.6*b*. For $\Delta z \to 0$, the mode coupling within the curved section is equal to that within the tilt of Fig. 2.6*c*, if the tilt angle is

$$\varphi = \Delta z c, \tag{2.5.2}$$

and hence is the same angle by which the z-axis changes it direction in Fig. 2.6*b*.

For the nth mode with amplitude $A_n^{(p)}(z_1)$, if it would propagate undisturbed through the waveguide tilt, the field in the adjoining waveguides would be given by

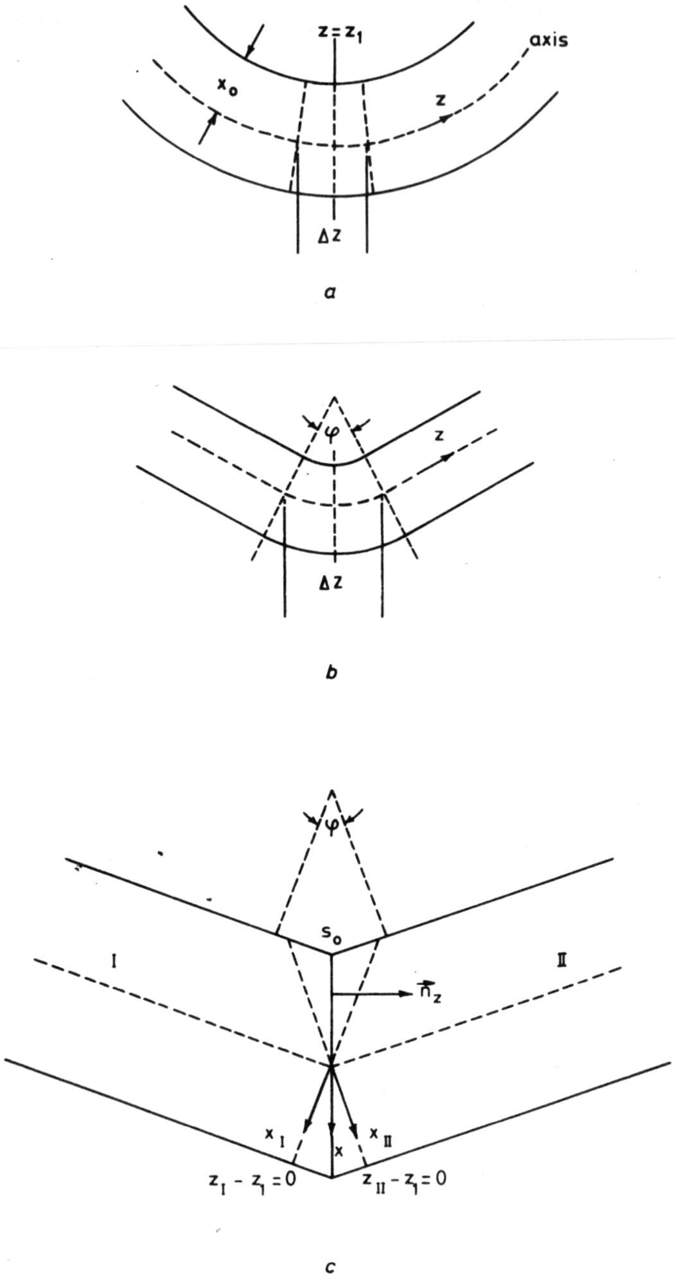

Fig. 2.6. *Curved section with constant cross-section*

$$E_{\mathrm{I}} = A_n^{(p)}(z_1)E_n^{(p)}(x_{\mathrm{I}}, y) \exp[-p\gamma_n(z_{\mathrm{I}} - z_1)] \quad (2.5.3a)$$

$$H_{\mathrm{I}} = A_n^{(p)}(z_1)H_n^{(p)}(x_{\mathrm{I}}, y) \exp[-p\gamma_n(z_{\mathrm{I}} - z_1)] \quad (2.5.3b)$$

$$E_{\mathrm{II}} = A_n^{(p)}(z_1)E_n^{(p)}(x_{\mathrm{II}}, y) \exp[-p\gamma_n(z_{\mathrm{II}} - z_1)] \quad (2.5.3c)$$

$$H_{\mathrm{II}} = A_n^{(p)}(z_1)H_n^{(p)}(x_{\mathrm{II}}, y) \exp[-p\gamma_n(z_{\mathrm{II}} - z_1)] \quad (2.5.3d)$$

$E_{\mathrm{I}}, H_{\mathrm{I}}$ and $E_{\mathrm{II}}, H_{\mathrm{II}}$ denote the fields in waveguide I and II, respectively, according to Fig. 2.6*c*. The co-ordinates $x_{\mathrm{I}}, z_{\mathrm{I}}$ and $x_{\mathrm{II}}, z_{\mathrm{II}}$ refer likewise to these waveguides. With n_z as the unit vector perpendicular to the centre plane S_0 of the tilt, and x and y as the transverse co-ordinates in this plane, the surface currents $-J_S, -M_S$, which are needed to compensate for the field discontinuity on S_0 according to eqns. 2.5.3, can be written as

$$\begin{aligned}-J_S &= n_z \times (H_{\mathrm{II}} - H_{\mathrm{I}}) \\ &= A_n^{(p)}(z_1)c\Delta z n_z \times (pH_{tn}^{(p)} x\gamma_n - H_{zn}^{(p)} n_x) + O(\varphi^2)\end{aligned} \quad (2.5.4a)$$

$$\begin{aligned}-M_S &= (E_{\mathrm{II}} - E_{\mathrm{I}}) \times n_z \\ &= A_n^{(p)}(z_1)c\Delta z(pE_{tn}^{(p)} x\gamma_n - E_{zn}^{(p)} n_x) \times n_z + O(\varphi^2)\end{aligned} \quad (2.5.4b)$$

n_x is the unit vector in the x-direction. The subscript t on E and H denotes the transverse components of the undisturbed field while the subscript z designates its z-components. Note that in eqns. 2.5.4 only terms proportional to φ and Δz, respectively, are included. The higher-order terms disappear in the limit of $\Delta z \to 0$.

To compensate for the sources $-J_S, -M_S$, according to eqns. 2.5.4, additional sources of magnitude J_S, M_S are introduced on S_0 which are then regarded as the sources which generate spurious modes. These spurious modes are again calculated by using eqn. 2.2.7, which for $\Delta z \to 0$ leads to coupled wave equations of the form of eqns. 2.3.5 and 2.3.6 with coupling coefficients according to

$$\begin{aligned}\kappa_{mn}^{(q,p)} = \frac{q}{2R_m} c \iint_{x,y} &[(pH_{tn}^{(p)} x\gamma_n - H_{zn}^{(p)} n_x) \times E_m^{(-q)} \\ &- H_m^{(-q)} \times (pE_{tn}^{(p)} x\gamma_n - E_{zn}^{(p)} n_x)] \cdot n_z dS\end{aligned} \quad (2.5.5)$$

For these coupling coefficients the relations 2.4.6 and 2.4.9 can again be verified. To this end one has to apply certain identities for integrals of field expressions which may be obtained using eqns. 2.2.11 or 2.2.2*a*.

Taking the results of Section 2.3 and the present Section together, the mode interaction characteristics of any curved and tapered waveguide section can in principle be calculated when the normal mode fields of the corresponding local cross-sections are known and when eqn. 2.5.1 is satisfied. If different types of nonuniformities occur simultaneously the corresponding coupling coefficients must all be added in the coupled wave equations. In such a situation the individual nonuniformities of a short nonuniform section can be regarded as being connected in series in one section of the same length. Note that any change of the transverse co-ordinates along the waveguide has to be taken into account in the evaluation of

$$\frac{d}{dz} H_{tn}(x,y), \quad \frac{d}{dz} E_{tn}(x,y)$$

in eqns. 2.3.7 or 2.3.9. In general, however, for waveguides with more complicated nonuniformities, it is quite difficult to find a co-ordinate system, which is suitable for calculating the coupling coefficients.

2.6 Waveguides filled with homogeneous, isotropic and lossless matter

In this Section the previous results will be applied to nonuniform waveguide sections which contain homogeneous, isotropic and lossless matter, the wave number of which is

$$k = \omega\sqrt{\mu\epsilon} \tag{2.6.1}$$

Coupled wave equations for such nonuniform waveguides have been derived in detail elsewhere.[4] We will hence list only those results regarding wave propagation and coupling, which we need in the following chapters.

The fields are represented in terms of modes of cylindrical waveguides, which in cross-sectional form and size correspond to the local cross-section of the nonuniform waveguide. The fields of these cylindrical waveguide modes, such as for example in the rectangular, round, or coaxial waveguides of Fig. 2.7, are derived from vector potentials, which have only a longitudinal component. The dependence of the vector potential component for any particular mode of index i on the transverse co-ordinates of the waveguide cross-section is given by its eigenfunction $T_i = T_i(x, y)$. This eigenfunction satisfies the transverse wave equation

$$\Delta_t T + k_{ci}^2 T = 0 \tag{2.6.2}$$

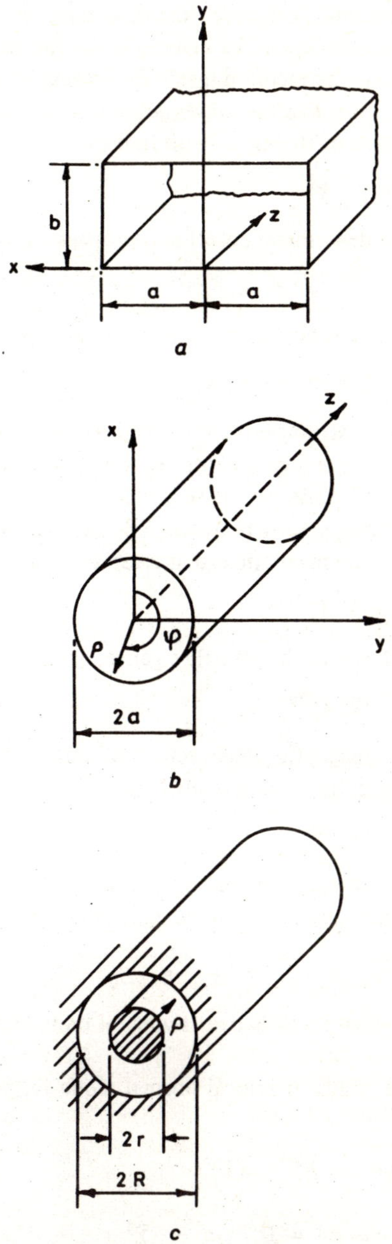

Fig. 2.7. *Waveguide containing homogeneous matter*
a rectangular waveguide
b circular waveguide
c coaxial line

where the index t on Δ denotes the Laplacian with respect to the transverse co-ordinates. k_{ci} is the eigenvalue of the ith mode, and also the wave number of the medium inside the waveguide at cutoff for this particular mode. For lossless waveguides it is a real quantity. The propagation constant of the mode with index i is given by

$$\gamma_i = \sqrt{k_{ci}^2 - k^2}$$

For $k > k_{ci}$ the mode is above cutoff and propagates with

$$\gamma_i = j\beta_i = jk\sqrt{1-(k_{ci}/k)^2} \tag{2.6.3a}$$

For $k < k_{ci}$ the mode is below cutoff and decays exponentially with

$$\gamma_i = \alpha_i = k_{ci}\sqrt{1-(k/k_{ci})^2} \tag{2.6.3b}$$

The cross-sectional distributions for the transverse electric and magnetic fields are given by the eigenvectors $\boldsymbol{e}_i = \boldsymbol{e}_i(x, y)$ and $\boldsymbol{h}_i = \boldsymbol{h}_i(x, y)$, respectively. When we derive these eigenvectors from the respective eigenfunction we distinguish between TE- and TM-waves. For TE- or H-waves we obtain the magnetic eigenvector according to

$$\boldsymbol{h}_i^{(H)} = -\nabla_t T_i^{(H)} \tag{2.6.4a}$$

while for TM- or E-waves the electric eigenvector follows as

$$\boldsymbol{e}_i^{(E)} = -\nabla_t T_i^{(E)} \tag{2.6.4b}$$

The electric and magnetic eigenvectors of any particular mode are perpendicular to each other and related by

$$\boldsymbol{h}_i = \boldsymbol{n}_z \times \boldsymbol{e}_i \tag{2.6.4c}$$

The components of each eigenvector are chosen to be real and so normalised that they obey the following orthonormality conditions:

$$\iint \boldsymbol{e}_m \cdot \boldsymbol{e}_n \, dS_0 = \iint \boldsymbol{h}_m \cdot \boldsymbol{h}_n \, dS_0 = \delta_{mn} \tag{2.6.4d}$$

where the integration extends over the total cross-section of the waveguide.

The transverse fields of a local normal mode follow from its eigenvectors according to

$$\boldsymbol{E}_{ti} = \sqrt{Z_i}\boldsymbol{e}_i = \boldsymbol{E}_{ti}^{(+)} = \boldsymbol{E}_{ti}^{(-)} \tag{2.6.5a}$$

$$\boldsymbol{H}_{ti} = \frac{1}{\sqrt{Z_i}}\boldsymbol{h}_i = \boldsymbol{H}_{ti}^{(+)} = -\boldsymbol{H}_{ti}^{(-)} \tag{2.6.5b}$$

Z_i is the characteristic or wave impedance of the normal mode. For a H-mode it is given by

$$Z_i^{(H)} = j\omega\mu/\gamma_i \tag{2.6.6a}$$

and for an E-mode by

$$Z_i^{(E)} = \gamma_i/(j\omega\epsilon) \tag{2.6.6b}$$

The characteristic impedance relates both transverse fields of a normal mode to each other according to

$$Z_i \boldsymbol{H}_{ti} = \boldsymbol{n}_z \times \boldsymbol{E}_{ti} \tag{2.6.7}$$

The longitudinal components E_{zi} or H_{zi} of the modal fields are derived by introducing the relations 2.6.5 into Maxwell's equations.

$$\text{H-mode:}\quad H_{zi}^{(+)} = H_{zi}^{(-)} = -\frac{1}{\gamma_i\sqrt{Z_i}}\nabla_t\cdot\boldsymbol{h}_i = \frac{k_{ci}^2}{\gamma_i\sqrt{Z_i}}T_i \tag{2.6.5c}$$

$$E_{zi}^{(H)} = 0$$

$$\text{E-mode:}\quad E_{zi}^{(+)} = -E_{zi}^{(-)} = -\frac{\sqrt{Z_i}}{\gamma_i}\nabla_t\cdot\boldsymbol{e}_i = k_{ci}^2\frac{\sqrt{Z_i}}{\gamma_i}T_i \tag{2.6.5d}$$

$$H_{zi}^{(E)} = 0$$

From the normalization of the fields in eqns. 2.6.5 we have

$$R_i = 1 \tag{2.6.8a}$$

and the transmitted power for $k > k_{ci}$ is

$$P_i = 1 \tag{2.6.8b}$$

while for $k < k_{ci}$ it is

$$P_i = \pm j \tag{2.6.8c}$$

It is not always adequate to describe the field in a waveguide in terms of waves which travel in the forward and backward z-direction. In particular, for modes near cutoff ($k \approx k_{ci}$) it may be more appropriate to combine the forward and backward travelling components of a mode and express its total transverse fields in the following form:

$$\begin{aligned}\boldsymbol{E}_t &= \boldsymbol{e}_i V_i(z)\\ &= \boldsymbol{E}_{ti}\,[A_i^{(+)}(z_0)\exp(-\gamma_i(z-z_0))\\ &\quad + A_i^{(-)}(z_0)\exp(\gamma_i(z-z_0))]\end{aligned} \tag{2.6.9a}$$

$$\begin{aligned}\boldsymbol{H}_t &= \boldsymbol{h}_i I_i(z)\\ &= \boldsymbol{H}_{ti}[A_i^{(+)}(z_0)\exp(-\gamma_i(z-z_0))\\ &\quad - A_i^{(-)}(z_0)\exp(\gamma_i(z-z_0))]\end{aligned} \tag{2.6.9b}$$

The factors V_i and I_i in this representation take account of the z-dependence and are given by

$$V_i(z) = \sqrt{Z_i}\,[A_i^{(+)}(z) + A_i^{(-)}(z)] \tag{2.6.10a}$$

$$I_i(z) = \frac{1}{\sqrt{Z_i}}\,[A_i^{(+)}(z) - A_i^{(-)}(z)] \tag{2.6.10b}$$

V_i and I_i are called mode voltage and mode current, respectively. In cylindrical guides without any mode interaction they satisfy the following coupled pair of ordinary transmission-line equations

$$\frac{dV_i}{dz} = -\gamma_i Z_i I_i$$

$$\frac{dI_i}{dz} = -\gamma_i \frac{1}{Z_i} V_i$$

To supplement the above results the eigenfunctions and eigenvalues for the modes of the rectangular waveguide according to Fig. 2.7*a* and the circular waveguide according to Fig. 2.7*b* are listed in Table 2.1. In addition the eigenfunction of the TEM-mode of the coaxial line according to Fig. 2.7*c* is also included in Table 2.1. This TEM-mode derives from eqn. 2.6.2 as the *E*-mode with zero eigenvalue.

The special characteristics of the TE- and TM-modes of cylindrical waveguides with isotropic and homogeneous filling allow us to further evaluate the coupling coefficients from eqns. 2.3.9 and 2.5.5. Thus, when the permittivity and the permeability of the material inside the nonuniform waveguides are scalar and constant quantities throughout a particular local cross-section, the coupling coefficients from eqns. 2.3.9 reduce to much simpler expressions.

With

$$\frac{d\boldsymbol{H}_{tn}}{dz} = \frac{1}{\sqrt{Z_n}}\frac{d\boldsymbol{h}_n}{dz} - \frac{1}{2Z_n^{3/2}}\frac{dZ_n}{dz}\boldsymbol{h}_n$$

according to eqn. 2.6.5*b*, and with eqns. 2.6.5*a* and 2.6.4*c*, the coupling coefficients according to eqns. 2.3.9 can be written as

$$\kappa_{mn}^{(q,p)} = \frac{1}{2}\left(C_{mn}\sqrt{\frac{Z_n}{Z_m}} - qpC_{nm}\sqrt{\frac{Z_m}{Z_n}}\right) \tag{2.6.11a}$$

for $m \neq n$ and as

$$\kappa_{mm}^{(q,q)} = 0, \quad \kappa_{mm}^{(-q,q)} = C_{mm} - \frac{d\sqrt{Z_m}}{dz}\frac{1}{\sqrt{Z_m}} \tag{2.6.11b}$$

Table 2.1 *Eigenvalues and eigenfunctions of modes in rectangular and circular waveguides and in the coaxial line*

Waveguide	Mode	Eigenfunction and eigenvalue
Rectangular guide according to Fig. 2.7*a*	H_{mn}	$T_i = \frac{1}{\pi}\left\{\frac{2ab\epsilon_m\epsilon_n}{(mb)^2+(n2a)^2}\right\}^{1/2}\cos\left(\frac{n\pi}{b}y\right)\begin{cases}(-1)^{(m-1)/2}\sin\left(\frac{m\pi}{2a}x\right) & \text{for } m \text{ odd}\\ (-1)^{m/2}\cos\left(\frac{m\pi}{2a}x\right) & \text{for } m \text{ even}\end{cases}$ $k_{ci}^2 = \left(\frac{m\pi}{2a}\right)^2+\left(\frac{n\pi}{b}\right)^2 > 0$
	E_{mn}	$T_i = \frac{2}{\pi}\left\{\frac{2ab}{(mb)^2+(n2a)^2}\right\}^{1/2}\sin\left(\frac{n\pi}{b}y\right)\begin{cases}\cos\left(\frac{m\pi}{2a}x\right) & \text{for } m \text{ odd}\\ \sin\left(\frac{m\pi}{2a}x\right) & \text{for } m \text{ even}\end{cases}$ $k_{ci}^2 = \left(\frac{m\pi}{2a}\right)^2+\left(\frac{n\pi}{b}\right)^2, \quad m>0, \quad n>0$

Table 2.1 *Continued*

Waveguide	Mode	Eigenfunction and eigenvalue
Circular guide according to Fig. 2.7*b*	H_{np}	$T_i = \left\{\dfrac{\epsilon_n}{[(x'_{np})^2 - n^2]\,\pi}\right\}^{1/2} \dfrac{J_n(x'_{np}\rho/a)}{J_n(x'_{np})} \begin{cases}\sin(n\varphi)^* \\ \cos(n\varphi)\end{cases}$
		$k_{ci} = \dfrac{x'_{np}}{a}; \quad x'_{np} > 0 \quad p\text{th zero of } J'_n(x)^\dagger$
	E_{np}	$T_i = \sqrt{\epsilon_n/\pi}\,\dfrac{J_n(x_{np}\rho/a)}{x_{np}J_{n-1}(x_{np})} \begin{cases}\sin(n\varphi)^* \\ \cos(n\varphi)\end{cases}$
		$k_{ci} = \dfrac{x_{np}}{a}; \quad x_{np} > 0 \quad p\text{th zero of } J_n(x)^\dagger$
Coaxial line according to Fig. 2.7*c*	TEM	$T_i = \ln\rho/\{2\pi\ln(R/r)\}^{1/2}; \quad e_i = \nabla_t T_i$
		$k_{ci} = 0$

* $\epsilon_n = 1$ if $n = 0$; $\epsilon_n = 2$ if $n \neq 0$

† $J_n(x)$ = Bessel function; $J'_n(x) = dJ_n(x)/dx$

for $m = n$, where the coupling factors C_{mn} are given by

$$C_{mn} = \iint \frac{de_m}{dz} \cdot e_n dS_0 \tag{2.6.12}$$

By introducing eqns. 2.6.9 and 2.6.10 into the coupled wave eqns. 2.3.5 we convert to a system of coupled differential equations for the mode voltages and currents in the following form

$$\frac{d}{dz}\begin{bmatrix} V_1 \\ V_2 \\ \cdot \\ \cdot \\ I_2 \\ I_2 \\ \cdot \\ \cdot \end{bmatrix} = \begin{bmatrix} C_{11} & C_{12} & \ldots & -\gamma_1 Z_1 & 0 & \ldots \\ C_{21} & C_{22} & \ldots & 0 & -\gamma_2 Z_2 & \ldots \\ \cdot & \cdot & & \cdot & \cdot & \\ \cdot & \cdot & & \cdot & \cdot & \\ -\gamma_1/Z_1 & 0 & \ldots & -C_{11} & -C_{21} & \ldots \\ 0 & -\gamma_2/Z_2 & \ldots & -C_{12} & -C_{22} & \ldots \\ \cdot & \cdot & & \cdot & \cdot & \\ \cdot & \cdot & & \cdot & \cdot & \end{bmatrix} \begin{bmatrix} V_1 \\ V_2 \\ \cdot \\ \cdot \\ I_1 \\ I_2 \\ \cdot \\ \cdot \end{bmatrix} \tag{2.6.13}$$

For a uniform waveguide without any coupling between modes these equations reduce to coupled pairs of ordinary transmission-line equations. There is one such pair for each mode. Eqns. 2.6.13 thus represent a generalisation of the ordinary transmission-line equations and are hence referred to as generalised transmission-line equations. They prove especially useful to analyse mode interaction when one of the modes under consideration is at or near cutoff.

At cutoff, for any particular mode n we have $k_{cn} = k$ and the coupling coefficients κ_{mn} for travelling wave amplitudes become infinite in the cutoff cross-section. The coupling factors C_{mn} for mode voltages and currents remain, however, finite and even small, if the nonuniform waveguide is tapered smoothly enough. Thus, although there is only finite coupling between modal fields in their representation with mode voltages and currents, the coupling between travelling wave components of these fields becomes infinite. In such cases the field presentation in terms of forward and backward travelling waves is clearly unsuitable and one has to resort to a representation in terms of mode voltages and currents. Furthermore special methods of analysis must be applied in this case to solve the overall mode coupling in a tapered section, as will be shown in Section 6.5.

In this section the dielectric material filling the nonuniform waveguide is assumed to be homogeneous. Any coupling between modes occurs only due to the deviation of the conducting walls from a cylindrical geometry. It seems therefore appropriate to express the coupling factor of eqns. 2.6.11 in terms of the slope ϕ of the conducting boundary in the z-direction. Fig. 2.8 shows the unit vectors normal and tangential to the boundary contour and the co-ordinates which we apply here. The unit vectors $\boldsymbol{n}_S$ and $\boldsymbol{l}_S$ are defined as perpendicular to the z-co-ordinate while $\boldsymbol{n}_0$ is the unit vector normal to the boundary wall. $\boldsymbol{n}_0$ and $\boldsymbol{n}_S$ point outward from the waveguide. $\boldsymbol{l}_S$ is taken tangential to the wall, so that we have

$$\boldsymbol{n}_S \times \boldsymbol{l}_S = \boldsymbol{n}_z \tag{2.6.14a}$$

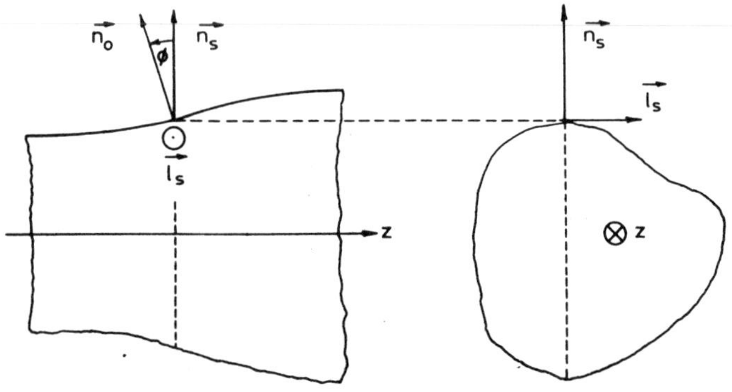

Fig. 2.8 *Nonuniform waveguide section*

ϕ is measured in the plane perpendicular to $\boldsymbol{l}_S$ according to

$$(\boldsymbol{n}_S \times \boldsymbol{n}_0)\cdot\boldsymbol{l}_S = \sin\phi \tag{2.6.14b}$$

By applying Green's identity in two dimensions

$$\iint [(\nabla_t\cdot u)(\nabla_t\cdot v) + u\Delta v]\, dS_0 = \oint u(\nabla_t\cdot v)\cdot\boldsymbol{n}_S\cdot d\boldsymbol{l}_S \tag{2.6.14c}$$

to the waveguide cross-section S_0, such that $\oint dl_S$ denotes integration along the boundary of the local cross-section, and by utilising eqn. 2.6.2 with the various boundary conditions for the eigenfunctions and the orthogonality relations 2.6.4d, we obtain formulas for the coupling factors as they are listed in Table 2.2. A detailed derivation of these formulas is given in Reference 5.

Before we conclude this Section we evaluate the coupling factors for the typical nonuniformities in Fig. 2.9. We restrict this evaluation to coupling between those modes that under practical circumstances are excited in the respective waveguide. The resulting coupling formulas are needed in the following chapters of the book.

Table 2.2 *Coupling factors of a nonuniform waveguide with homogeneous dielectric according to Fig. 2.8*

Index n, m denoting an E-, H-mode	Coupling factor
m E-mode n E-mode $m \neq n$	$C_{mn} = \dfrac{k_{cm}^2}{k_{cn}^2 - k_{cm}^2} \oint_L \tan\phi \dfrac{\partial T_m}{\partial n_s} \dfrac{\partial T_n}{\partial n_s} dl_s$
$m = n$ E-mode	$C_{mm} = -\dfrac{1}{2} \oint_L \tan\phi \left(\dfrac{\partial T_m}{\partial n_s}\right)^2 dl_s$
m H-mode n H-mode $m \neq n$	$C_{mn} = \dfrac{k_{cn}^2}{k_{cm}^2 - k_{cn}^2} \oint_L \tan\phi\, T_n \dfrac{\partial^2 T_m}{\partial n_s^2} dl_s$
$m = n$ H-mode	$C_{mm} = -\dfrac{1}{2} \oint_L \tan\phi \left(\dfrac{\partial T_m}{\partial l_s}\right)^2 dl_s$
m E-mode n H-mode	$C_{mn} = -\oint_L \tan\phi \dfrac{\partial T_m}{\partial n_s} \dfrac{\partial T_n}{\partial l_s} dl_s$ $C_{nm} = 0$

$\oint_L dl_s$ denotes integration along the boundary of the cross-section

$$\frac{\partial T}{\partial n_s} = (\nabla_t T)\cdot n_s; \quad \frac{\partial T}{\partial l_s} = (\nabla_t T)\cdot l_s; \quad n_s \times l_s = n_z; \quad \sin\phi = (n_s \times n_o)\cdot l_s$$

A coaxial line is ordinarily excited only in its TEM-mode, and only this mode propagates while all higher-order modes are below cutoff and evanescent. We therefore consider interaction only between TEM-fields and neglect any excitation of higher-order evanescent modes. The coupling factor between the forward and backward TEM-waves are calculated from eqn. 2.6.11 and given by

$$\kappa^{(+-)} = \kappa^{(-+)} = \frac{\dfrac{1}{r}\dfrac{dr}{dz} - \dfrac{1}{R}\dfrac{dR}{dz}}{2\ln(R/r)}; \quad \kappa^{(++)} = \kappa^{(--)} = 0 \quad (2.6.15)$$

R and r define the cross-section of the coaxial line according to Fig. 2.9*a*. As will be shown in Section 5.1, this result can also be obtained by converting the well known transmission-line equations for the current and voltage of the TEM-mode to the coupled wave equations in terms of forward and backward wave amplitudes of the TEM-mode.

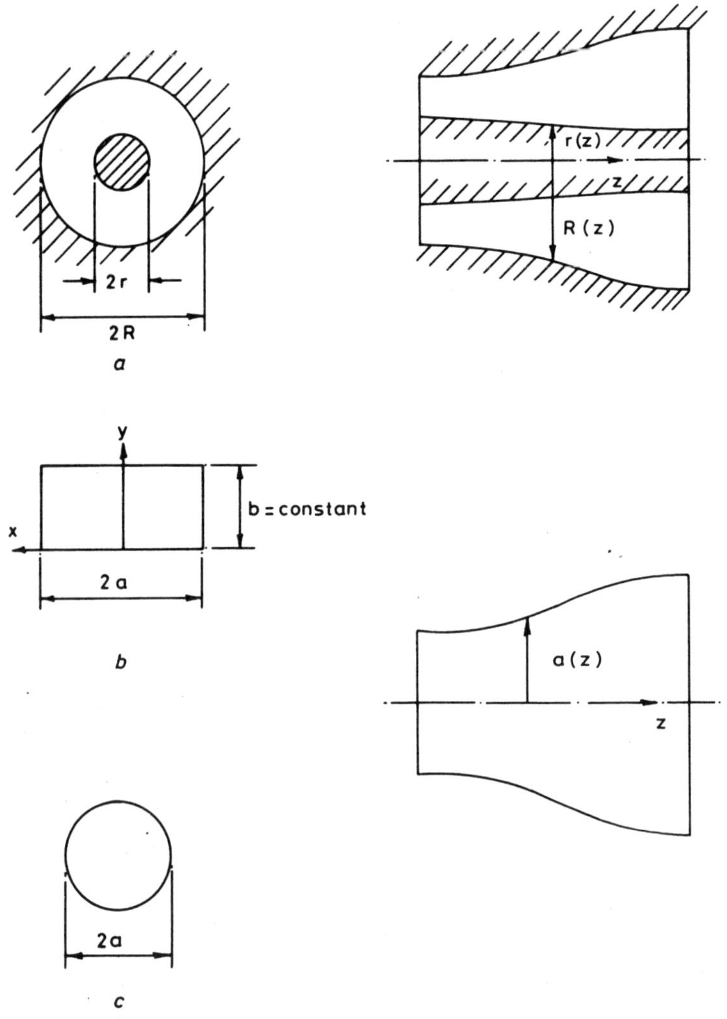

Fig. 2.9 *Examples for nonuniform waveguide sections*

The rectangular transition section of Fig. 2.9*b* is assumed to be symmetrical with respect to its centre plane $x = 0$ and its height b is constant. Thus all coupling effects are due to the change of the width $2a$ of the broad side as a function of z:

$$\frac{da}{dz} = a' \neq 0$$

In the following we shall consider only fields which are excited when the fundamental (H_{10})-mode is incident. This mode excites only H_{30}, H_{50} . . .-modes. Thus only the coupling factors for these H_{n0}-modes with $n = 1, 3, 5$ . . . need to be evaluated. With

$$x_n = n\pi/2 \tag{2.6.16}$$

we obtain from the results of Tables 2.1 and 2.2

$$C_{nn} = 0 \tag{2.6.17a}$$

$$C_{mn} = \frac{2x_m x_n}{x_n^2 - x_m^2}\frac{a'}{a} \tag{2.6.17b}$$

The coupling in the circular transitions according to Fig. 2.9*c* is due to the change of the radius a as a function of z

$$\frac{da}{dz} = a' \neq 0$$

In subsequent chapters we shall be interested only in those fields which are excited by the H_{01}-mode and which contain only H_{0n}-modes with $n = 1, 2, 3$ . . . .

With

$$x_n = x'_{0n}; \quad J'_0(x'_{0n}) = 0 \tag{2.6.18}$$

according to Table 2.1 we obtain with Tables 2.1 and 2.2 the coupling factor C_{mn} of eqns. 2.6.17.

2.7 Multihole couplers

In the preceding Sections mode-coupling structures were considered for which complete orthogonal systems of normal modes with their corresponding normal-mode fields for all cross-sections along the nonuniform waveguide are either known, or assumed to be known.

There are, however, waveguide applications where we lack this knowledge, and have not even approximations for normal modes available. In this Section as well as in the following we investigate two such examples, and calculate the coupled wave equations for them. The overall transmission properties of these coupling structures will be analysed in a later chapter.

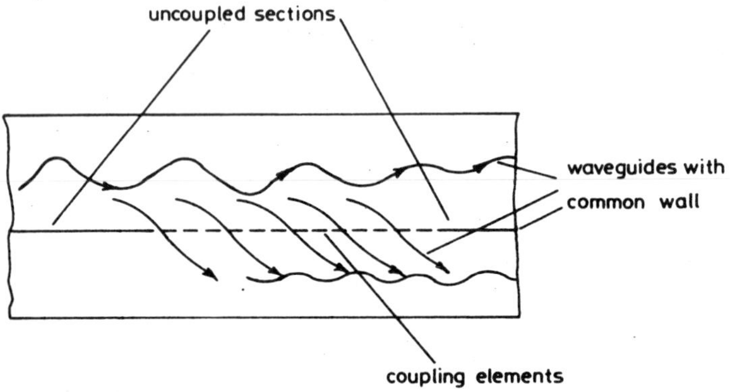

Fig. 2.10 *Longitudinal cut through a coupled waveguide line*

First, let us look at the directional coupler of Fig. 2.10. Two waveguides with constant (not necessarily equal) cross-sections are coupled through apertures in the common wall. Let us, at the moment, assume that the coupling element is a longitudinal slot with its width varying slowly as a function of the axial distance z. We can regard these two coupled waveguides as just one waveguide with its cross-section consisting of two parts connected by the slot region. The normal modes of this waveguide are not readily known to us. At the ends of the coupling section, where the slot is closed, the total fields can be described in terms of normal modes of the separate waveguide ports. To calculate the overall coupling of these modes through the slot it is best to describe the field in the coupled cross-sections also in terms of the normal modes of the separate waveguides.

In the coupling region, and in particular within the slot region, this set of normal modes is, however, not complete in the sense that any possible field distribution can be described in terms of fields of these normal modes. The normal modes of the separate waveguide modes have in particular no electric field component parallel to the conducting walls of the undisturbed waveguide. At the slot they hence cannot account for the electric field component in the y-direction of Fig. 2.11. Thus we need, in addition, slot waves to account for this field component and to avoid convergence problems.

The individual field of any of these modes in either one of the coupled waveguides, or of the slot waves does not, however, in general satisfy Maxwell's equations under the existing boundary conditions of the coupled cross-sections. This is the reason why coupling between these modes must be introduced, even when the coupling structure is uniform with the cross-section shown in Fig. 2.11 not changing with z. To calculate this mode coupling we can again introduce fictitious sources which allow a single mode to propagate undisturbed. The compensating sources of opposite sign constitute sources which generate spurious mode. It is, however, difficult to determine the spurious mode generation because the fields of the individual modes are not orthogonal with respect to each other. Eqn. 2.2.7 cannot therefore be applied directly.

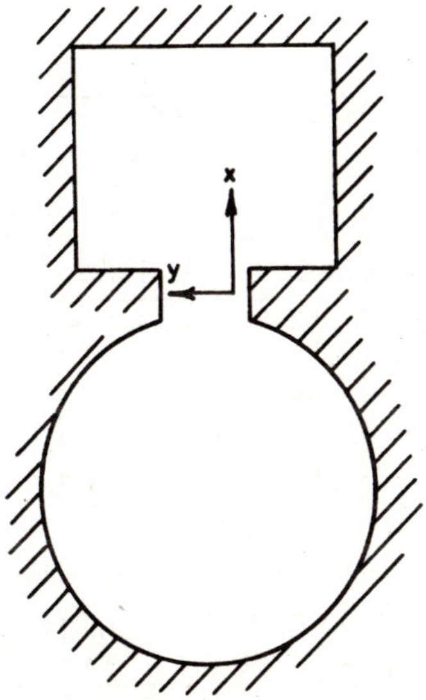

Fig. 2.11 *Cross-section of coupled waveguide line*

After this brief discussion of the difficulties which are encountered when coupled mode propagation with nonorthogonal fields is analysed, we shall simplify our present problem. In practice, waveguide coupling by continuous long slots has several disadvantages. The main problem occurs due to the reflection of the fundamental slot wave at the ends of the coupling sections. Such reflections degrade the directivity and the wideband performance of the coupler. Therefore the long coupling slot is usually replaced by a number of coupling apertures which do not allow any aperture waves to propagate. In most cases the coupling is accomplished by a series of discrete holes. Accordingly, we shall investigate the coupling properties of an array of round coupling holes with small spacings between them and with the overall coupling per unit length varying slowly as a function of z.

Let us at first look at a single hole with diameter $2r$ in the ideally conducting wall of thickness t between waveguides I and II at $z = z_1$ as shown in Fig. 2.12. The coupling effects of small elliptic holes in thin walls is treated in detail for example in Reference 2. To simplify the notation for the coupled modes we denote any mode in the undisturbed waveguide I and II by a different subscript i which is positive for waves travelling in the positive z-direction and negative for waves travelling in the negative z-direction. In addition we choose all values of R_i of eqn. 2.2.2a equal to unity

$$R_i = 1 \tag{2.7.1}$$

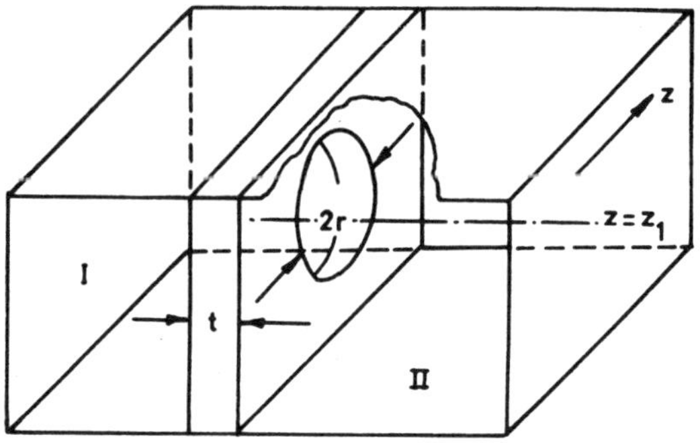

Fig. 2.12 *Coupling hole in the common thick wall between two waveguides*

Suppose now that only the field of the normal mode n is present in the region $z < z_1 - r$ of the structure in Fig. 2.12. If the hole were closed this mode would have the following field components on the waveguide wall near the centre of the hole:

$$E = A_n(z_1)E_{hn} \tag{2.7.2a}$$

$$H = A_n(z_1)H_{hn} \tag{2.7.2b}$$

$A_n(z_1)$ represents the amplitude of mode n and E_{hn}, H_{hn} are its normalised fields at the centre of the hole. The effect of the hole on the excitation of fields in the region $z > z_1 + r$ can be described by fictitious dipole sources on the waveguide wall, where the centre of the hole had been, before it was closed.[2] The moments of these electric and magnetic dipoles $\boldsymbol{m}$ and $\boldsymbol{i}$ respectively, are related to the incident field of eqns. 2.7.2 as follows:

Dipoles in the waveguide in which mode n is incident have

$$\boldsymbol{i}_e = -j\omega\epsilon p_{ie}\boldsymbol{E}_{hn}A_n(z_1) \tag{2.7.3a}$$

$$\boldsymbol{m}_e = j\omega\mu p_{me}\boldsymbol{H}_{hn}A_n(z_1) \tag{2.7.3b}$$

Dipoles in the other waveguide have

$$\boldsymbol{i}_0 = -j\omega\epsilon p_{io}\boldsymbol{E}_{hn}A_n(z_1) \tag{2.7.3c}$$

$$\boldsymbol{m}_0 = j\omega\mu p_{mo}\boldsymbol{H}_{hn}A_n(z_1) \tag{2.7.3d}$$

The following relations hold for the real and positive scalar factors $p_{ie}, p_{me}, p_{io}, p_{mo}$ of these dipole formulas

$$p_{ie} = \tfrac{2}{3}r^3R_IK_{Ie} \tag{2.7.4a}$$

$$p_{me} = \tfrac{4}{3}r^3R_MK_{Me} \tag{2.7.4b}$$

$$p_{io} = -\tfrac{2}{3}r^3R_IK_{Io} \tag{2.7.4c}$$

$$p_{mo} = -\tfrac{4}{3}r^3R_MK_{Mo} \tag{2.7.4d}$$

For small holes in thin walls the values of p follow from the static polarisability of small dielectric or magnetic discs as shown for example in Reference 2. In this case the factors R and K are equal to unity. For larger holes in thick walls the factors R and K are listed in Table 2.3. The formulas of this table have been obtained by applying the results of Reference 6 for rectangular coupling holes to circular holes. They offer sufficient accuracy as long as the holes are small enough for

$$kr \leqslant 1$$

to be satisfied.

To determine the mode excitation caused by the dipoles of eqns. 2.7.3 we apply again eqns. 2.2.7, and obtain for the amplitude $\Delta A_m(z_1)$, with which mode m is excited:

$$\Delta A_m(z_1) = -(\boldsymbol{i}\cdot\boldsymbol{E}_{h(-m)} - \boldsymbol{m}\cdot\boldsymbol{H}_{h(-m)})\frac{\text{sign}\,(m)}{2} \tag{2.7.5}$$

The cross-section at $z = z_1$ represents the plane of reference and

$$\text{sign}\,(m) = 1 \text{ holds for } m > 0, \tag{2.7.6a}$$

while

$$\text{sign}\,(m) = -1 \text{ is valid for } m < 0 \tag{2.7.6b}$$

$\boldsymbol{E}_{h(-m)}$, $\boldsymbol{H}_{h(-m)}$ represent the normalised field components of mode $-m$ on the waveguide wall at the centre of the closed hole. If the subscripts

n and m denote modes in the same waveguide the dipoles $\boldsymbol{i}_e$ and $\boldsymbol{m}_e$ of eqns. 2.7.3*a*, b have to be introduced into eqn. 2.7.5. Otherwise the dipoles $\boldsymbol{i}_0, \boldsymbol{m}_0$ must be taken.

Table 2.3 *Coefficients of the polarisability for circular holes*

$$R_M = \frac{1}{1-0{\cdot}4(kr)^2}$$

$$R_I = \frac{1}{1+0{\cdot}4(kr)^2}$$

$$K_{Me} = 1-(0{\cdot}35+0{\cdot}15(kr)^2)\,[1-\exp(-2\gamma^{(H)}t)]$$

$$K_{Mo} = \exp(-\gamma^{(H)}t)\,[1-\{0{\cdot}1+(0{\cdot}3+0{\cdot}15(kr)^2)^2\}\{1-\exp(-2\gamma^{(H)}t)\}]$$

with

$$\gamma^{(H)} = \left\{\left(\frac{1{\cdot}84}{r}\right)^2 - k^2\right\}^{\frac{1}{2}}$$

$$K_{Ie} = 1-0{\cdot}13\exp(-2\gamma^{(E)}t)$$

$$K_{Io} = \exp(-\gamma^{(E)}t)\,[1-0{\cdot}14\{1-\exp(-2\gamma^{(E)}t)\}]$$

with

$$\gamma^{(E)} = \left\{\left(\frac{2{\cdot}405}{r}\right)^2 - k^2\right\}^{\frac{1}{2}}$$

The preceding results of hole coupling can be summarised by the following vector equation:

$$\boldsymbol{\Delta} A(z_1) = \mathbf{T}A(z_1) \tag{2.7.7a}$$

$\boldsymbol{\Delta}A(z_1)$ is a vector which has the individual wave amplitudes $\Delta A_m(z_1)$ of the excited modes as its components; the vector $A(z_1)$ has as its components the amplitudes $A_n(z_1)$ of the incident modes. The elements T_{mn} of the matrix $\mathbf{T}$ follow from eqns. 2.7.3 and 2.7.5; they are given by

$$\text{sign}\,(m)T_{mn} = \frac{j\omega\epsilon}{2}p_i\boldsymbol{E}_{hn}\cdot\boldsymbol{E}_{h(-m)} + \frac{j\omega\mu}{2}p_m\boldsymbol{H}_{hn}\cdot\boldsymbol{H}_{h(-m)} \tag{2.7.7b}$$

$p_i = p_{ie}$ and $p_m = p_{me}$ must be taken if both modes are normal modes of the same waveguide. Else $p_i = p_{io}$ and $p_m = p_{mo}$ have to be applied.

We proceed now to consider a coupling structure which consists of an array of many coupling elements with the overall coupling characteristics varying slowly as a function of the axial distance z. The overall coupling of this structure is conveniently analysed with the help of its coupled wave equations. To set up these equations the discrete coupling according to eqn. 2.7.7 must be replaced by an equivalent distributed coupling along the coupling region.

To accomplish this replacement we consider a short section of coupled lines between $z = z_1 - s/2$ and $z = z_1 + s/2$ and assume its coupled wave eqns. 2.4.1b to have the form

$$\frac{d}{dz}A = (-\gamma + \kappa)A \tag{2.7.8}$$

The elements $\gamma_{nn} = \gamma_n$ and κ_{mn} of the coupling matrix are assumed to be constant along this line section. In addition we assume that the overall coupling between the individual modes in this section is small. Under these assumptions we can solve eqns. 2.7.8 using an iterative approach. As a first approximation we neglect any mode coupling and consider only a mode n to be incident in one waveguide and propagate through the short coupling section according to

$$A_n(z) = A_n(z_1)\exp[-\gamma_n(z - z_1)] \tag{2.7.9a}$$

To calculate the coupling of this nth mode to the mth mode we integrate the coupled wave equation

$$\frac{d}{dz}A_m = -\gamma_m A_m + \kappa_{mn}A_n \tag{2.7.9b}$$

between $z = z_1 - s/2$ and $z = z_1 + s/2$. We thereby obtain the amplitude ΔA_m with which mode m is excited.

$$\Delta A_m(z_1) = \int_{z_1 - s/2}^{z_1 + s/2} A_n(z)\kappa_{mn}\exp[-\gamma_m(z_1 - z)]\,dz \tag{2.7.9c}$$

$z = z_1$ is taken here as a reference cross-section. When we replace $\gamma_{m,n}$ by

$$\gamma_m = j\beta_m; \quad \gamma_n = j\beta_n \tag{2.7.10}$$

and introduce eqn. 2.7.9a into eqn. 2.7.9c we obtain

$$\Delta A_m(z_1) = A_n(z_1)\kappa_{mn}s\,\frac{\sin[(\beta_m \quad \beta_n)s/2]}{(\beta_m - \beta_n)s/2} \tag{2.7.9d}$$

Comparing now eqn. 2.7.9*d* with eqn. 2.7.7 the coupling coefficients κ can be calculated for any array of equal coupling holes with spacing s between the holes.

$$\kappa_{mn} = \frac{T_{mn}}{s} \frac{(\beta_m - \beta_n)s/2}{\sin[(\beta_m - \beta_n)s/2]} \tag{2.7.10b}$$

This result can be generalised and also applied to coupling arrays with slowly varying properties of the coupling elements as a function of the axial distance z.

As an example for multihole couplers we treat the coupling between a rectangular and a round waveguide in Fig. 2.13. Only those field distributions are considered which are excited by the fundamental mode of the rectangular guide. Because of the symmetry of the coupling arrangement in Fig. 2.13 these field distributions do not contain any E_x- or H_y-field components in the plane $y = 0$. Thus the respective waveguide modes are coupled only by the components H_{zi} of their undisturbed normal mode fields at the holes. For the fundamental H_{10}-mode of the rectangular guide, and according to Section 2.6, this field component is given by

$$H_{zi} = H_{z(-i)} = \frac{k_{ci}}{(j\omega\mu\gamma_{|i|})^{1/2}} \frac{1}{\sqrt{\hat{a}b}} = \frac{h_i}{(j\omega\mu\gamma_{|i|})^{1/2}} \tag{2.7.11a}$$

For the H_{np}-mode of the circular guide it is given by

$$H_{zi} = H_{z(-i)} = \frac{k_{ci}}{(j\omega\mu\gamma_{|i|})^{1/2}} \sqrt{\frac{\epsilon_n}{\pi a^2}} \frac{1}{\{1-(n/k_{ci}a)^2\}^{1/2}} = \frac{h_i}{\{j\omega\mu\gamma_{|i|}\}^{1/2}} \tag{2.7.11b}$$

k_{ci} is the cutoff wave number of the mode i, according to Table 2.1 with i positive. The propagation constant of this mode is

$$-\gamma_{(-i)} = \gamma_i = (k_{ci}^2 - k^2)^{1/2} = j(k^2 - k_{ci}^2)^{1/2} = j\beta_i$$
$$= -j\beta_{(-i)} \tag{2.7.11c}$$

ϵ_n represents Neumann's number.

$\hat{a}$, b and a denote width, height and diameter, respectively, of the coupled waveguides.

If in eqn. 2.7.10*b* we have

$$(\beta_m - \beta_n)s \ll \pi \tag{2.7.11d}$$

then κ_{mn} obtains as

$$\kappa_{mn} = \pm\frac{2}{3}\frac{r^3}{s}R_M K_M h_{|m|}h_{|n|}\,\text{sign}\,(m)\frac{1}{(\gamma_{|n|}\gamma_{|m|})^{1/2}} \tag{2.7.11e}$$

where the plus sign and $K_M = K_{Me}$ must be taken for interaction between waves of the same waveguide, and else the minus sign and $K_M = K_{Mo}$ is valid.

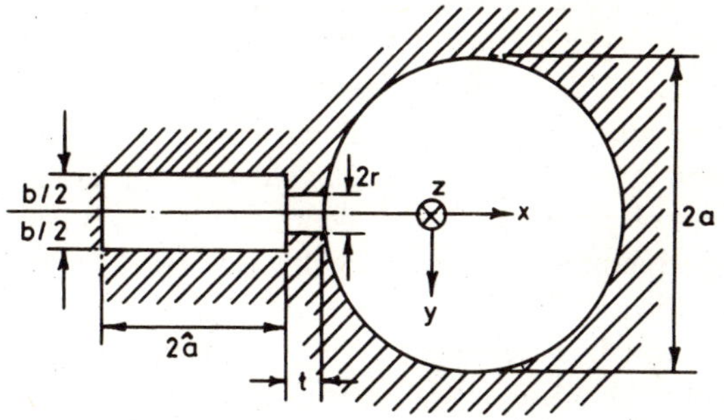

Fig. 2.13 *Cross-section of multihole coupler between rectangular and round waveguide*

The real, positive factors R_M and K_{Me}, K_{Mo} are given in Table 2.3. $h_{|m|}$ and $h_{|n|}$ are positive real constants according to eqns. 2.7.11*a* and 2.7.11*b*. If the two coupled modes are above cutoff, then

$$(\gamma_{|m|}\gamma_{|n|})^{1/2} = j(\beta_{|m|}\beta_{|n|})^{1/2}$$

must be substituted in eqn. 2.7.11*e*, with real and positive phase constants $\beta_{|m|}$ and $\beta_{|n|}$.

2.8 Coupled dielectric waveguides

Dielectric waveguides serve as guiding structures mainly for electromagnetic waves at optical frequencies. Quite often in these applications two or more such optical waveguides run parallel to each other and interact with one another by way of their transversely evanescent fields.

The interaction leads to power transfer from a mode on one of the parallel waveguides to a mode on another waveguide. In optical circuits or cables such power transfer may degrade the desired performance. It then needs to be reduced to a low enough value by spacing parallel waveguides sufficiently apart or by otherwise reducing the interaction.

In some applications, however, waveguides in optical circuits are brought close together and parallel to each other to intentionally transfer power in a specific way between particular modes of both guides. Such waveguide couplers are used to monitor the power in one waveguide by way of a secondary waveguide, or, to launch a specific mode into a waveguide from the mode of another waveguide. These represent but two examples for the numerous and varied applications which waveguide couplers find in optical circuits.

To avoid parasitic coupling between parallel waveguides or to design waveguide couplers we need to analyse the interaction between any two modes of such guides and find its dependence on waveguide and mode parameters. For this purpose we consider two parallel dielectric waveguides, and begin by defining first the individual optical waveguide and its normal modes of propagation. This definition may not include all possible guiding structures of practical significance for light waves. All relevant characteristics of mode propagation and mode coupling in optical waveguides will, however, be described by this model.

We shall regard as a dielectric optical waveguide a suitable cylindrical distribution of linear, isotropic matter with permittivity $\epsilon_{\mathrm{I}}(x, y)$ which is a function of the transverse co-ordinates x, y only, and which does not change as a function of the axial distance z. This guide is assumed to be confined to a certain region of x and y. Outside this region the permittivity is assumed to be constant and equal to ϵ. Normally we have $\epsilon_{\mathrm{I}}(x,y) \geqslant \epsilon$. To obtain a waveguide model, that complies with requirements in previous sections of this book, we enclose the actual optical waveguide in cylindrical walls of perfect conductivity. These opaque walls are placed at a sufficient transverse distance from the light guiding region of optical waveguide with $\epsilon_{\mathrm{I}} > \epsilon$ so that most of the transversely evanescent waves which are guided within the cross-section with $\epsilon_{\mathrm{I}} > \epsilon$ have only extremely low fields at the walls. The permeability μ is assumed to be constant over the guide cross-section. The guide is assumed to be lossless.

Fig. 2.14 shows the cross-section of this dielectric waveguide model. In it, source-free field distributions can again be described by the fields of normal modes, which propagate independently of one another.

$$\boldsymbol{E}(x,y,z) = \sum_i \boldsymbol{E}_i(x,y) A_i(z_1) \exp\left[-\gamma_i(z-z_1)\right] \qquad (2.8.1a)$$

$$\boldsymbol{H}(x,y,z) = \sum_i \boldsymbol{H}_i(x,y) A_i(z_1) \exp\left[-\gamma_i(z-z_1)\right] \qquad (2.8.1b)$$

For convenience, waves which propagate in the positive z-direction are denoted by positive subscripts i while the respective negative subscripts

$-i$ designate waves propagating in the negative z-direction. Thus, if the subscript i represents the forward travelling mode i, the subscript $-i$ represents its replica travelling in opposite direction.

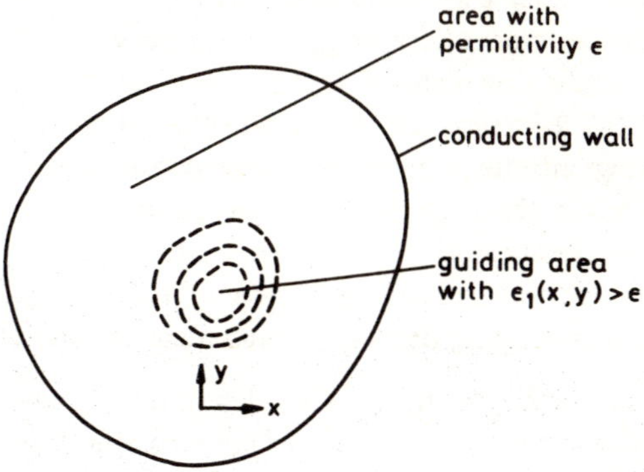

Fig. 2.14 *Cross-section of an optical guide model surrounded by conducting walls*

By a suitable normalisation of the eigenfields the following orthonormality relation can be satisfied:

$$\frac{1}{2}\iint (\boldsymbol{E}_m \times \boldsymbol{H}_n + \boldsymbol{E}_n \times \boldsymbol{H}_m)\cdot\boldsymbol{n}_z\, dS_0 = \delta_{mn}\ \text{sign}\ (m) \tag{2.8.2a}$$

Here the integration $\iint dS_0$ extends over the total cross-section and δ_{mn} denotes the Kronecker symbol. The value of sign (m) is given by

$$\begin{aligned} \text{sign}\ (m) &= 1 \qquad \text{for} \quad m > 0 \\ \text{sign}\ (m) &= -1 \qquad \text{for} \quad m < 0 \end{aligned} \tag{2.8.2b}$$

According to eqns. 2.2.9 of Section 2.2, the transverse normal mode fields $\boldsymbol{E}_{ti}$, $\boldsymbol{H}_{ti}$ and the longitudinal field components E_{zi}, H_{zi} of the forward and backward waves are related by

$$\begin{aligned} \boldsymbol{E}_{ti} &= \boldsymbol{E}_{t(-i)} \qquad \boldsymbol{H}_{ti} = -\boldsymbol{H}_{t(-i)} \\ E_{zi} &= -E_{z(-i)} \quad\ H_{zi} = H_{z(-i)} \end{aligned} \tag{2.8.2c}$$

Referring to the normal mode fields at the conducting walls far away from the guiding area with $\epsilon_I > \epsilon$, we can distinguish two types of normal modes of the waveguide model:

(*a*) There exists a limited number of normal modes which have fields that decrease exponentially as a function of the distance from the guiding area. The field energy of these modes is concentrated in or near the guiding area with $\epsilon_I > \epsilon$. The propagation constant γ_i of these modes is purely imaginary for lossless guides.
(*b*) All the other modes have fields extending over the full cross-section out to the conducting walls. A finite number of them are propagating modes and, in the lossless case, have purely imaginary values of γ_i, while the remaining infinite number is below cutoff and evanescent with purely real γ_is in the lossless case. If the distance between the guiding area and the conducting wall is extended to infinity, then this set of modes forms the continuous spectrum of space or radiating modes, with which any radiating fields are represented (Reference 7).

Note that, in general, for the rigorous description of an electromagnetic field distribution the normal modes of type (*a*) as well as the infinite set of modes of type (*b*) are needed.

We shall denote the complete waveguide model as defined by $\epsilon_I(x, y)$ and including the conducting wall as waveguide I. In the following we shall assume that the complete set of modes of this guide is known. Let us now define a waveguide II with the same conducting wall in the identical location as for waveguide I, but with a different distribution of matter with permittivity $\epsilon_{II}(x, y)$. The part of the cross-section of waveguide I with $\epsilon_I(x, y) \neq \epsilon$ and that of waveguide II with $\epsilon_{II}(x, y) \neq \epsilon$ are assumed to be separated from each other inside the same conducting walls. The complete set of normal modes of waveguide II is supposed to be known as well.

Let any subscript i denote only one mode of either waveguide I or waveguide II. Its field, which according to eqn. 2.8.1 is written as

$$\boldsymbol{E}_i = \boldsymbol{E}_i(x, y) \exp[-\gamma_i(z - z_1)] A_i(z_1) \qquad (2.8.3a)$$

$$\boldsymbol{H}_i = \boldsymbol{H}_i(x, y) \exp[-\gamma_i(z - z_1)] A_i(z_1) \qquad (2.8.3b)$$

will by itself satisfy Maxwell's equation in the corresponding waveguide I or II

$$\nabla \times \boldsymbol{H}_i = j\omega\epsilon_i(x, y)\boldsymbol{E}_i \qquad (2.8.3c)$$

$$-\nabla \times \boldsymbol{E}_i = j\omega\mu\boldsymbol{H}_i \qquad (2.8.3d)$$

where $\epsilon_i(x, y)$ is defined by

$$\epsilon_i(x, y) = \epsilon_I(x, y) \qquad (2.8.4a)$$

if the ith mode is a mode of waveguide I and by

$$\epsilon_i(x, y) = \epsilon_{II}(x, y) \tag{2.8.4b}$$

if this mode belongs to waveguide II.

These definitions and notation will now be applied to the cylindrical structure of which Fig. 2.15 shows the cross-section. This structure encloses both dielectric waveguides I and II within a common wall of perfect conductivity.

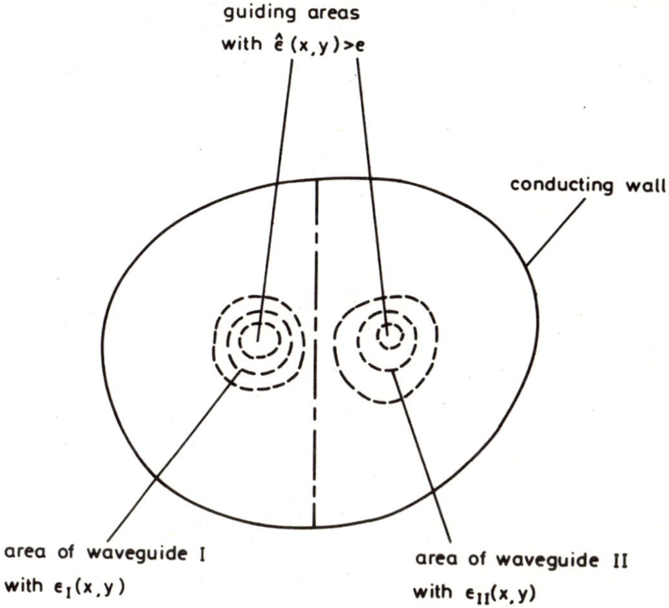

Fig. 2.15 *Cross-section of waveguide I and waveguide II laid in parallel*

The parallel arrangement of both waveguides within the common wall is described by a cross-sectional distribution $\hat{\epsilon}(x, y)$ of isotropic permittivity with

$$\hat{\epsilon}(x, y) = \epsilon_I(x, y) \tag{2.8.4c}$$

in the region of waveguide I, and

$$\hat{\epsilon}(x, y) = \epsilon_{II}(x, y) \tag{2.8.4d}$$

in the region of waveguide II, but

$$\hat{\epsilon}(x, y) = \epsilon \tag{2.8.4e}$$

over the remaining part of the cross-section.

The two parallel dielectric waveguides together with their enclosure of perfect conductivity form a uniform cylindrical structure. A complete set of normal modes exists for this structure which propagate independently of one another.

Usually, however, these modes are not readily known to us. But we also do not really need them. In the structure of Fig. 2.15 both waveguides I and II couple with each other and power converts between modes of the separate waveguides I and II. It is therefore, more appropriate to describe the fields in the structure of Fig. 2.15 by the superposition of the normal modes of waveguides I and II. In the following this superposition will be needed for the transverse fields in any given cross-section with $z = z_1$:

$$\boldsymbol{E}_t = \sum_i \boldsymbol{E}_{ti}(x,y)A_i(z_1) \tag{2.8.5a}$$

$$\boldsymbol{H}_t = \sum_i \boldsymbol{H}_{ti}(x,y)A_i(z_1) \tag{2.8.5b}$$

Note that the summations include terms with positive and negative subscripts i representing modes, which propagate in the positive and negative z-direction, respectively.

Although this set of normal mode fields is not orthogonal, it should be recognised that these fields are nevertheless well suited to describe the field in a given cross-section and that also the series in eqn. 2.8.5 will converge quite rapidly for any actual field $\boldsymbol{E}_t, \boldsymbol{H}_t$.

The individual normal mode fields by themselves do not satisfy Maxwell's equations in the guide of Fig. 2.15. We therefore have coupling between these modes. To determine this coupling we resort to the concept of coupled wave equations. By writing these coupled wave equations in the form of eqn. 2.4.1b we obtain

$$\frac{d}{dz}A = (-\gamma + \kappa)A \tag{2.8.6}$$

where the elements of the diagonal matrix γ represent the undisturbed propagation constants of the modes in waveguide I or II. The corresponding coupling coefficients κ_{mn} are calculated using the principle of effective sources.

We can make any mode with subscript n to propagate undisturbed if we impress electric currents of density $-\boldsymbol{i}_n = -\boldsymbol{i}_n(x, y, z)$ which is so chosen, that Maxwell's equations in the medium with $\hat{\epsilon}$ are satisfied. Considering eqn. 2.8.3c we obtain

$$\nabla \times \boldsymbol{H}_n = j\omega\epsilon_n(x,y)\boldsymbol{E}_n = j\omega\hat{\epsilon}(x,y)\boldsymbol{E}_n - \boldsymbol{i}_n,$$

and the effective source is given by

$$-\boldsymbol{i}_n = j\omega[\epsilon_n(x,y) - \hat{\epsilon}(x,y)]\,\boldsymbol{E}_n \tag{2.8.7}$$

This is the only current source which has to be impressed to make the nth mode to propagate independently. Since μ is unchanged the second Maxwell's eqns. 2.8.3d is satisfied in any case.

For source-free field distributions the current density $+\boldsymbol{i}_n$ can be regarded as the origin of mode coupling. After introducing eqn. 2.8.3a into eqn. 2.8.7 the value of $\boldsymbol{i}_n$ at z is obtained as

$$\boldsymbol{i}_n(x,y) = -j\omega\Delta\epsilon_n(x,y)\boldsymbol{E}_n(x,y)A_n(z) \tag{2.8.8a}$$

with

$$\Delta\epsilon_n(x,y) = \epsilon_n(x,y) - \hat{\epsilon}(x,y) \tag{2.8.9}$$

We have

$$\Delta\epsilon_n(x,y) = \epsilon - \epsilon_{\mathrm{I}}(x,y) \text{ if } n \text{ denotes a mode of waveguide II}$$

$$\Delta\epsilon_n(x,y) = \epsilon - \epsilon_{\mathrm{II}}(x,y) \text{ if } n \text{ denotes a mode of waveguide I}$$

We now consider sources $\boldsymbol{i}_n$ according to eqn. 2.8.8a in the short region $z_1 - \Delta z < z < z_1$, and the transverse components $\boldsymbol{E}'_{tn}$ and $\boldsymbol{H}'_{tn}$ of the fields which these sources generate in the cross-section at $z = z_1$. In our coupled wave representation these transverse fields $\boldsymbol{E}'_{tn}$ and $\boldsymbol{H}'_{tn}$ are formed by the superposition of the fields of all normal modes i which, according to the coupled wave equations, are generated by their interaction with the nth mode in the region $z_1 - \Delta z < z < z_1$. In the limit of $\Delta z \to 0$ the coupled wave equations yield for these transverse fields

$$\boldsymbol{E}'_{tn} = \sum_i \boldsymbol{E}_{ti}\,\Delta z \kappa_{in} A_n(z_1) \tag{2.8.8b}$$

$$\boldsymbol{H}'_{tn} = \sum_i \boldsymbol{H}_{ti}\,\Delta z \kappa_{in} A_n(z_1) \tag{2.8.8c}$$

Note that $\boldsymbol{i}_n$ is assumed to radiate only in the positive z-direction. This can in general only be achieved by introducing additional sources S_{na} in the region $z > z_1$ which compensate for the overall field terms propagating into the region $z < z_1 - \Delta z$.

To calculate the coupling from the nth mode to the mth mode we introduce at $z = z_1$ surface currents of magnitude

$$\boldsymbol{J}_{(-m)}(x,y) = -\boldsymbol{n}_z \times \boldsymbol{H}_{t(-m)}(x,y) \tag{2.8.10a}$$

$$\boldsymbol{M}_{(-m)}(x,y) = \boldsymbol{n}_z \times \boldsymbol{E}_{t(-m)}(x,y) \tag{2.8.10b}$$

These sources are assumed to radiate into the negative z-direction only. To prevent them from radiating into the region $z > z_1$ it may be necessary to impress additional sources $S_{(-m)a}$ at $z < z_1 - \Delta z$. The overall arrangement of sources and fields is shown in Fig. 2.16. To determine the field $E'_{(-m)}$, $H'_{(-m)}$ which the sources of eqns. 2.8.10 excite in the region $z_1 - \Delta z \leqslant z \leqslant z_1$ in the limit $\Delta z \to 0$ we consider eqn. 2.1.2 and find that the transverse components of this field are equal to the transverse field of the $(-m)$th mode.

$$E'_{(-m)} \times n_z = E_{t(-m)}(x,y) \times n_z \qquad (2.8.10c)$$

$$H'_{(-m)} \times n_z = H_{t(-m)}(x,y) \times n_z \qquad (2.8.10d)$$

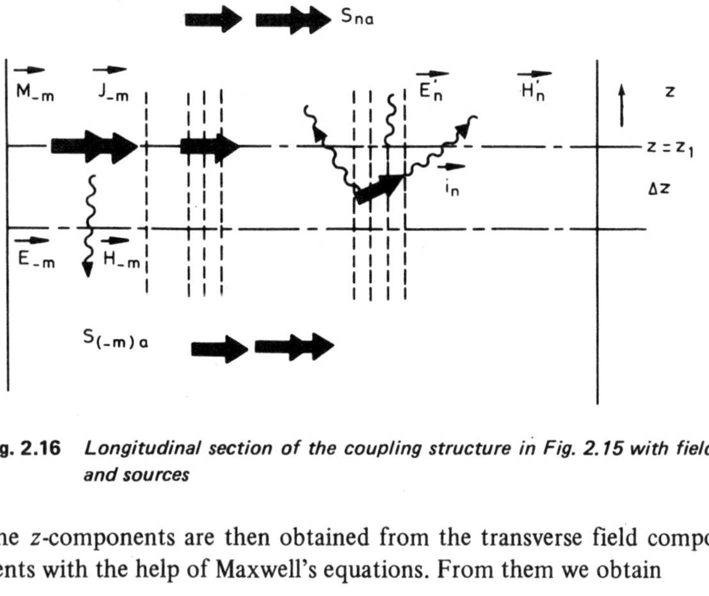

Fig. 2.16 ***Longitudinal section of the coupling structure in Fig. 2.15 with fields and sources***

The z-components are then obtained from the transverse field components with the help of Maxwell's equations. From them we obtain

$$j\omega\hat{\epsilon}(x,y)(E_{(-m)}\cdot n_z) = n_z \cdot [\nabla \times H_{t(-m)}(x,y)]$$

$$j\omega\mu(H_{(-m)}\cdot n_z) = -n_z \cdot [\nabla \times E_{t(-m)}(x,y)]$$

Comparing these equations with eqns. 2.8.3c and 2.8.3d leads to

$$E'_{(-m)}\cdot n_z = \frac{\epsilon_m(x,y)}{\hat{\epsilon}(x,y)} E_{z(-m)}(x,y) \qquad (2.8.10e)$$

$$H'_{(-m)}\cdot n_z = H_{z(-m)}(x,y), \qquad (2.8.10f)$$

where $E_{z(-m)}(x, y)$ and $H_{z(-m)}(x, y)$ denote the z-components of the field of the $(-m)$th mode.

To calculate the coupling coefficients κ_{mn} we can now apply the reciprocity theorem according to eqns. 2.1.4 to the sources of eqns. 2.8.8 and 2.8.10 and to their fields. The additional sources S_{na} and $S_{(-m)a}$ can be neglected since the fields which have to be multiplied by these sources are zero in the corresponding regions. We thus obtain for $\Delta z \to 0$

$$\Delta z \iint i_n(x,y)\cdot E'_{(-m)} dS_0 =$$

$$\iint [J_{(-m)}(x,y)\cdot E'_{tn} - M_{(-m)}(x,y)\cdot H'_{tn}]\, dS_0 \qquad (2.8.11)$$

Introducing for the fields and sources the results of eqns. 2.8.8 and 2.8.10 and evaluating the resulting expressions leads to

$$\kappa'_{mn} = \sum_i W_{mi}\kappa_{in} \qquad (2.8.12a)$$

with

$$\kappa'_{mn} = \frac{j\omega}{2}\iint (\epsilon_n - \hat{\epsilon})(E_{tn}\cdot E_{t(-m)} + \frac{\epsilon_m}{\hat{\epsilon}} E_{zn}E_{z(-m)})dS_0$$

$$= \frac{j\omega}{2}\iint (\epsilon_n - \hat{\epsilon})(E_{tn}\cdot E_{tm} - \frac{\epsilon_m}{\hat{\epsilon}} E_{zn}E_{zm})dS_0$$

$$(2.8.12b)$$

$$W_{mi} = \frac{1}{2}\iint [E_{t(-m)} \times H_{ti} - E_{ti} \times H_{t(-m)}]\cdot n_z dS_0$$

$$= \frac{1}{2}\iint [E_{tm} \times H_{ti} + E_{ti} \times H_{tm}]\cdot n_z dS_0 \qquad (2.8.12c)$$

Eqn. 2.8.12*a* when written in matrix form appears as

$$\boldsymbol{\kappa}' = \mathbf{W}\boldsymbol{\kappa} \qquad (2.8.13a)$$

The elements of $\boldsymbol{\kappa}'$ are given by eqn. 2.8.12*b* and those of $\mathbf{W}$ follow from eqn. 2.8.12*c*. They can be determined if the normal-mode fields are known. The element of $\boldsymbol{\kappa}$ can then also be evaluated from

$$\boldsymbol{\kappa} = \mathbf{W}^{-1}\boldsymbol{\kappa}' \qquad (2.8.13b)$$

The results of eqn. 2.8.13 together with eqns. 2.8.12*b* and 2.8.12*c* are rigorous. Their evaluation, however, is very involved. To calculate the coupling between two specific modes it is necessary to consider all modes of type a and b. Thus, usually only an approximate solution can be achieved, which, however, suffices for all practical purposes if only the conducting boundaries are far enough removed from the guiding regions.

To find this approximation it is helpful to derive a more suitable relation for the elements of the matrix **W**. For this purpose we calculate the real part Re (P) of the total power which is carried by a field distribution with an arbitrary set of mode amplitudes A_i. Only the propagating modes contribute to this real part of transmitted power. From

$$\mathrm{Re}\,(P) = \mathrm{Re}\left[\iint (\boldsymbol{E} \times \boldsymbol{H}^*)\cdot \boldsymbol{n}_z\, dS_0\right.$$

$$= \mathrm{Re}\left\{\iint \left[\left(\sum_m A_m \cdot \boldsymbol{E}_{tm}\right) \times \left(\sum_n A_n^* \boldsymbol{H}_{tn}^*\right)\right]\cdot \boldsymbol{n}_z\, dS_0\right\}$$

$$= \sum_m \sum_n \frac{1}{2} A_m A_n^* \left[\iint (\boldsymbol{E}_{tm} \times \boldsymbol{H}_{tn} + \boldsymbol{E}_{tn} \times \boldsymbol{H}_{tm})\cdot \boldsymbol{n}_z\, dS_0\right]$$

we find

$$\mathrm{Re}\,(P) = \sum_m \sum_n A_m A_n^* W_{mn}$$

which in matrix notation appears as

$$\mathrm{Re}\,(P) = \boldsymbol{A}^T \mathbf{W} \boldsymbol{A}^* \tag{2.8.14}$$

Note that the components of $\boldsymbol{E}_{tm}$ and $\boldsymbol{H}_{tm}$ are assumed to be real quantities. With the field normalisations of eqns. 2.8.2 this is possible only for propagating modes. Thus in eqn. 2.8.14 we assume that the real power is carried by the fields of propagating modes only. Consequently, in the column vector $\boldsymbol{A}$ only the amplitudes of propagating modes are considered. The elements of $\boldsymbol{A}^*$ denote the conjugate complex elements of $\boldsymbol{A}$. $\boldsymbol{A}^T$ is the transposed (row) vector of $\boldsymbol{A}$. Since the guiding structure is lossless, the value of Re (P) must be constant along the coupled waveguides, which requires

$$\frac{d}{dz}\,\mathrm{Re}\,(P) = \left(\frac{d}{dz}\boldsymbol{A}^T\right)\mathbf{W}\boldsymbol{A}^* + \boldsymbol{A}^T\mathbf{W}\left(\frac{d}{dz}\boldsymbol{A}^*\right) = 0$$

In this equation we substitute for the rate of change of $\boldsymbol{A}$ with z from eqn. 2.8.6 and eqn. 2.8.13 according to

$$\frac{d}{dz}A = [-\gamma + \mathbf{W}^{-1}\,\kappa']A \qquad (2.8.15)$$

We note that for all propagating modes in the lossless structure the diagonal matrix has only imaginary elements,

$$\gamma = \gamma^T = -\gamma^*, \qquad (2.8.16a)$$

and that for all modes which propagate in the same direction the matrix **W** according to eqn. 2.8.12*c* is symmetrical, and that all its elements are real

$$\mathbf{W} = \mathbf{W}^T = \mathbf{W}^* \qquad (2.8.16b)$$

This latter assumption is always justified if a wave travelling in one direction interacts significantly only with waves travelling in the same direction, so that any interaction between waves travelling in opposite directions may be neglected.

Utilising all the above relations we obtain the following condition for the conservation of power along the coupled waveguides

$$\begin{aligned} 0 &= A^T[-\gamma + \kappa'^T\mathbf{W}^{-1}]\mathbf{W}A^* + A^T\mathbf{W}[-\gamma^* + \mathbf{W}^{-1}\,\kappa'^*]A^* \\ &= A^T[-\gamma\mathbf{W} + \mathbf{W}\gamma + \kappa'^T + \kappa'^*]A^* \end{aligned}$$

To satisfy this condition for any combination of wave amplitudes, we must require that the expression within the brackets vanishes, thus

$$-\mathbf{W}\gamma + \gamma\mathbf{W} = \kappa'^* + \kappa'^T \qquad (2.8.17)$$

holds for all elements m, n with m and n denoting the subscripts of propagating modes, only. In addition, all modes under consideration are assumed to propagate in positive z-direction, so that only positive subscripts m and n are taken.

For $m = n$ we obtain with

$$W_{nn} = 1 \qquad (2.8.18a)$$

the condition

$$\text{Re}\,(\kappa'_{nn}) = 0$$

This condition may also be inferred from eqn. 2.8.12*b* when the appropriate normalisation of fields in eqns. 2.8.2 is taken into consideration. For any propagating mode this normalisation implies

$$E_{tn} = \text{Re}\,(E_{tn})$$

and

$$E_{zn} = j\,\text{Im}\,(E_{zn})$$

Eqn. 2.8.17 yields for the off-diagonal terms of **W** with $m \neq n$:

$$W_{mn} = \frac{\kappa'^{*}_{mn} + \kappa'_{nm}}{\gamma_m - \gamma_n} \tag{2.8.18b}$$

This result allows us to very conveniently estimate the interaction terms in the coupled wave equations 2.8.15. Considering only propagating modes we find that the absolute value of W_{mn} will be the smaller the larger the value of $|\gamma_m - \gamma_n|$ is. Usually one is interested only in those fields which are excited by the modes of type *a* as described above. The phase constant $\beta_n = \gamma_n/j > k = \omega\sqrt{\mu\epsilon}$ of these modes is the greater the further the particular mode is above its cutoff frequency, while for the propagating modes of type *b* the phase constants β_n are always smaller than k. Hence, if only the coupling between the modes of type *a* is to be analysed then usually only the elements of **W** which follow from interaction between just these modes need to be considered.

According to eqn. 2.8.2

$$W_{mn} = \delta_{mn}$$

holds in any case in which n and m denote modes of the same dielectric waveguide. If n denotes a mode of one of the two waveguides I or II and m a mode of the other waveguide then the value of W_{mn} will in general not vanish. In any case, however, its absolute value is smaller than 1. If the modes m and n are of the previously described type *a*, then the absolute value of W_{mn} is the smaller the larger the distance between the two guiding areas is and the higher the modes are above cutoff. In this case the fields of modes n and m overlap only little.

Thus, if we assume that the coupling between the modes of type *a* is small, then we can neglect the off-diagonal elements of **W** and obtain

$$\kappa_{mn} = \kappa'_{mn} \tag{2.8.19}$$

This simple approximation is usually applied when interaction between modes of parallel dielectric waveguides is analysed. It is also obtained if the field excitation of the current i_n is determined by using eqn. 2.2.7. In the preceding sections this equation was always used for similar purposes. Eqn. 2.2.7 was, however, derived subject to the condition that the fields of the modes are orthogonal. This condition is only approximately satisfied in the present case.

Results similar to eqn. 2.8.19 with eqn. 2.8.12*b* have been derived previously, for example in References 8 and 9 by applying different methods. These methods do not, however, lead to the rigorous formulation of eqn. 2.8.13*b*.

As an example to illustrate the coupled wave theory for parallel dielectric waveguides we consider the case of the two parallel dielectric slabs in Fig. 2.17. Single dielectric slabs represent the simplest form of optical waveguides, they also serve as a model for other optical waveguides such as transparent films on transparent substrates or transparent strips, which are deposited on or embedded in transparent substrates. By the same token the two parallel dielectric slabs in Fig. 2.17 represent a simple form of coupled optical waveguides, and serve well as a model for other optical waveguide couplers.

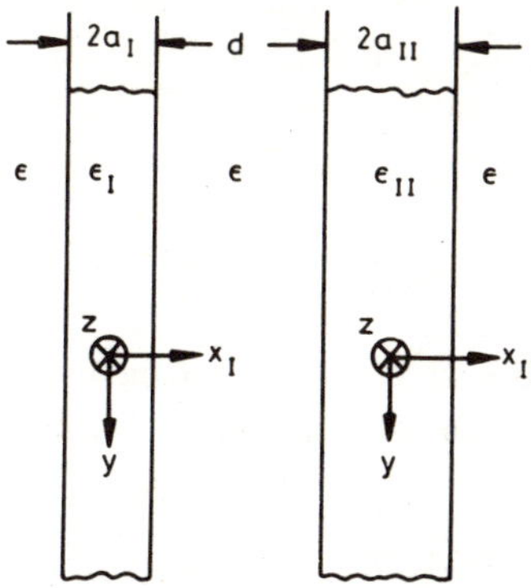

Fig. 2.17 *Two coupled dielectric slab guides*

The slabs in Fig. 2.17 are assumed to extend to infinity in the y- and z-direction, and we take the z-co-ordinate for the direction in which slab waves propagate. Fig. 2.17 indicates the dimensions and permittivities of the five-layer structure which forms the coupled slabs. The conducting enclosure of the guiding structure which our present coupled wave theory requires is here assumed to be removed to infinity in the $\pm x$-directions as well as in the $\pm y$-direction.

The guided modes of slab I with permittivity ϵ_I will be represented in terms of the transverse co-ordinate x_I and that of slab II with ϵ_{II} in terms of x_{II}. To unify the representation we use the notation x_i and ϵ_i with $i = I$ or II, respectively. All fields are constant in the y-direction, and hence independent of the y-co-ordinate. We will consider here the coupling between TE- and H-waves of the slab which have only a y-component for their electric field

$$\mathbf{E}_i = \mathbf{n}_y E_i \tag{2.8.20a}$$

The index i on the fields and on their propagation parameters indicates that they belong to the mode i in slab i.

To evaluate the coupling coefficients according to eqn. 2.8.12 we need first to determine the field distribution of guided modes in each of the slabs without the other slab present. Each of the two slabs by itself represents a symmetrical structure with its centre plane $x_i = 0$ as the plane of symmetry. Its guided modes have hence either an even or an odd field distribution about this centre plane and are therefore called even and odd modes, respectively.

The fundamental mode of the symmetrical slab is its even mode. Actually this mode has zero cutoff. Since this mode is of most practical concern we will consider only coupling between even TE- or H-modes of the slabs, and list only their field distributions.

From Maxwell's equations follows the wave equation for the only electric-field component of TE-modes according to

$$\Delta E_i + \frac{\epsilon_i}{\epsilon} k^2 E_i = 0 \tag{2.8.21}$$

with $k = \omega\sqrt{\mu\epsilon}$ as the wave number for the medium outside the slabs. Its plane wave solution with an even distribution about x_i is for $|x_i| < a_i$

$$E_i = N_i \cos(p_i x_i / a_i) \exp(-j\beta_i z) \tag{2.8.20b}$$

where p_i follows from the separation condition for eqn. 2.8.21 inside the slab

$$(p_i/a_i)^2 = k^2 \epsilon_i/\epsilon - \beta_i^2 \tag{2.8.22a}$$

Plane wave solutions for $|x_i| > a_i$, which match the field in eqn. 2.8.20*b* at $|x_i| = a_i$, are

$$E_i = N_i \cos(p_i) \exp[-q_i(|x_i/a_i| - 1)] \exp(-j\beta_i z) \tag{2.8.20c}$$

where q_i follows from the separation condition for eqn. 2.8.21 outside the slab

$$(q_i/a_i)^2 = \beta_i^2 - k^2 \tag{2.8.22b}$$

One of Maxwell's equations, namely

$$-\nabla \times \boldsymbol{E}_i = j\omega\mu \boldsymbol{H}_i$$

allows to derive the magnetic field from the above electric field. Its only components are

$$H_{zi} = \frac{j}{\omega\mu} \frac{\partial}{\partial x_i} E_i \tag{2.8.20d}$$

$$H_{xi} = -\frac{\beta_{|i|}}{\omega\mu} E_i \,\mathrm{sign}\,(i) \tag{2.8.20e}$$

Matching the tangential components H_z at the slab surface leads to the characteristic equation

$$p_i \tan(p_i) = q_i \qquad (2.8.23a)$$

We furthermore obtain from both separation conditions 2.8.22*a* and 2.8.22*b*

$$p_i^2 + q_i^2 = a_i^2 k^2(\epsilon_i/\epsilon - 1) \equiv V_i \qquad (2.8.23b)$$

To normalise the guided mode fields we require it to carry unit power in the z-direction per unit width of the slab in the y-direction

$$\int_{-\infty}^{\infty} (\boldsymbol{E}_i \times \boldsymbol{H}_i^*)\cdot \boldsymbol{n}_z dx_i = -\int_{-\infty}^{\infty} E_i H_{xi} dx_i = \text{sign}(i) \qquad (2.8.24a)$$

The normalisation factor follows from this condition as

$$N_i = \left[\frac{\beta_{|i|} a_i}{\omega\mu}\left(1 + \frac{1}{q_i}\right)\right]^{-1/2} \qquad (2.8.24b)$$

This normalisation corresponds to the procedure which eqn. 2.8.2*a* prescribes. Note that the present structure extends to infinity in the y-direction but that the fields are independent of y. When normalising the fields, and also when evaluating the coupling coefficients we need to integrate only over the transverse x-co-ordinate.

Eqns. 2.8.20, 2.8.23 and 2.8.24*b* provide us with all the expressions that we require to evaluate the coupling between any of the even TE-modes of two parallel slabs. For a given value of the slab parameter V_i in eqn. 2.8.23*b* the respective slab guides a finite and definite number of such modes. For $V_i \leqslant \pi$ it guides only the fundamental TE-mode.

We designate a TE-mode in one slab with index m and a TE-mode in the other slab with index n. By evaluating eqn. 2.8.12*b* for the fields of these two modes we find for the coupling coefficient between them

$$\kappa'_{mn} = j\sqrt{\frac{\epsilon_m - \epsilon}{\epsilon_n - \epsilon}} \frac{1/2}{(\beta_m a_m \beta_n a_n)^{1/2}} \frac{p_n/a_n}{\{1 + (1/q_n)\}^{1/2}}$$

$$\frac{p_m/a_m}{\{1 + (1/q_m)\}^{1/2}} \frac{\exp(-q_n d/a_n)}{(q_n/a_n)^2 + (p_m/a_m)^2}$$

$$\left[\left(\frac{q_m}{a_m} - \frac{q_n}{a_n}\right)\exp(-2q_n a_m/a_n) + \frac{q_m}{a_m} + \frac{q_n}{a_n}\right] \qquad (2.8.25)$$

This formula shows clearly the exponential decay of coupling strength with increasing values of d and q_n/a_n. The two modes couple via their evanescent fields in the gap of width d. Therefore, if d or if the transverse attenuation q_n/a_n of fields increase the coupling strength decreases exponentially.

The coupling coefficients of eqn. 2.8.25 do not obey the exprs. 2.4.11 and 2.4.12 which follow from reciprocity and when the coupling structure is lossless. The reason for this discrepancy is that, because of eqn. 2.8.13b, the formula eqn. 2.8.12b only approximates the actual coupling coefficient, and that the exprs. 2.4.11 and 2.4.12 assume the coupled modes to be orthogonal to each other. In the present situation, however, the guided modes of the single slabs are not orthogonal to each other in the composite structure.

The same result as given by eqn. 2.8.19 together with eqn. 2.8.25 has also been obtained in Reference 9. Reference 10 investigates in detail under what conditions these approximate results remain valid. It is found there, that eqn. 2.8.19 is rigorous when the two coupled modes m and n are degenerate. The simplification eqn. 2.8.19 may therefore be utilised whenever coupled modes do not differ greatly in their propagation constants.

2.9 Dielectric waveguide transitions

Dielectric waveguides, in contrast to metallic waveguides, have no conducting walls; they guide waves by total reflection at interfaces of different dielectric media or by continuous focusing with graded dielectric media. Transitions which connect dielectric waveguides of different cross-sectional sizes and forms not only couple their guided modes with one another, they also convert guided-mode energy to the radiation field.

So far in this book we have not concerned ourselves with radiation, but only with guided modes. Metallic waveguide transitions have conducting walls. For them an infinite but complete set of discrete modes, either propagating or evanescent, exists in any of its cross-sections. The fields may be expanded in terms of these local normal modes and transmission through the transition involves interaction between these modes. All other waveguides considered so far were enclosed by an opaque screen, to obtain a complete set of functions in terms of the discrete modes with which to represent any fields along these waveguides. These modes are orthogonal to each other and, by way of eqn. 2.2.17, were so normalised that the squared magnitude of their amplitudes $A_i^{(\pm)}$ represents the complex power which they transmit.

In the previous Section the wave coupling between two parallel dielectric waveguides was analysed by also enclosing them in an opaque screen and thus obtaining a discrete set of propagating and evanescent modes for the complete field representation. For the evaluation of coupling coefficients in a particular example we let the screen become infinitely large in cross-section, and we then used the fields of guided modes on the open dielectric waveguide. In that case we were interested only in coupling between guided modes of the dielectric waveguides and did not have any need for their radiation field. If at all, radiation in such dielectric waveguide couplers occurs only at transitions from the coupling section to the connecting waveguides.

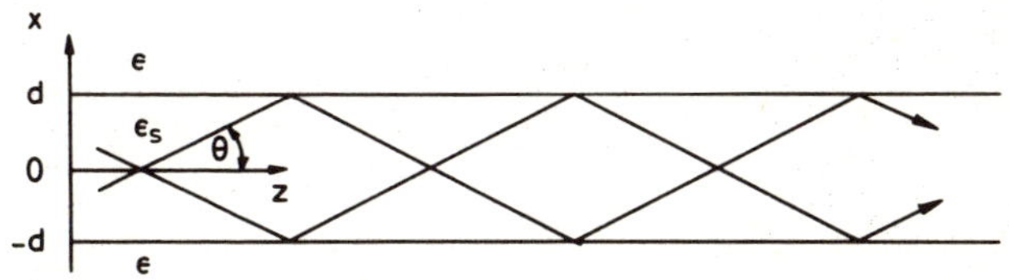

Fig. 2.18 *Ray picture of guided modes of a dielectric slab*

In dielectric waveguide transitions which more or less gradually change their cross-sectional form or size or even both, we are very often, and sometimes primarily, concerned with the radiation.

To adopt a suitable presentation of the radiation field we will first consider as a simple but representative example the symmetrical dielectric slab in Fig. 2.18 of relative permittivity ϵ_s embedded in a medium with relative permittivity ϵ. Its guided modes are not only the even TE-modes, for which in the previous Section we evaluated the coupling coefficients between two parallel such slabs, but also odd TE-modes, as well as even and odd TM-modes. These four different classes of guided modes of the symmetrical slab may, according to Fig. 2.18, be regarded as uniform plane waves which propagate inside the slab at an angle θ with respect to the longitudinal z-direction, and experience total reflection at both slab interfaces. If the uniform plane wave is polarised with its electric-field vector parallel to the slab we obtain TE-modes; if it is polarised with its magnetic-field vector parallel to the slab we obtain TM-modes. The uniform plane waves in propagating on their zig-zag path superimpose upon themselves and in the transverse direction perpendicular to the slab form standing waves. But for these standing waves to exist, the phase delay $2\sqrt{\epsilon_s}k_0 d \sin \theta$ of the uniform plane wave in this transverse direction between both slab interfaces, plus the phase delay $-\phi$ of total reflection, must equal an integer multiple m of π. We hence obtain the following phase condition for TE- or TM-modes to be guided modes of the slab:

$$2\sqrt{\epsilon_s}k_0 d \sin\theta - \phi = m\pi \tag{2.9.1}$$

The polarisation that leads to TE-modes suffers the phase shift

$$\phi = 2\arctan\frac{\{\cos^2\theta - (\epsilon/\epsilon_s)\}^{1/2}}{\sin\theta} \tag{2.9.2}$$

upon total reflection, the polarisation for TM-modes suffers

$$\phi = 2\arctan\frac{\epsilon_s\{\cos^2\theta - (\epsilon/\epsilon_s)\}^{1/2}}{\epsilon\sin\theta} \tag{2.9.3}$$

Because of these two different phase shifts each positive integer m, including $m = 0$, can have associated with it two different solutions of eqn. 2.9.1, one for a TE-mode with ϕ from eqn. 2.9.2 and the other for a TM-mode with ϕ from eqn. 2.9.3. These solutions exist, of course, only in the range of total reflection

$$\theta < \theta_c = \text{arc}\cos\sqrt{\epsilon/\epsilon_s} \tag{2.9.4}$$

where real phase shifts ϕ follow from eqns. 2.9.2 and 2.9.3. In the limit of $\theta = \theta_c$ we have $\phi = 0$ and eqn. 2.9.1 reduces to

$$2k_0 d\sqrt{\epsilon_s - \epsilon} = m\pi \tag{2.9.5}$$

This equation represents the cut-off condition for guided modes. For any given value of $k_0 d\sqrt{\epsilon_s - \epsilon}$ we have only a finite number of guided modes equal to twice the next integer larger than

$$2k_0 d\sqrt{\epsilon_s - \epsilon}/\pi.$$

For m even the E-field in the case of TE-modes and the H-field in the case of TM-modes have an even distribution about the centreplane $x = 0$ of the slab. They are therefore called even modes of the slab. For m odd the distribution of these transverse fields is odd about the centreplane and the modes are called odd modes of the slab.

To obtain simple expressions for field and power of guided modes we introduce the phase constant β for their propagation along the slab

$$\beta = \sqrt{\epsilon_s}k_0 \cos\theta \tag{2.9.6}$$

as well as their transverse phase constant inside the slab

$$\tau_s = \sqrt{\epsilon_s}k_0 \sin\theta \tag{2.9.7}$$

and their transverse attenuation constant α outside the slab

$$\alpha = k_0\sqrt{\epsilon_s\cos^2\theta - \epsilon}. \tag{2.9.8}$$

In terms of these parameters the characteristic eqn. 2.9.1 of even TE-modes reduces to

$$\tau_s \tan(\tau_s d) = \alpha \tag{2.9.9}$$

and their transverse electric field is given by

$$E_y = E_e \cos(\tau_s x) \quad |x|<d$$

$$E_y = E_e \cos(\tau_s d) \exp[-\alpha(|x|-d)] \quad |x|>d \tag{2.9.10}$$

For odd TE-modes eqn. 2.9.1 reduces to

$$\tau_s \cot(\tau_s d) = -\alpha \tag{2.9.11}$$

and the electric field has the distribution

$$E_y = E_0 \sin(\tau_s x) \quad |x|<d$$

$$E_y = E_0 \sin(\tau_s d) \exp[-\alpha(x-d)] \quad x>d$$

$$E_y = -E_0 \sin(\tau_s d) \exp[+\alpha(x+d)] \quad x<-d \tag{2.9.12}$$

The only components of the magnetic field of TE-waves follow from their transverse electric field according to

$$H_x = -(\beta/\omega\mu_0)E_y, \quad H_z = (j/\omega\mu_0)\partial E_y/\partial x \tag{2.9.13}$$

With these field distributions the modes are not yet normalised for power.

To achieve such a normalisation we calculate the power, which a TE-mode guides per unit width of the slab

$$P = -\int_{-\infty}^{\infty} E_y H_x^* dx = \frac{\beta}{\omega\mu_0}\left(d+\frac{1}{\alpha}\right)|E_{o,e}|^2 \tag{2.9.14}$$

A TE-mode with

$$E_{o,e} = \left[\frac{\beta}{\omega\mu_0}\left(d+\frac{1}{\alpha}\right)\right]^{-1/2} \tag{2.9.15}$$

will therefore guide unit power per unit width of the slab.

For even TM-modes the characteristic eqn. 2.9.1 reduces to

$$\tau_s \tan(\tau_s d) = (\epsilon_s/\epsilon)\alpha \tag{2.9.16}$$

and their magnetic field is given by

$$H_y = H_e \cos(\tau_s x) \quad |x|<d$$

$$H_y = H_e \cos(\tau_s d) \exp[-\alpha(|x|-d)] \quad |x|>d \tag{2.9.17}$$

Odd TM-modes have

$$\tau_s \cot(\tau_s d) = -(\epsilon_s/\epsilon)\alpha \tag{2.9.18}$$

as their characteristic equation and a magnetic field according to

$$H_y = H_0 \sin(\tau_s x) \quad |x|<d$$

$$H_y = H_0 \sin(\tau_s d) \exp[-\alpha(x-d)] \quad x>d$$

$$H_y = -H_0 \sin(\tau_s d) \exp[\alpha(x+d)] \quad x<-d \tag{2.9.19}$$

The electric field of both even and odd TM-modes inside the slab follows from

$$E_x = \frac{\beta}{\omega\epsilon_s\epsilon_0} H_y; \quad E_z = \frac{-j}{\omega\epsilon_s\epsilon_0}\frac{\partial H_y}{\partial z} \tag{2.9.20a}$$

while outside the slab it is given by

$$E_x = \frac{\beta}{\omega\epsilon\epsilon_0} H_y \; ; \; E_z = \frac{-j}{\omega\epsilon\epsilon_0}\frac{\partial H_y}{\partial x} \tag{2.9.20b}$$

For the power which TM-modes guide per unit width of the slab we calculate

$$P = \int_{-\infty}^{\infty} E_x H_y^* dx = \frac{\beta}{\omega\epsilon_0}\left[\frac{d}{\epsilon_s} + \frac{\epsilon(\epsilon_s-\epsilon)k_0^2}{(\epsilon^2\tau_s^2 + \epsilon_s^2\alpha^2)\alpha}\right]|H_{e,o}|^2 \tag{2.9.21}$$

so that with

$$H_{e,o} = \left[\frac{\beta}{\omega\epsilon_0}\left(\frac{d}{\epsilon_s} + \frac{\epsilon(\epsilon_s-\epsilon)k_0^2}{(\epsilon^2\tau_s^2 + \epsilon_s^2\alpha^2)\alpha}\right)\right]^{-1/2} \tag{2.9.22}$$

they guide unit power per unit width of the slab.

All guided modes of the slab have propagation angles θ and phase constants β in the ranges

$$0<\theta<\theta_c \tag{2.9.23}$$

$$\sqrt{\epsilon_s}\,k_0 > \beta > \sqrt{\epsilon}\,k_0 \tag{2.9.24}$$

For $\theta > \theta_c$ the phase constant becomes smaller than $\sqrt{\epsilon}\,k_0$ and the uniform plane waves do not any longer reflect totally at the slab interfaces; they rather radiate into the outside regions.

To obtain a complete representation for the radiation field we let plane uniform waves be incident onto the slab and consider all angles of incidence with planes of incidence parallel to the z-axis. When such waves are incident at equal angles and with equal amplitude and phase from both sides of the slab as in Fig. 2.19, they superimpose to an even field distribution with respect to $x = 0$. When they have opposite phase but are otherwise equal they superimpose to an odd field distribution. For the electric field polarised in the plane of incidence the total field is transverse magnetic with respect to z; for the perpendicular polarisation it is transverse electric. We thus obtain for a complete representation even and odd TE- and TM-fields. Since they are solutions of Maxwell's equations for the radiation field under the boundary conditions of the guiding structure they are referred to as radiation modes of the waveguide. In contrast to the guided modes their fields are not confined to the guide with only evanescent fields outside; they rather extend to infinity in the transverse direction. Also these modes are not limited to discrete values of θ and β; they rather form a continuum with respect to θ and hence also with respect to β. For radiation modes of the even TE-type we start with the following set up for the standing waves of the electric field

$$E_y = E_{es} \cos(\tau_s x) \quad |x| < d$$

$$E_y = E_e \cos[\tau(|x| - d) + \phi] \quad |x| > d \tag{2.9.25}$$

with

$$\tau_s = (\epsilon_s k_0^2 - \beta^2)^{1/2}, \quad \tau = (\epsilon k_0^2 - \beta^2)^{1/2} \tag{2.9.26}$$

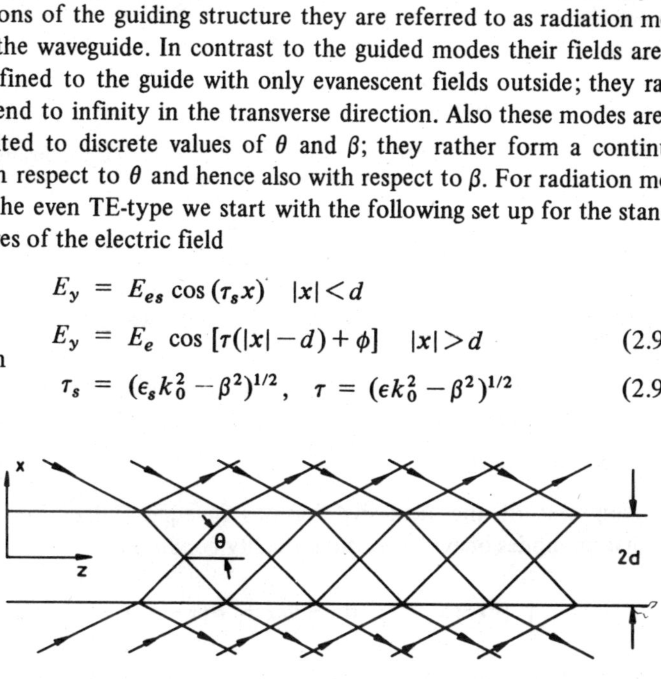

Fig. 2.19 *Ray picture of radiation modes of a dielectric slab*

Their magnetic field follows from eqn. 2.9.13, and to match tangential fields at $x = \pm d$ the quantities ϕ and E_e must be chosen according to

$$\tan\phi = \tau_s \tan(\tau_s d)$$

$$E_e^2 = E_{es}^2 [\cos^2(\tau_s d) + (\tau_s/\tau)^2 \sin^2(\tau_s d)] \tag{2.9.27}$$

The power which this radiation mode transmits is infinite for any finite field E_e. We arrive nevertheless at a suitable normalisation with respect to power by considering the cross product $E_y(\tau)H_x^*(\tau')$ of the electric field of an even TE-mode with transverse phase constant τ and the transverse magnetic field of an even TE-mode with τ'. Its integral over the cross-section of the guide is referred to as crosspower. Accordingly

$$\bar{P} = -\int_{-\infty}^{\infty} E_y(\tau)H_x^*(\tau')dx \tag{2.9.28}$$

represents the cross-power per unit slab width for TE-modes. This cross-power vanishes for $\tau' \neq \tau$ indicating orthogonality also for radiation modes. If now, in addition, we integrate $\bar{P}$ over τ including $\tau = \tau'$ in the range of integration we obtain with E_y from eqn. 2.9.10 and H_x from eqn. 2.9.13

$$\int_{-\infty}^{\infty} \bar{P}d\tau = \frac{\pi\beta}{\omega\mu_0} E_e^2 \tag{2.9.29}$$

This result allows us to write the cross-power of even TE-modes in the form

$$\bar{P} = \frac{\pi\beta}{\omega\mu_0} E_e^2\, \delta(\tau-\tau') \tag{2.9.30}$$

where $\delta(\tau-\tau')$ represents Dirac's delta function. We should keep in mind that this delta function is defined only when it appears as part of an integrand as in eqn. 2.9.29. When we now normalise the fields in eqn. 2.9.25 according to

$$E_e^2 = \omega\mu_0/(\pi\beta) \tag{2.9.31}$$

the crosspower of the field distributions in eqns. 2.9.25 for even TE-radiation modes obeys the orthonormality condition

$$\bar{P} = \int_{-\infty}^{\infty} [\boldsymbol{E}(\tau) \times \boldsymbol{H}^*(\tau')] \cdot \boldsymbol{n}_z dx = \delta(\tau-\tau') \tag{2.9.32}$$

It corresponds to the orthogonality condition eqn. 2.2.2*a* for guided modes and allows us to incorporate radiation modes of the even TE-type in a more general field representation.

TE-radiation modes with odd field distributions about the centre-plane of the slab consist of plane uniform waves which as in Fig. 2.19 are incident onto the slab from both sides with equal angles of incidence and with equal amplitudes, but with opposite phase. Their incident and reflected components superimpose and form standing waves with an electric field distribution

$$E_y = E_{os} \sin(\tau_s x) \quad |x| < d$$

$$E_y = E_o \sin[\tau(x-d)+\phi] \quad x > d$$

$$E_y = E_o \sin[\tau(x+d)-\phi] \quad x<-d \tag{2.9.33}$$

Their magnetic field follows also from eqn. 2.9.13, and in order to match all tangential fields at $x = \pm d$ the phase shift ϕ and E_0 must be chosen according to

$$\tau \cot\phi = \tau_s \cot(\tau_s d)$$
$$E_0^2 = E_{os}^2[\sin^2(\tau_s d) + (\tau_s/\tau)^2 \cos^2(\tau_s d)] \tag{2.9.34}$$

For the cross-power between these odd TE-modes the same expression 2.9.29 results from eqn. 2.9.28, only with E_e replaced by E_0. If therefore we let

$$E_0^2 = \omega\mu_0/(\pi\beta) \tag{2.9.35}$$

the odd TE-radiation modes will also obey the orthonormality condition 2.9.32.

For radiation modes of the even TM-type we set up the following transverse magnetic-field distribution

$$H_y = H_{se}\cos(\tau_s x) \quad |x|<d$$
$$H_y = H_e \cos[\tau(|x|-d)+\phi] \quad |x|>d \tag{2.9.36}$$

with transverse phase constants τ_s and τ as in eqn. 2.9.26. The TM-mode electric field for this particular magnetic field derives from eqns. 2.9.20. When tangential field components are matched at the slab interfaces we obtain the following equations for ϕ and H_e

$$\epsilon_s\tau \tan\phi = \epsilon\tau_s \tan(\tau_s d)$$
$$H_e^2 = H_{se}^2\left[\cos^2(\tau_s d) + \left(\frac{\epsilon\tau_s}{\epsilon_s\tau}\right)^2 \sin^2(\tau_s d)\right] \tag{2.9.37}$$

For odd TM-radiation modes we set up the magnetic field according to

$$H_y = H_{so}\sin(\tau_s x) \quad |x|<d$$
$$H_y = H_0 \sin[\tau(x-d)+\phi] \quad x>d$$
$$H_y = H_0 \sin[\tau(x+d)-\phi] \quad x<-d \tag{2.9.38}$$

and from matching field components at the interfaces obtain

$$\epsilon_s\tau \cot\phi = \epsilon\tau_s \cot(\tau_s d)$$
$$H_0^2 = H_{so}^2\left[\sin^2(\tau_s d) + \left(\frac{\epsilon\tau_s}{\epsilon_s\tau}\right)^2 \cos^2(\tau_s d)\right] \tag{2.9.39}$$

as conditions for ϕ and H_0.

To satisfy the orthonormality condition of eqn. 2.9.32 for TM-radiation modes we need to choose

$$H_{e,o}^2 = \omega\epsilon_0/(\pi\beta) \tag{2.9.40}$$

Orthogonality holds also between even and odd radiation modes as well as between TE- and TM-modes.

The integration over τ in eqn. 2.9.29 extends from $-\infty$ to $+\infty$. Actually, however, radiation modes of the slab occur only for $|\tau| < \sqrt{\epsilon}k_0$, because only for τ values smaller in absolute value than this limit has the longitudinal phase constant β according to $\beta = (\epsilon k_0^2 - \tau^2)^{1/2}$ real values. For $\tau > \sqrt{\epsilon}k_0$ the phase constant β becomes imaginary and thus turns into an attenuation constant. In this range of τ the fields that are described by the radiation-mode formulas decay exponentially in propagation direction without any change in phase. The modes no longer propagate in z-direction but are evanescent. These evanescent modes form a continuous spectrum with τ ranging from $\pm\sqrt{\epsilon}k_0$ to $\pm\infty$ and with $-j\beta$ ranging from zero to $\pm\infty$. For a complete representation of fields on the dielectric waveguide they must also be taken into account. But in dielectric waveguide transitions with only gradual changes in cross-section the evanescent modes are hardly ever excited to any significant extent so that we do not have any need for them in this context.

After this discussion of guided and radiation modes as well as evanescent modes, for the dielectric slab as an example of a uniform dielectric waveguide, let us now consider the nonuniform case of a dielectric waveguide transition. Quite generally such a transition consists of a permittivity distribution $\epsilon = \epsilon(x, y, z)$ which not only depends on the transverse co-ordinates x and y but also changes in longitudinal direction with z. By analysing wave propagation in dielectric waveguide transitions in terms of guided and radiation modes we follow a representation that was first developed by Marcuse[7] and applied to nonuniform dielectric films and fibres.

In Section 2.6 we found a representation in terms of mode voltages and currents with eigenvectors for the transverse fields and obtained generalised transmission-line equations. To derive such equations for the dielectric waveguide transition we adopt the corresponding representation, and first decompose the field vectors into their transverse and longitudinal parts

$$\boldsymbol{E} = \boldsymbol{E}_t + \boldsymbol{n}_z E_z, \quad \boldsymbol{H} = \boldsymbol{H}_t + \boldsymbol{n}_z H_z \tag{2.9.41}$$

Instead of introducing equivalent sources to account for changes of the guide in the longitudinal direction we will derive generalised transmission-line equations, as well as coupled wave equations, directly from Maxwell's equations:

$$\nabla \times \boldsymbol{E} = -j\omega\mu_0 \boldsymbol{H} \quad \nabla \times \boldsymbol{H} = j\omega\epsilon\epsilon_0 \boldsymbol{E} \tag{2.9.42}$$

The generalised transmission-line equations in Section 2.6 were obtained for the mode voltages and mode currents of the transverse field. We will arrive at corresponding equations for dielectric waveguide transitions when we eliminate the longitudinal field components from Maxwell's equations. It will lead us to differential equations for only the transverse fields. To this end we decompose the curl operator according to

$$\nabla \times = \nabla_t \times + \boldsymbol{n}_z \times \frac{\partial}{\partial z} \tag{2.9.43}$$

where $\nabla_t \times$ contains all the partial derivatives with respect to transverse co-ordinates. The vector $\nabla \times \boldsymbol{E}$ then has the transverse part

$$(\nabla \times \boldsymbol{E})_t = \nabla_t \times (\boldsymbol{n}_z E_z) + \boldsymbol{n}_z \times \frac{\partial \boldsymbol{E}_t}{\partial z} \tag{2.9.44}$$

and the longitudinal component

$$(\nabla \times \boldsymbol{E})\cdot\boldsymbol{n}_z = \nabla\cdot(\boldsymbol{E}_t \times \boldsymbol{n}_z) \tag{2.9.45}$$

With these relations we separate both of Maxwell's equations 2.9.42 into their respective transverse parts

$$(\nabla_t E_z) \times \boldsymbol{n}_z + \boldsymbol{n}_z \times \frac{\partial \boldsymbol{E}_t}{\partial z} = -j\omega\mu_0 \boldsymbol{H}_t$$

$$(\nabla_t H_z) \times \boldsymbol{n}_z + \boldsymbol{n}_z \times \frac{\partial \boldsymbol{H}_t}{\partial z} = j\omega\epsilon\epsilon_0 \boldsymbol{E}_t \tag{2.9.46}$$

and their longitudinal components

$$\nabla\cdot(\boldsymbol{n}_z \times \boldsymbol{E}_t) = j\omega\mu_0 H_z$$

$$\nabla\cdot(\boldsymbol{H}_t \times \boldsymbol{n}_z) = j\omega\epsilon\epsilon_0 E_z \tag{2.9.47}$$

In the transverse equations we can now substitute for E_z and H_z from the longitudinal equations and obtain the desired relations for just the transverse field vectors

$$-\frac{\partial \boldsymbol{E}_t}{\partial z} = j\omega\mu_0 \boldsymbol{H}_t \times \boldsymbol{n}_z - \frac{1}{j\omega\epsilon_0}\nabla_t\left[\frac{1}{\epsilon}\nabla\cdot(\boldsymbol{H}_t \times \boldsymbol{n}_z)\right]$$

$$-\frac{\partial \boldsymbol{H}_t}{\partial z} = j\omega\epsilon\epsilon_0 \boldsymbol{n}_z \times \boldsymbol{E}_t - \frac{1}{j\omega\mu_0}\nabla_t[\nabla\cdot(\boldsymbol{n}_z \times \boldsymbol{E}_t)] \tag{2.9.48}$$

Let us now expand the transverse fields in terms of local normal modes along the transition. To be complete this expansion must include all guided as well as all radiation modes. It hence consists of a finite sum of the discrete terms for guided modes and an integral over the continuous spectrum of radiation modes

$$E_t = \sum_{i=1}^{M} v_i E_{ti} + \int v(\tau) E_t(\tau) d\tau$$

$$H_t = \sum_{i=1}^{M} i_i H_{ti} + \int i(\tau) H_t(\tau) d\tau \tag{2.9.49}$$

We use here for the mode voltages and currents the lower case letters v_i and i_i to indicate that they are meant to be dimensionless quantities. For any particular mode which in a uniform waveguide propagates with β_i along z, they are both equal to $\exp(-j\beta_i z)$. The summations in eqns. 2.9.49 include all guided modes. In case of the slab they can be even and odd TE- and TM-modes. For two dimensional dielectric waveguides such as rectangular and round guides we have two labels for each guided mode, designating the order of transverse field distributions in each of the two transverse co-ordinates. In this case the single summation in eqn. 2.9.49 stands as a symbol for the double summation. The same is true for the integrals over the radiation modes in eqn. 2.9.49. They also should include all possible kinds of radiation modes and should extend over the full range of the respective transverse phase parameter for which the particular mode represents a radiation mode. Despite all the instructions that eqns. 2.9.49 contain we will go even further in simplifying the notation and write

$$E_t = \sum_i v_i E_{ti} \qquad H_t = \sum_i i_i H_{ti} \tag{2.9.50}$$

for eqns. 2.9.49, where we designate the summation of guided modes as well as the integration over radiation modes by just one summation. E_{ti} and H_{ti} denote the transverse field vectors of either guided or radiation modes. For any mode propagating with β_i in positive z-direction in a uniform waveguide with a z-independent permittivity distribution $\epsilon = \epsilon(x, y)$ the transverse field equations 2.9.48 reduce to

$$j\beta_i E_{ti} = j\omega\mu_0 H_{ti} \times n_z - \frac{1}{j\omega\epsilon_0} \nabla_t \left[\frac{1}{\epsilon} \nabla \cdot (H_{ti} \times n_z) \right]$$

$$j\beta_i H_{ti} = j\omega\epsilon\epsilon_0 n_z \times E_{ti} - \frac{1}{j\omega\mu_0} \nabla_t [\nabla \cdot (n_z \times E_{ti})] \tag{2.9.51}$$

All of these modes are orthogonal to each other with respect to their cross-power and are assumed to be so normalised that they obey the following orthonormality condition

$$\iint \boldsymbol{n}_z \cdot (\boldsymbol{E}_i \times \boldsymbol{H}_j^*) dxdy = \delta_{ij} \tag{2.9.52}$$

For radiation modes this orthonormality condition is understood as a generalisation of eqn. 2.9.32.

For a dielectric waveguide transition with its permittivity distribution changing with z we interpret the guided and radiation modes of the transverse field expansion as local normal modes, which satisfy boundary conditions and propagate independently from each other only in the uniform waveguide that corresponds to the respective local cross-section of the transition. Since actually the waveguide changes along the transition these local normal modes not only depend on z, but they also do not satisfy the field equations and boundary conditions individually. However, they still represent a complete enough set of functions to serve for the field expansion in eqn. 2.9.50.

When we substitute for the transverse fields from eqn. 2.9.50 into eqns. 2.9.48 and take into account that $\boldsymbol{E}_{ti}$ and $\boldsymbol{H}_{ti}$ depend on z, but that locally they satisfy eqns. 2.9.51, we obtain the following two equations:

$$\sum_i \left[\left(\frac{dv_i}{dz} + j\beta_i i_i \right) \boldsymbol{E}_{ti} + v_i \frac{\partial \boldsymbol{E}_{ti}}{\partial z} \right] = 0$$

$$\sum_i \left[\left(\frac{di_i}{dz} + j\beta_i v_i \right) \boldsymbol{H}_{ti} + i_i \frac{\partial \boldsymbol{H}_{ti}}{\partial z} \right] = 0 \tag{2.9.53}$$

We now form the vector product of the first of these two equations with $\boldsymbol{H}_{tk}^*$ and of the second equation with $\boldsymbol{E}_{tk}^*$, integrate the resulting equations for the z-directed vectors over the local cross-section of the transition, and take advantage of the orthonormality condition 2.9.52. The result may then be written in the following form:

$$\frac{dv_k}{dz} + j\beta_k i_k = \sum_i C_{ki} v_i$$

$$\frac{di_k}{dz} + j\beta_k v_k = \sum_i D_{ki} i_i \tag{2.9.54}$$

where we abbreviate by C_{ki} and D_{ki} the coupling coefficients

$$C_{ki} = -\iint n_z \cdot \left(\frac{\partial E_i}{\partial z} \times H_k^* \right) dS$$

$$D_{ki} = -\iint n_z \cdot \left(E_k^* \times \frac{\partial H_i}{\partial z} \right) dS \tag{2.9.55}$$

These coupling coefficients are written with cross products of the complete vectors of the local normal mode fields, but actually only the transverse parts of these field vectors contribute to the z-component of the cross products in eqn. 2.9.55.

For a uniform waveguide we have $\partial E_i/\partial z = \partial H_i/\partial z = 0$ and all the coefficients C_{ki} and D_{ki} vanish. Eqns. 2.9.54 reduce under these conditions to the ordinary transmission-line equations for voltages and currents, one independent pair for each mode of the guide. In the nonuniform waveguide of the transition we have finite coefficients C_{ki} and D_{ki} which couple the modes with each other. Eqns. 2.9.54 then represent generalised transmission-line equations in which the phase constants β_k of local normal modes as well as the coupling coefficients C_{ki} and D_{ki} depend on z.

Due to the changes in waveguide cross-section different modes couple with each other. They interact most strongly when they travel in the same direction with nearly the same phase constant. To analyse this interaction we transform from the mode voltages v_i and currents i_i to the amplitudes $A_i^{(+)}$ and $A_i^{(-)}$ of forward and backward travelling waves

$$v_i = A_i^{(+)} + A_i^{(-)}, \quad i_i = A_i^{(+)} - A_i^{(-)} \tag{2.9.56}$$

For a uniform waveguide these forward and backward wave amplitudes depend on z according to

$$A_i^{(+)} \sim \exp(-j\beta_i z), \quad A_i^{(-)} \sim \exp(j\beta_i z)$$

where the phase constant β_i of the particular mode is independent of z. In the nonuniform regions of waveguide transitions the phase constants β_i of local normal modes change with z. Disregarding their interaction, their amplitudes would depend on z as

$$A_i^{(+)} = a_i^{(+)} \exp\left(-j \int_0^z \beta_i dz\right), \quad A^{(-)} = a_i^{(-)} \exp\left(j \int_0^z \beta_i dz\right) \tag{2.9.57}$$

where $a_i^{(+)}$ and $a_i^{(-)}$ are the initial values for these amplitudes at $z = 0$. To account for the interaction we let $a_i^{(+)}$ and $a_i^{(-)}$ be functions of z

which we can expect to vary much less with z than the amplitudes $A_i^{(+)}$ and $A_i^{(-)}$. We therefore transform to these slowly varying mode amplitudes by substituting for $A_i^{(+)}$ and $A_i^{(-)}$ in eqns. 2.9.56 from eqns. 2.9.57, and by introducing these in turn into the generalised transmission-line equations. When we add and subtract suitable pairs of eqns. 2.9.54 we can solve for $da_k^{(+)}/dz$ and $da_k^{(-)}/dz$ and obtain the following system of coupled wave equations:

$$\frac{da_k^{(+)}}{dz} =$$
$$\sum_i \left[\kappa_{ki}^{(+,+)} a_i^{(+)} \exp\left(j \int_0^z (\beta_k - \beta_i) dz \right) + \kappa_{ki}^{(+,-)} a_i^{(-)} \exp\left(j \int_0^z (\beta_k + \beta_i) dz \right) \right]$$

$$\frac{da_k^{(-)}}{dz} =$$
$$\sum_i \left[\kappa_{ki}^{(-,+)} a_i^{(+)} \exp\left(-j \int_0^z (\beta_k + \beta_i) dz \right) + \kappa_{ki}^{(-,-)} a_i^{(-)} \exp\left(-j \int_0^z (\beta_k - \beta_i) dz \right) \right] \quad (2.9.58)$$

The coupling coefficients are given in terms of C_{ki} and D_{ki} by

$$\kappa_{ki}^{(p,q)} = \tfrac{1}{2}(C_{ki} + pqD_{ki}) \quad (2.9.59)$$

where the factors p and q are $+1$ or -1 depending on whether the superscripts of κ are $+$ or $-$. We use eqns. 2.9.55 to express these coupling coefficients directly in terms of the integrals of cross-products of local normal mode fields

$$\kappa_{ki}^{(p,q)} = -\frac{p}{2} \iint n_z \cdot \left[\frac{\partial E_i^{(q)}}{\partial z} \times (H_k^{(p)})^* + (E_k^{(p)})^* \times \frac{\partial H_i^{(q)}}{\partial z} \right] dS \quad (2.9.60)$$

The fields have been furnished here with superscripts, which correspond to those on κ. They are meant to indicate in which direction the local normal mode is travelling for which we insert the particular field in eqn. 2.9.60. From eqns. 2.2.9 we have

$$E_{ti}^{(-)} = E_{ti}^{(+)}$$

and would not need a superscript for the electric field but for

$$H_{ti}^{(-)} = -H_{ti}^{(+)}$$

the superscript indicates a change in sign between the transverse magnetic field of forward and backward travelling modes.

The coupling coefficients according to eqn. 2.9.60 contain in the integrand the derivatives with respect to z of local normal mode fields, they are hence not so convenient to evaluate. We arrive at a simpler although approximate expression for κ by considering Maxwell's eqns. 2.9.42 and inserting into them the fields of local normal modes. When we decompose the curl operator according to eqn. 2.9.43 and take account of the exponential z-dependence of normal mode field we obtain

$$\nabla_t \times \boldsymbol{E}_k^{(p)} - jp\beta_k (\boldsymbol{n}_z \times \boldsymbol{E}_k^{(p)}) = j\omega\mu_0 \boldsymbol{H}_k^{(p)} \tag{2.9.61}$$

$$\nabla_t \times \boldsymbol{H}_k^{(p)} - jp\beta_k (\boldsymbol{n}_z \times \boldsymbol{H}_k^{(p)}) = j\omega\epsilon\epsilon_0 \boldsymbol{E}_k^{(p)} \tag{2.9.62}$$

The local normal mode fields by themselves do not exactly satisfy these field equations, but for gradual transitions they obey these equations to a rather good approximation. Also for such gradual transitions the coupling coefficients are quite small and any error which we make by evaluating them with the help of eqns. 2.9.61 and 2.9.62 remains secondary in nature.

To obtain an expression as it appears in the integrand of eqn. 2.9.60 we take the scalar product of the conjugate complex of eqn. 2.9.61 with $\partial \boldsymbol{E}_i^{(q)}/\partial z$ and add to it the scalar product of the conjugate complex of eqn. 2.9.62 with $\partial \boldsymbol{H}_i^{(q)}/\partial z$. When we solve the sum equation for the two terms that appear in eqn. 2.9.60 we obtain

$$-jp\beta_k \left[\frac{\partial \boldsymbol{E}_i^{(q)}}{\partial z} \times (\boldsymbol{H}_k^{(p)})^* + (\boldsymbol{E}_k^{(p)})^* \times \frac{\partial \boldsymbol{H}_i^{(q)}}{\partial z}\right] \cdot \boldsymbol{n}_z$$

$$= -\frac{\partial \boldsymbol{E}_i^{(q)}}{\partial z} \cdot \nabla_t \times (\boldsymbol{H}_k^{(p)})^* + \frac{\partial \boldsymbol{H}_i^{(q)}}{\partial z} \cdot \nabla_t \times (\boldsymbol{E}_k^{(p)})^* - j\omega\epsilon\epsilon_0 (\boldsymbol{E}_k^{(p)})^* \cdot \frac{\partial \boldsymbol{E}_i^{(q)}}{\partial z}$$

$$-j\omega\mu_0 (\boldsymbol{H}_k^{(p)})^* \cdot \frac{\partial \boldsymbol{H}_i^{(p)}}{\partial z} \tag{2.9.63}$$

Next we write both eqns. 2.9.61 and 2.9.62 for the mode i and differentiate them partially with respect to z. Then we take the scalar product of the result from eqn. 2.9.61 with $(\boldsymbol{E}_k^{(p)})^*$ and add to it the scalar product of the result from eqn. 2.9.62 with $(\boldsymbol{H}_k^{(p)})^*$. We solve again for the two terms that appear in eqns. 2.9.60 and obtain

$$jq\beta_i \left[\frac{\partial \boldsymbol{E}_i^{(q)}}{\partial z} \times (\boldsymbol{H}_k^{(p)})^* + (\boldsymbol{E}_k^{(p)})^* \times \frac{\partial \boldsymbol{H}_i^{(q)}}{\partial z}\right] \cdot \boldsymbol{n}_z$$

$$= (\boldsymbol{H}_k^{(p)})^* \cdot \nabla_t \times \frac{\partial \boldsymbol{E}_i^{(q)}}{\partial z} - (\boldsymbol{E}_k^{(p)})^* \cdot \nabla_t \times \frac{\partial \boldsymbol{H}_i^{(q)}}{\partial z} + j\omega\epsilon\epsilon_0 (\boldsymbol{E}_k^{(p)})^* \cdot \frac{\partial \boldsymbol{E}_i^{(q)}}{\partial z}$$

$$+j\omega\mu_0 (H_k^{(p)})^* \cdot \frac{\partial H_i^{(q)}}{\partial z} + j\omega\epsilon_0 \frac{\partial \epsilon}{\partial z} E_i^{(q)} (E_k^{(p)})^*$$

$$-jq\frac{\partial \beta_i}{\partial z} n_z [E_i^{(q)} \times (H_k^{(p)})^* + (E_k^{(p)})^* \times H_i^{(q)}] \tag{2.9.64}$$

Adding both these equations and integrating the result over the waveguide cross-section leads to

$$j(q\beta_i - p\beta_k) \iint \left[\frac{\partial E_i^{(q)}}{\partial z} \times (H_k^{(p)})^* + (E_k^{(p)})^* \times \frac{\partial H_i^{(q)}}{\partial z}\right] \cdot n_z dS$$

$$= \iint \nabla_t \cdot \left[\frac{\partial E_i^{(q)}}{\partial z} \times (H_k^{(p)})^* - (E_k^{(p)})^* \times \frac{\partial H_i^{(q)}}{\partial z}\right] dS$$

$$-j\frac{\partial \beta_i}{\partial z} \iint n_z \cdot [E_i^{(q)} \times (H_k^{(p)})^* + (E_k^{(p)})^* \times H_i^{(q)}]\, dS$$

$$+j\omega\epsilon_0 \iint \frac{\partial \epsilon}{\partial z} E_i^{(q)} \cdot (E_k^{(p)})^* dS \tag{2.9.65}$$

The first term on the right-hand side with the surface integral of a divergence converts to a line integral of the vector products around the open waveguide at an infinite distance. This line integral in turn vanishes for guided modes, but also for radiation modes as long as $i \neq k$. The second term on the right-hand side vanishes for guided and radiation modes by virtue of their orthogonality, if only $i \neq k$. All that remains on the right-hand side for $i \neq k$ is the last term. With it the mutual coupling coefficients for $i \neq k$ are given by

$$\kappa_{ki}^{(p,q)} = \frac{p\omega\epsilon_0}{2(p\beta_k - q\beta_i)} \iint \frac{\partial \epsilon}{\partial z} (E_k^{(p)})^* \cdot E_i^{(q)} dS \tag{2.9.66}$$

For $i = k$ not only this last term but also the first two terms on the right-hand side of eqn. 2.9.65 contribute to the coefficient $\kappa_{ii}^{(p,q)}$. Any coefficient $\kappa_{ii}^{(p,p)}$, however, represents only a modification of the propagation constant $j\beta_i$ of that particular mode. This modification remains small, and hence insignificant for gradual transitions. A coefficient $\kappa_{ii}^{(p,q)}$ with different signs p and q couples forward and backward waves of one and the same mode with each other. From this interaction a mode travelling in one direction will reflect part of its power. For gradual transitions this reflection within one mode remains small compared to the power conversion between modes which travel in the same direction. We therefore have no need to consider it here.

The actual coupling coefficients in eqn. 2.9.66 depend in their magnitude on the rate with which the permittivity distribution over the cross-section of the dielectric waveguide changes in axial direction. For low rates of change in gradual transitions the coupling is rather weak. The coupling coefficients depend also on the scalar product of electric fields in the regions of permittivity change. If the electric field of one or both modes is small in these regions or if both fields are polarised at right angles with respect to each other the coupling will also remain weak. Finally the axial phase constants of coupled modes appear in the denominator of the coupling coefficients. If both modes travel in the same direction so that p and q have the same sign, this denominator consists of the difference of the phase constants of both modes and may render the coupling coefficient relatively large in magnitude, in particular when both modes are close together in their phase constants.

Alternatively, if the two modes travel in opposite directions so that p and q are of opposite sign, the sum of the coupling coefficients appears in the denominator of eqn. 2.9.66, and the coupling remains usually quite weak. For the same reason a guided mode couples more strongly only to those radiation modes that propagate at small angles θ in nearly the same direction as the guided mode. It scatters power into the radiation field mainly in forward direction. For larger propagation angles the phase constant β_l of the radiation mode becomes so small and then differs so much from the β_k of guided modes that the large denominator of eqn. 2.9.66 renders the coupling quite weak. Guided modes in gradual transitions scatter hardly any power sideways or in backward direction.

To further evaluate the mode coupling according to eqn. 2.9.66 we need to know the cross-sectional permittivity distribution of the dielectric waveguide, and its change in axial direction along the transition. In the case of dielectric slabs, films or strips and rods the permittivity changes more or less abruptly between an inner region of high permittivity and an outer region of lower permittivity. For a uniform dielectric waveguide of this kind the boundary between these two regions may be specified in terms of suitable cross-section co-ordinates x and y as a function $x = f(y)$ which is independent of z. For the non-uniform case of a waveguide transition however, this function depends also on z

$$x = f(y, z) \tag{2.9.67}$$

We restrict attention here to waveguide transitions in which the boundary may locally be specified by either $x = f(z)$ or $y = f(z)$, and have in mind transitions for slabs and films or dielectric strips of rectangular geometry, as well as round waveguide transitions with their radius $x = b(z)$ changing axially but being independent of the angle φ of cylinder co-ordinates. Fig. 2.20 shows a longitudinal section of such a waveguide transition. In this section $b(z)$ defines the boundary and $\boldsymbol{n}$ is the unit vector normal to the boundary. To evaluate the integral in eqn. 2.9.66 for the abrupt permittivity transition from ϵ_1 to ϵ_2 at this boundary we express $\partial\epsilon/\partial z$ by the derivative of ϵ normal to the boundary

$$\frac{\partial \epsilon}{\partial z} = \frac{\partial \epsilon}{\partial n}\frac{\partial n}{\partial z} = -\frac{\partial \epsilon}{\partial n}\sin\phi \tag{2.9.68}$$

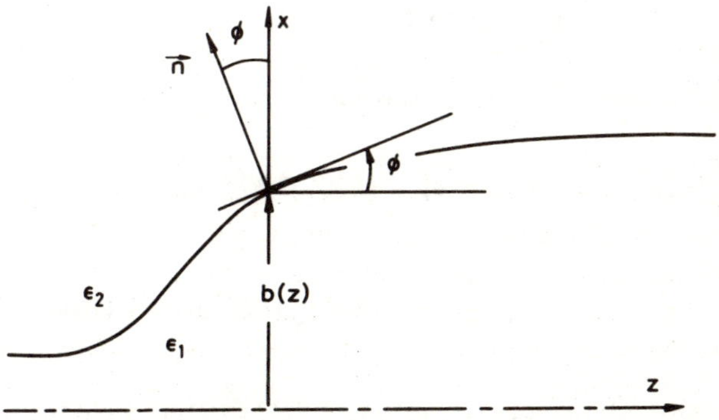

Fig. 2.20 *Longitudinal section of the boundary between regions of different permittivity in a dielectric waveguide transition*

where ϕ is the angle that the longitudinal tangent to the boundary makes with the z-direction. For the abrupt transition the normal derivative $\partial\epsilon/\partial n$ does not really exist, but we can consider this abrupt transition as the limiting case of a gradual transition, that changes permittivity in a very narrow interval. The electric-field components in the integrand of eqn. 2.9.66 which are tangential to the boundary do not change within this narrow interval. Their contribution to the coupling coefficient is hence obtained from

$$\iint \frac{\partial \epsilon}{\partial z}(E_{k\,\tan}^{(p)})^* \cdot E_{i\,\tan}^{(q)}\,dS =$$

$$-\int_c (E_{k\,\tan}^{(p)})^* \cdot E_{i\,\tan}^{(q)} \sin\phi\,dc \int \frac{\partial \epsilon}{\partial n}dx \tag{2.9.69}$$

where the integration $\int_c dc$ extends along the boundary of the local cross-section and where the integral across the boundary may, with $dx = dn/\cos\phi$, be evaluated as

$$-\sin\phi \int \frac{\partial\epsilon}{\partial n} dx = -\tan\phi \int \frac{\partial\epsilon}{\partial n} dn = (\epsilon_1 - \epsilon_2)\tan\phi \tag{2.9.70}$$

The electric-field components which at the boundary are normal to it change across the boundary according to

$$E_n = \frac{\epsilon_1}{\epsilon(n)} E_{n1} \tag{2.9.71}$$

where E_{n1} is the normal electric field component on that side of the boundary with $\epsilon = \epsilon_1$. These normal electric-field components contribute to the integral in eqn. 2.9.66 with

$$\iint \frac{\partial\epsilon}{\partial z}(E_{kn}^{(p)})^* E_{in}^{(q)} dS = -\epsilon_1^2 \int_c (E_{kn1}^{(p)})^* E_{in1}^{(q)} \tan\phi dc \int_{\epsilon_1}^{\epsilon_2} \frac{d\epsilon}{\epsilon^2}$$

$$= \frac{\epsilon_1}{\epsilon_2}(\epsilon_1 - \epsilon_2) \int_c (E_{kn1}^{(p)})^* E_{in1}^{(q)} \tan\phi dc \tag{2.9.72}$$

Both contributions eqns. 2.9.69 and 2.9.72 together yield the following expression for the coupling coefficients

$$\kappa_{ki}^{(p,q)} = \frac{p\omega\epsilon_0(\epsilon_1 - \epsilon_2)}{2(p\beta_k - q\beta_i)} \int_c \left[(E_{k\tan}^{(p)})^* \cdot E_{i\tan}^{(q)} + \frac{\epsilon_1}{\epsilon_2}(E_{kn1}^{(p)})^* E_{in1}^{(q)} \right] \frac{\partial b}{\partial z} dc \tag{2.9.73}$$

The tangent of the angle ϕ equals the rate at which the waveguide dimension b changes with z, and in eqn. 2.9.73 has been replaced accordingly.

To illustrate this rather general result we return to the simple example of the dielectric slab, and consider the transition between two slabs of different thickness. To avoid any interaction between even and odd modes we let the transition be symmetrical with respect to the centre plane $x = 0$ of the slabs as in Fig. 2.21. For slab modes propagating in z-direction the fields do not change in the transverse direction parallel to the slabs. To evaluate the coupling coefficients according to eqn. 2.9.73 we only need to insert the appropriate field quantities at the boundaries but no integration along the boundary is necessary. Of the guided modes of the slab both the lowest order even TE- and TM-mode have zero cutoff. But the lowest-order even TE-modes is under all conditions best confined to the slab. It is hence the fundamental mode of the slab. If this even TE-mode is incident on the slab transition it couples to all other even TE-modes, both guided and radiation modes. For the interaction between guided modes of the even TE-type we obtain, with the electric field in eqn. 2.9.10 normalised according to eqn. 2.9.15, the following coupling coefficient

$$K_{ki}^{(p,q)} = \frac{pk_0^2(\epsilon_s - \epsilon)\cos(\tau_{sk}\,d)\cos(\tau_{si}d)}{(p\beta_k - q\beta_i)\left[\beta_k\beta_i\left(d+\frac{1}{\alpha_k}\right)\left(d+\frac{1}{\alpha_i}\right)\right]^{1/2}}\,\frac{d}{dz}(d) \tag{2.9.74}$$

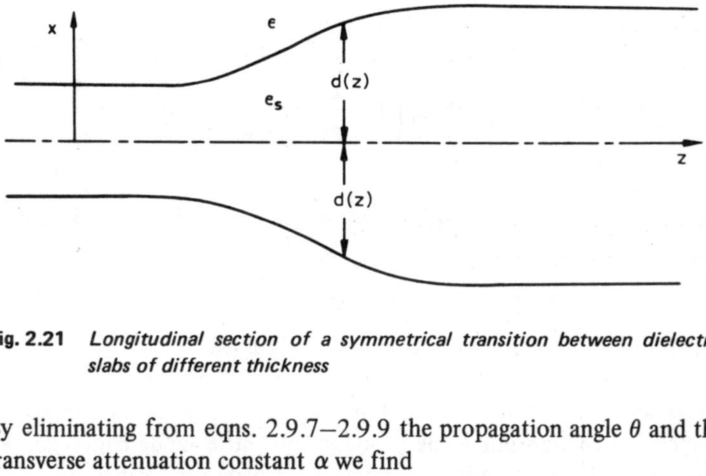

Fig. 2.21 *Longitudinal section of a symmetrical transition between dielectric slabs of different thickness*

By eliminating from eqns. 2.9.7–2.9.9 the propagation angle θ and the transverse attenuation constant α we find

$$k_0\sqrt{\epsilon_s - \epsilon}\;\cos(\tau_s d) = \tau_s \tag{2.9.75}$$

with which the coupling coefficient between guided modes of the even TE-type reduces to

$$K_{ki}^{(p,q)} = \frac{p\tau_{sk}\tau_{si}}{(p\beta_k - q\beta_i)\left[\beta_k\beta_i\left(d+\frac{1}{\alpha_k}\right)\left(d+\frac{1}{\alpha_i}\right)\right]^{1/2}}\,\frac{d}{dz}(d) \tag{2.9.76}$$

For interaction between the guided mode k of the even TE-type and the continuous spectrum of even TE-radiation modes we obtain the coupling coefficient by substituting for the field of radiation modes from eqn. 2.9.25 and by normalising this field according to eqn. 2.9.31. For the field of guided modes we substitute as before from eqns. 2.9.10 and 2.9.15. The result appears as

$$\kappa_k^{(p,q)}(\tau) = \frac{pk_0^2(\epsilon_s - \epsilon)\cos(\tau_{sk}d)\cos\phi}{(p\beta_k - q\beta_\tau)\left[\pi\beta_k\beta_\tau\left(d + \frac{1}{\alpha_k}\right)\right]^{1/2}} \frac{d}{dz}(d) \quad (2.9.77)$$

The phase shift ϕ for the outside field of the radiation mode follows from eqn. 2.9.27, while with eqn. 2.9.75 we may replace $\cos(\tau_{sk}d)$ in eqn. 2.9.77. The coupling between guided and radiation modes of the even TE-type is then given by

$$\kappa_k^{(p,q)}(\tau) = \frac{pk_0\tau_k\tau}{p\beta_k - q\beta_\tau} \left[\frac{\epsilon_s - \epsilon}{\pi\beta_k\beta_\tau\left(d + \frac{1}{\alpha_k}\right)(\tau^2 + \tau_s^2\tan^2(\tau_s d))}\right]^{1/2} \frac{d}{dz}(d)$$

(2.9.78)

It is a function of the continuous variable τ, with τ_s and β_τ also depending on τ according to

$$\beta_\tau = \sqrt{\epsilon k_0^2 - \tau^2} \qquad \tau_s = (\epsilon_s - \epsilon)k_0^2 + \tau^2 \quad (2.9.79)$$

In addition to the interaction among guided modes and between guided and radiation modes there is also mutual interaction between radiation modes with different parameters τ. For gradual transitions which are to transmit guided modes with little mode conversion any interaction between radiation modes is of little significance, and we do not need it in the analysis and the design of these transitions. We also do not need to consider the interaction between waves which travel in opposite directions because, due to the large factor $(\beta_k + \beta_i)$ in the denominator of the corresponding coupling coefficients, this interaction stays quite weak. All that remains under these restrictions, of the general form of coupled wave eqns. 2.9.58 is one set of equations for the rate of change of guided wave amplitudes

$$\frac{da_k^{(+)}}{dz} = \sum_i \kappa_{ki}^{(+,+)} a_i^{(+)} \exp\left(j\int_0^z (\beta_k - \beta_i)dz\right)$$

$$+\int_0^{\epsilon k_0} \kappa_k^{(+,+)}(\tau)a^{(+)}(\tau)\exp\left(j\int_0^z(\beta_k-\beta_\tau)dz\right)d\tau \tag{2.9.80}$$

and one equation for the rate of change of the amplitude function $a^{(+)}(\tau)$ of radiation modes

$$\frac{\partial a^{(+)}(\tau)}{\partial z}=\sum_i \kappa_i^{(+)}(\tau)a_i^{(+)}\exp\left(j\int_0^z(\beta_\tau-\beta_i)dz\right) \tag{2.9.81}$$

The summation in these expressions extends over the finite number of guided modes. The last eqn. 2.9.81 gives the rate of change of radiation mode amplitudes only to the extent that they are excited by guided modes.

References

1 HARRINGTON, R. F.: 'Time-harmonic electromagnetic fields' (McGraw-Hill, 1961)
2 COLLIN, R. E.: 'Field theory of guided waves' (McGraw-Hill, 1960)
3 SCHELKUNOFF, S. A.: 'Conversion of Maxwell's Equations into generalized telegraphist's equations', *Bell Syst. Tech. J.*, 1955, **34**, pp. 995–1043
4 REITER, G.: 'Generalized telegraphist's equation for waveguides of varying cross-section', *Proc. IEE,* 1959, **106B**, pp. 54–57
5 SOLYMAR, L.: 'Spurious mode generation in non-uniform waveguide', *IEEE Trans.*, 1959, **MTT-7**, pp. 379–383
6 SPORLEDER, F.: 'Erweiterte Theorie der Lochkopplung'. Dr.-Ing.-Thesis, Technische, Universität Braunschweig, Germany, 1976
7 MARCUSE, D.: 'Theory of dielectric optical waveguides' (Academic Press, New York and London, 1974)
8 MARCUSE, D.: 'The coupling of degenerate modes in two parallel dielectric waveguides', *Bell Syst. Tech. J.*, 1971, **50**, pp. 1791–1816
9 SNYDER, A. W.: 'Coupled mode theory for optical fibers', *J. Opt. Soc. Amer.* 1972, **62**, pp. 1267–1277
10 KUESTER, E. F., and CHANG, D. C.: 'Nondegenerate surface-wave mode coupling between dielectric waveguides', *IEEE Trans.*, 1975, **MTT-23**, pp. 877–882

Chapter 3

Taper analysis with horn modes

In Chapter 2 the coupled wave equations have been derived for a number of transmission-line arrangements which exhibit mode coupling. Usually, uniform cylindrical waveguides are connected to these coupling sections. Therefore it seems appropriate to represent the fields in the coupling sections in terms of cylindrical waveguide modes which at the ends of the coupling sections are identical with the normal modes of the adjoining uniform guides. In principle, however, many different sets of modes can be used to represent the fields in the coupling sections. If the corresponding coupled wave equations are determined properly and if they are solved accurately all these presentations will render identical overall transmission properties of the coupling structures at the respective ports.

Exact solutions of the coupled wave equations cannot, however, in general be found. Thus approximate solutions which consider only a limited number of the infinite set of coupled modes with sometimes only approximate values of the coupling coefficients must be used. The results obtained by applying these approximate calculation methods become more accurate, if by a suitable choice of modes the mode coupling per unit length in the coupling structure can be kept weak. If the overall transmission characteristics are to be calculated to a high degree of accuracy the modes in the inhomogeneous or coupling section should be chosen such that any of the modal fields by themselves represent good approximations to the solution of the coupled wave equations under the respective boundary conditions for any particular point within the inhomogeneous or coupling section.

Such modes will then show small interaction per unit length, and therefore any solution of the coupled wave equations will converge rapidly.[1–3]

To investigate these convergence properties in more detail we choose the two examples of tapered waveguide sections of Fig. 2.9*b* and 2.9*c* which have already been treated in Section 2.6. There, the field has been represented by normal modes of a cylindrical waveguide, corresponding in cross-section to the respective local cross-section of the taper.

In contrast with this presentation in terms of cylindrical waveguide modes we will now use modes which are the solutions of the wave equations of uniform horns. These normal modes of uniform horns we refer to as horn modes.

The sector horn and round horn shown in Fig. 3.1 are two examples of uniform horns. Quite generally we call a horn uniform when its metallic walls coincide with co-ordinate surfaces of co-ordinate systems in which the wave equations can be solved by the separation of variables.

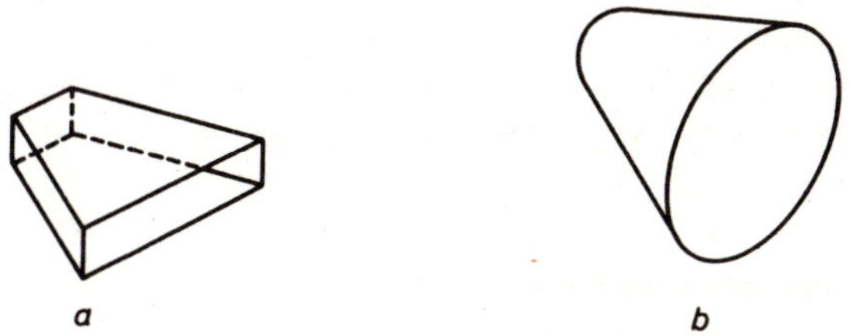

Fig. 3.1 *Uniform horns*
a sector horn
b round horn

In the case of the sector horn the walls are co-ordinate surfaces of the co-ordinates of the circular cylinder and normal modes are derived from cylindrical wave functions. With the round horn the wall is a co-ordinate surface of spherical co-ordinates and modes are derived from spherical wave functions. These modes propagate independently of each other along the uniform horn. Horns which do not fit into a system of separable co-ordinates are called nonuniform. Such nonuniform horns occur as waveguide tapers to join waveguides of different sizes as well as transitions between different uniform horns.

When deriving the coupled wave equations for horn modes two distinct difficulties arise. First, the propagation characteristics of horn modes are significantly more complex than those of the cylindrical waveguide modes. The second difficulty concerns the natural co-ordinates of the nonuniform horn. Natural co-ordinates of the horn are those co-ordinates that have the walls of the horn as co-ordinate surfaces. These natural co-ordinates are in general nonorthogonal, and therefore give rise to off-diagonal terms in their matrix of metrical coefficients.

The present analysis investigates these points in detail, and leads in principle to accurate solutions also for such singular situations where local normal modes are at cutoff either within the nonuniform horn or at the cylindrical ends of the waveguide taper.

The solution which takes full account of the nonorthogonality of natural co-ordinates will be derived here, by way of example, for the relatively simple structure of the sector horn. For this sector-horn solution a method will be developed which bears close resemblance to the method of effective sources which we used before. Subsequently a solution for the round horn will be derived directly from Maxwell's equations. For both cases the degree of nonorthogonality of the natural co-ordinates will be inspected. For axially-symmetric fields in the round horn wave propagation will be analysed with one formulation for the solution above cutoff, and another near and below cutoff.

The theory of coupled horn modes presented in this Chapter is mainly based on the results of References 4 and 5.

3.1 Nonuniform sector horn

The waveguide structure to be analysed is shown in Fig. 3.2*a* with Fig. 3.2*b* as a longitudinal section through this structure. The overall arrangement is assumed symmetrical with respect to the plane $x = 0$. The waveguide height b is constant all through the transition section. Any point P within the ideally conducting boundaries can be specified by the natural co-ordinates y, η, z. They have the boundaries $y = 0, b$ and $\varphi = \pm\phi$ and are locally circularly cylindrical, with y as the axial distance of the local circular cylinder, and $\eta = \varphi/\phi$ as the local azimuth angle relative to half the local flare angle of the horn. The co-ordinate z of point P obtains as the distance along the ζ-axis at which the circular cylinder cuts through the taper contour at right angles. z therefore measures the axial distance along the taper contour. The local taper with

$$a = \rho \sin\phi \tag{3.1.1}$$

changes according to

$$\frac{da}{dz} = a' = \tan\phi \tag{3.1.2}$$

with axial distance, while the flare angle changes according to

$$\frac{d\phi}{dz} = \phi' = \frac{a''}{1+(a')^2} \tag{3.1.3}$$

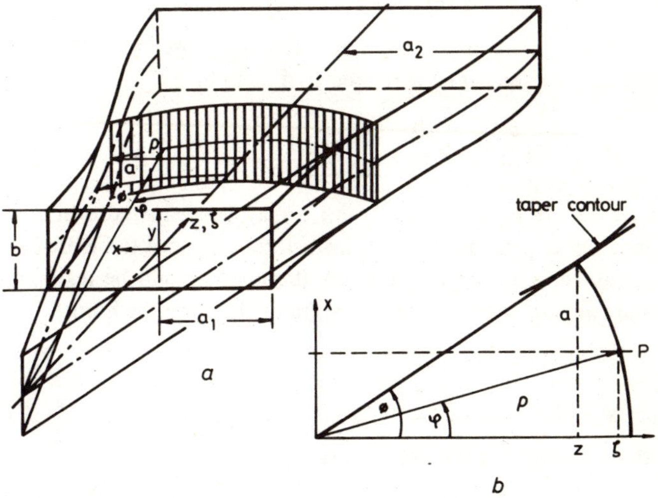

Fig. 3.2 *Sectoral horn with natural co-ordinates*
a perspective view
b longitudinal section

and the radius ρ of the local cylinder surface according to

$$\frac{d\rho}{dz} = \rho' = \frac{a'}{\sin\phi} - a\phi'\frac{\cos\phi}{\sin^2\phi} \tag{3.1.4}$$

Also the Cartesian co-ordinates x and ζ of point P are related to the natural co-ordinates η and z by

$$\zeta = z + \rho(\cos(\eta\phi) - \cos\phi) \tag{3.1.5a}$$

and

$$x = \rho\sin(\eta\phi) \tag{3.1.5b}$$

For later evaluation of the curl operation in these natural co-ordinates the differential distance vector

$$dr = \boldsymbol{u}_y dy + \boldsymbol{u}_\eta d\eta + \boldsymbol{u}_z dz \tag{3.1.6}$$

will be needed. $\boldsymbol{u}_y$ represents the unit vector in the y-direction, while $\boldsymbol{u}_\eta$ is given by

$$\boldsymbol{u}_\eta = \boldsymbol{n}_\eta \rho\phi = \boldsymbol{n}_\eta \frac{a\phi}{\sin\phi} = \boldsymbol{n}_\eta [1 + \phi^2/6 + O(\phi^4)]\, a \qquad (3.1.7)$$

with $\boldsymbol{n}_\eta = \boldsymbol{n}_\varphi$ as the unit vector in the φ-direction. $O(x)$ denotes a term proportional to x. Alternatively $\boldsymbol{u}_\eta$ is given by

$$\boldsymbol{u}_\eta = u_\eta [-\boldsymbol{n}_\zeta \sin(\eta\phi) + \boldsymbol{n}_x \cos(\eta\phi)] \qquad (3.1.8)$$

with $\boldsymbol{n}_x$ and $\boldsymbol{n}_\zeta$ as the unit vectors in the x- and ζ-direction, respectively. Furthermore, the vector $\boldsymbol{u}_z$ is related to $\boldsymbol{n}_x$ and $\boldsymbol{n}_\zeta$ by

$$\boldsymbol{u}_z = \boldsymbol{n}_\zeta \frac{d\zeta}{dz} + \boldsymbol{n}_x \frac{dx}{dz} \qquad (3.1.9)$$

which by employing eqns. 3.1.1–3.1.5 may be evaluated exactly. The resulting expressions are, however, much too involved to be used in the subsequent computations. Therefore the analysis will be restricted to small flare angles, which change only gradually along the horn:

$$\phi^2 \ll 1, \qquad (3.1.10a)$$

and

$$\phi' \ll 1/a \qquad (3.1.10b)$$

With these restrictions the approximations

$$\frac{d\zeta}{dz} \approx 1 + \frac{\phi^2}{2}(1-\eta^2) + \frac{\phi' a}{2}(1-\eta^2)$$

$$\frac{dx}{dz} \approx \eta\phi\left(1 + \frac{\phi^2}{2} - \eta^2\frac{\phi^2}{6}\right) + \frac{\phi' a}{3}(1-\eta^2)\phi\eta \qquad (3.1.11)$$

lead to

$$u_z = 1 + \frac{\phi^2}{2} + \frac{\phi' a}{2}(1-\eta^2) + O(\phi^4, \phi'\phi^2, (\phi')^2) \qquad (3.1.12)$$

$\boldsymbol{u}_z$ is not quite orthogonal to $\boldsymbol{u}_\eta$. With the angle between the vectors $\boldsymbol{u}_\eta$ and $\boldsymbol{u}_z$ expressed as

$$\measuredangle(\boldsymbol{u}_\eta, \boldsymbol{u}_z) = \frac{\pi}{2} - \alpha \qquad (3.1.13)$$

where

$$\sin\alpha = \frac{\boldsymbol{u}_\eta \cdot \boldsymbol{u}_z}{u_\eta u_z} \qquad (3.1.14)$$

the deviation α from orthogonality follows from inserting the expressions 3.1.8, 3.1.9, 3.1.11 and 3.1.12 into eqn. 3.1.14

$$\sin\alpha = -\frac{\eta\phi}{6}\phi' a(1-\eta^2)[1 + O(\phi^2, \phi')] \qquad (3.1.15)$$

The curl of any vector V may be evaluated in these co-ordinates according to Reference 6.

$$\nabla \times V = \frac{1}{u_z u_\eta \cos\alpha} \begin{vmatrix} u_z & u_\eta & u_y \\ \dfrac{\partial}{\partial z} & \dfrac{\partial}{\partial \eta} & \dfrac{\partial}{\partial y} \\ u_z \cdot V & u_\eta \cdot V & u_y \cdot V \end{vmatrix} \tag{3.1.16}$$

Attention will be restricted here to field distributions which are excited when the dominant mode of the rectangular waveguide with its electric field polarised in the y-direction is incident. For such distributions the electric-field vector remains parallel to the y-direction and the magnetic-field vector parallel to the $x\zeta$-plane throughout the horn. The latter will be split here into two orthogonal components of which one will be taken as the η-component parallel to u_η while the other will be called the ρ-component parallel to the unit vector

$$n_\rho = n_\eta \times n_y \tag{3.1.17}$$

It is our aim to represent the field distribution within the taper by local normal modes of the respective uniform sector horn, which change along the horn and interact with one another due to the changing flare angle. To determine the interaction between these horn modes we select a particular section of the given taper between $z_1 \leqslant z \leqslant z_1 + \Delta z$ and join uniform horns to it on both sides as shown in Fig. 3.3.

The electric field of a horn wave in a uniform horn is obtained as a solution of the wave equation in the system of cylindrical co-ordinates ρ, φ, y

$$E_n^{(\pm)} = n_y e_{yn}^{(\pm)} \tag{3.1.18}$$

with

$$e_{yn}^{(\pm)} = a_n^{(\pm)}(\rho_0) \cos\left(\frac{n\pi}{2}\frac{\varphi}{\phi}\right) \frac{H_\nu^{\overset{(2)}{(1)}}(k\rho)}{H_\nu^{\overset{(2)}{(1)}}(k\rho_0)} (-1)^{(n-1)/2} \tag{3.1.19a}$$

$a_n^{(\pm)}(\rho_0)$ represents the amplitude of the forward (+) and backward (−) travelling horn wave, respectively, at $\rho = \rho_0$. Only modes with $n = 1, 3, 5 \ldots$ are coupled to the fundamental mode which is normally incident from one of the connecting cylindrical waveguides.

The quantity k represents the wavenumber

$$k = \omega\sqrt{\mu\epsilon} \tag{3.1.19b}$$

inside the horn, with ω as the circular frequency of the steady-state oscillation and μ and ϵ as permeability and permittivity of the homogeneous and isotropic material filling the horn. $H_\nu^{(i)}(k\rho)$ designates the Hankel function of the ith kind. The order ν of the Hankel function for the nth mode obtains as

$$\nu = \frac{n\pi}{2\phi} \tag{3.1.19c}$$

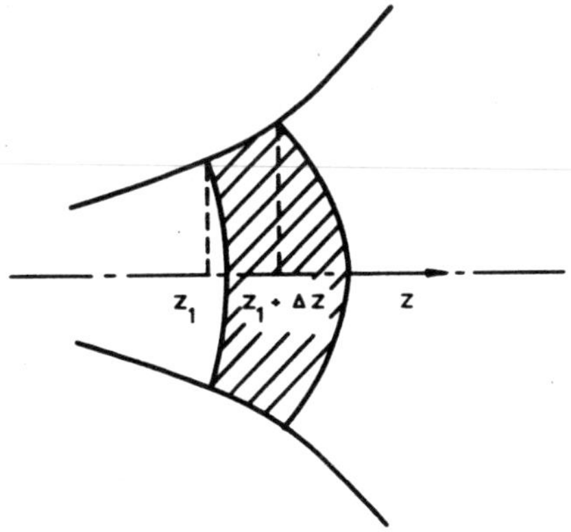

Fig. 3.3 *Short nonuniform section between uniform horns*

The factor $(-1)^{(n-1)/2}$ in eqn. 3.1.19*a* is introduced to simplify the coupling results.

The corresponding magnetic field is given by

$$H_n^{(\pm)} = \mathbf{n}_\eta h_{\eta n}^{(\pm)} + \mathbf{n}_\rho h_{\rho n}^{(\pm)} \tag{3.1.18b}$$

with

$$h_{\eta n}^{(\pm)} = \mp Y_n^{(\pm)} e_{yn}^{(\pm)} \tag{3.1.18c}$$

and

$$h_{\rho n}^{(\pm)} = -j \frac{n}{2\omega\mu a} \tan\left(\frac{n\pi}{2}\eta\right) e_{yn}^{(\pm)} \tag{3.1.18d}$$

$Y_n^{(\pm)}$ represents the wave admittance, which in general is complex and given by

$$Y_n^{(\pm)} = \pm jY_0 \frac{H_\nu^{\substack{(2)\\(1)}}(k\rho)'}{H_\nu^{\substack{(2)\\(1)}}(k\rho)} \tag{3.1.20a}$$

$$Y_0 = \sqrt{\epsilon/\mu} \tag{3.1.20b}$$

The prime at the Hankel function denotes differentiation with respect to the argument.

To utilise these local normal mode fields in the nonuniform horn section we take their components in the co-ordinate system y, η, z which is defined not only for uniform but for nonuniform horns as well. For this purpose eqn. 3.1.19*a* is written as

$$e_{yn}^{(\pm)} = a_n^{(\pm)}(z)(-1)^{(n-1)/2} \cos\left(\frac{n\pi}{2}\eta\right)$$

$$= a_n^{(\pm)}(z_1)(-1)^{(n-1)/2} \cos\left(\frac{n\pi}{2}\eta\right) \exp\left[\mp j \int_{z_1}^{z} \beta_n^{(\pm)} dz\right] \tag{3.1.21}$$

with

$$\beta_n^{(\pm)} = \frac{k}{\cos\phi} \frac{Y_n^{(\pm)}}{Y_0} \tag{3.1.20c}$$

The modal field according to eqn. 3.1.21 with eqns. 3.1.20 and 3.1.18 satisfies Maxwell's equations of course only in a uniform horn.

If such a horn mode is incident from either side on to the non-uniform section of Fig. 3.3 other forward and backward travelling horn modes will be excited. Excitation of such spurious modes can, however, be suppressed by introducing within the section from z_1 to $z_1 + \Delta z$ electric and magnetic current sources of such a distribution that the field of the incident horn mode alone together with these sources satisfies Maxwell's equations.

These fictitious sources of electric current $-\boldsymbol{i}_n^{(q)}$ as well as magnetic current $-\boldsymbol{m}_n^{(q)}$ will be calculated here such that a horn mode with index n travelling in the qz-direction passes through the nonuniform section without any disturbance. By not allowing any field discontinuities and the associated surface currents the fictitious current distributions follow from Maxwell's equations as

$$-\boldsymbol{i}_n^{(q)} = \nabla \times \boldsymbol{H}_n^{(q)} - j\omega\epsilon \boldsymbol{E}_n^{(q)} \tag{3.1.22a}$$

$$-\boldsymbol{m}_n^{(q)} = -\nabla \times \boldsymbol{E}_n^{(q)} - j\omega\mu \boldsymbol{H}_n^{(q)} \tag{3.1.22b}$$

These expressions may be evaluated by substituting for $\boldsymbol{E}_n^{(q)}$ and $\boldsymbol{H}_n^{(q)}$ from eqns. 3.1.18 and 3.1.21. With the restriction to small flare angles ϕ and a low rate of change ϕ' only those terms need to be taken into account which are proportional to ϕ' and $\phi'\phi$. Higher-order terms in ϕ' and ϕ can be neglected.

To this approximation the nonorthogonality factor $\cos\alpha$ in the curl operation (eqn. 3.1.16) remains insignificant. A contribution to the current sources which is due to nonorthogonality arises only when the rectangular field components are inserted in eqn. 3.1.16 and the rectangular components of $\boldsymbol{m}$ and $\boldsymbol{i}$ are taken from the oblique components of eqn. 3.1.22. Thus the η-component of the magnetic current density appears as

$$-m_\eta^{(q)} = \frac{1}{u_z u_\eta \cos\alpha}\left[u_\eta \frac{\partial}{\partial z} e_{yn}^{(q)} - \sin\alpha(u_z)\frac{\partial}{\partial\eta} e_{yn}^{(q)}\right] - j\omega\mu h_{\eta n}^{(q)} \tag{3.1.23a}$$

and the y-component of the electric current density as

$$-i_y^{(q)} = \frac{1}{u_z u_\eta \cos\alpha}\left[\frac{\partial}{\partial z}(u_\eta h_{\eta n}^{(q)}) - \frac{\partial}{\partial\eta}(u_z \cos\alpha(h_{\rho n}^{(q)}) + u_z \sin\alpha(h_{\eta n}^{(q)}))\right] - j\omega\epsilon e_{yn}^{(q)} \tag{3.1.23b}$$

The terms with $\sin\alpha$ are caused by the nonorthogonality of the natural co-ordinates.

Further evaluating these expressions leads to

$$-\boldsymbol{m}_n^{(q)} = \boldsymbol{n}_\eta m_\eta^{(q)} + \boldsymbol{n}_\rho O(\phi'^2\phi^2)$$

$$-\frac{m_\eta^{(q)}}{e_{yn}^{(q)}} = jq\beta_n^{(q)}\frac{a\phi'}{2}(1-\eta^2) + \frac{n}{12}\phi'\phi\tan\left(\frac{n\pi}{2}\eta\right)(\eta^3-\eta) + O(\phi'\phi^2, \phi'^2) \tag{3.1.24a}$$

$$-\boldsymbol{i}_n^{(q)} = -\boldsymbol{n}_y i_y^{(q)}$$

$$-\frac{i_y^{(q)}}{e_{yn}^{(q)}} = -\left(j\beta_n^{(q)} - kq\frac{\phi}{Y_n^{(q)}}\frac{\partial Y_n^{(q)}}{\partial ka}\right)Y_n^{(q)}\frac{a\phi'}{2}(1-\eta^2) - q\phi'\frac{\partial Y_n^{(q)}}{\partial\phi} - j\frac{n\pi}{2\omega\mu a}\tan\left(\frac{n\pi}{2}\eta\right)\phi'\eta - q\frac{n\pi}{12}\phi'\phi Y_n^{(q)}\tan\left(\frac{n\pi}{2}\eta\right)(\eta^3-\eta) + O(\phi'\phi^2, \phi'^2) \tag{3.1.24b}$$

To physically interpret the individual terms of the expressions for i_y and m_η, consider a corner in the taper contour as it occurs in the horn junction of Fig. 3.4. The fictitious currents account for the change of certain propagation characteristics of the field distribution for a horn wave which is incident from the left-hand side on to the surface S_1. With these currents the same horn wave continues to travel in the forward direction.

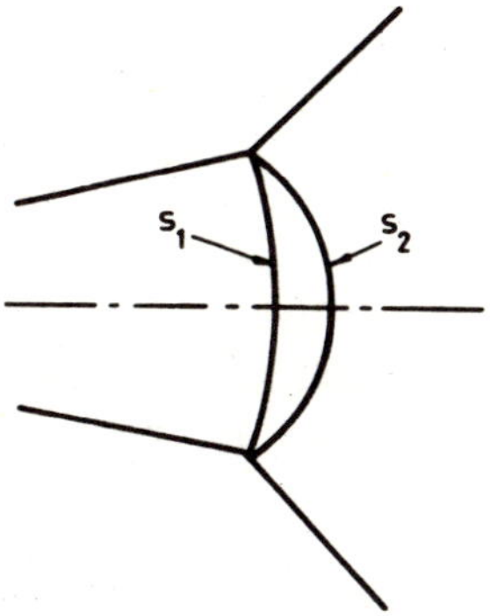

Fig. 3.4 *Horn junction*

The terms

$$-j\left(\beta_n^{(q)} + jk\frac{\phi}{Y_n^{(q)}}\frac{\partial Y_n^{(q)}}{\partial ka}\right)Y_n^{(q)}\frac{a\phi'}{2}(1-\eta^2) \quad \text{in} \quad \frac{i_y^{(q)}}{e_{yn}^{(q)}}$$

and

$$j\beta_n^{(q)}\frac{a\phi'}{2}(1-\eta^2) \quad \text{in} \quad \frac{m_\eta^{(q)}}{e_{yn}^{(q)}}$$

arise from the differences in path length between corresponding points on S_1 and S_2. The term $-\phi'(\partial Y_n^{(q)}/\partial\phi)$ takes account of the change in wave admittance between the horn wave on S_1 and the corresponding wave on S_2. The term

$$-j\frac{n\pi}{2\omega\mu a}\tan\left(\frac{n\pi}{2}\eta\right)\phi'\eta$$

arises from the rotation of the transverse direction with respect to the direction of propagation. Owing to this rotation part of the h_ρ-component from S_1 appears as h_η-component at S_2.

The terms which are proportional to $\phi'\phi$ essentially arise from the nonorthogonality of the natural co-ordinates as it was explained for eqn. 3.1.23. Physically they displace the Poynting vector somewhat when moving from S_1 to S_2. Quasioptically the horn wave would move perpendicular to S_1. The fictitious sources, however, force the wave to propagate obliquely along lines $\eta = \text{constant}$.

The fictitious sources which were introduced to have the horn wave propagate without disturbance through the inhomogeneous section from z_1 to $z_1 + \Delta z$, change the actual source-free problem to a fictitious problem with sources. To return again to the actual source-free problem we introduce another set of source distributions, equal in magnitude but of opposite sign, which compensate the fictitious sources. These sources $+\boldsymbol{m}_n^{(q)}$ and $+\boldsymbol{i}_n^{(q)}$ excite spurious horn modes. Physically these spurious modes represent mode conversion from the primary horn mode at the inhomogeneity of the horn.

To determine the spurious horn mode excitation we assume again that $\boldsymbol{m}_n^{(q)}$ and $\boldsymbol{i}_n^{(q)}$ radiate fields into the positive z-direction only. These fields consist of the fields of horn waves travelling in positive and negative z-direction, respectively, and we are interested here in the calculation of their amplitudes $\Delta a_m^{(p)}$ at $z = z_0 > z_1 + \Delta z$. For this purpose we can use the same principles which have been applied in Section 2.2 for cylindrical waveguides. In fact for $\Delta a_m^{(p)}$ we obtain again the expression 2.2.7 which in the limit of $\Delta z \to 0$ and $z_0 - z_1 \to 0$ may be written as

$$\Delta a_m^{(p)} = -p \frac{1}{2R_m} \iiint (\boldsymbol{E}_m^{(-p)} \cdot \boldsymbol{i}_n^{(q)} - \boldsymbol{H}_m^{(-p)} \cdot \boldsymbol{m}_n^{(q)}) dV \tag{3.1.25a}$$

The integration in this expression extends from the surface $z = z_1$ to $z = z_1 + \Delta z$, in which region the current sources have finite amplitudes.

In our case, where the guide connected to the region which contains the fictitious sources, is a horn, the normalisation factor R_m needs to be defined somewhat differently than the definition 2.2.2*a*, which was defined independently of z for plane cross-sections of cylindrical waveguides. The value of R_m in eqn 3.1.25*a* has to be calculated according to

$$R_m = -\frac{1}{2} \iint (\boldsymbol{E}_m^{(+)} \times \boldsymbol{H}_m^{(-)} - \boldsymbol{E}_m^{(-)} \times \boldsymbol{H}_m^{(+)}) \cdot \boldsymbol{n}_z dS_0 \tag{3.1.25b}$$

where $\iint dS_0$ denotes integration over the surface $z = z_0 = \text{constant}$, and where the fields of the normal mode with subscript m have to be taken according to eqns. 3.1.18 and 3.1.21 with $a_m^{(p)}(z_0) = 1$ at $z = z_0$.

In the limit of $\Delta z \to 0$ the amplitude of the spurious mode m, which the incident mode with amplitude $a_n^{(p)}(z_1)$ generates in the infinitesimally short region $z_1 < z < z_1 + \Delta z$, follows from substitution of eqns. 3.1.24 into eqn. 3.1.25 as

$$\lim_{\substack{\Delta z \to 0 \\ z_0 - z_1 \to 0}} \frac{\Delta a_m^{(p)}}{\Delta z} = \sqrt{Y_n^T / Y_m^T}\, \kappa_{mn}^{(p,q)} a_n^{(q)}(z_1)$$

$$= -\frac{p}{Y_m^T} \int_{-1}^{+1} (E_m^{(-p)} \cdot i_n^{(q)} - H_m^{(-p)} \cdot m_n^{(q)}) d\eta \quad (3.1.26)$$

with

$$Y_m^T = Y_m^{(+)} + Y_m^{(-)} \quad (3.1.27)$$

The factor $\sqrt{Y_n^T / Y_m^T}$ in eqn. 3.1.26 is introduced to arrive at a more suitable wave normalisation, which will be described in due course.

Evaluating the surface integral of eqn. 3.1.26, with fields of eqns. 3.1.18 and 3.1.21, at $z = z_0$, and with the current sources of eqn. 3.1.24, yields

$$\kappa_{mn}^{(p,q)} = \frac{pq\phi'}{\sqrt{y_m^T y_n^T}} \frac{16mn}{\pi^2 (m^2 - n^2)^2}$$

$$\left\{ jka \left(q y_n^{(q)} y_n^{(q)} + p y_n^{(q)} y_m^{(-p)} + j\phi \frac{\partial y_n^{(q)}}{\partial ka} \right) \right.$$

$$\left. + \frac{\phi}{2} \frac{m^2 + 3n^2}{m^2 - n^2} (y_n^{(q)} + pq y_m^{(-p)}) - jq \frac{(m^2 - n^2)\pi^2}{8ka} \right\}$$

$$+ O(\phi' \cdot \phi^2, \phi'^2) \quad (3.1.28a)$$

for $m \neq n$

and

$$\kappa_{mm}^{(p,q)} = \frac{pq\phi'}{y_m^T} \left\{ -\left(\frac{1}{3} + \frac{1}{m^2 \pi^2} \right) jka \right.$$

$$\left(q y_m^{(q)} y_m^{(q)} + p y_m^{(q)} y_m^{(-p)} + j\phi \frac{\partial y_m^{(q)}}{\partial ka} \right)$$

$$\left. - \frac{\partial y_m^{(q)}}{\partial \phi} + \frac{\phi}{2m^2 \pi^2} (y_m^{(q)} + pq y_m^{(-p)}) - jq \frac{1}{2ka} \right\}$$

$$+ O(\phi' \phi^2, \phi'^2) \quad (3.1.28b)$$

for $m = n$.

$y_n^{(p)}$ and y_n^T are the normalised wave admittances according to

$$y_n^{(p)} = Y_n^{(p)} / Y_0; \quad y_n^T = Y_n^T / Y_0 \quad (3.1.29)$$

If all horn modes incident from both sides as forward and backward waves are taken into consideration, their propagation through the nonuniform section, as well as the spurious mode generation by each of them, is accounted for by the following system of differential equations

$$
\frac{d}{dz}\begin{bmatrix} a_1^{(+)} \\ a_3^{(+)} \\ \vdots \\ a_1^{(-)} \\ a_3^{(-)} \\ \vdots \end{bmatrix} =
$$

$$
\begin{bmatrix}
-j\beta_1^{(+)} + \kappa_{11}^{(++)} & \kappa_{13}^{(++)}\sqrt{\dfrac{y_3^T}{y_1^T}} & \cdots & \kappa_{11}^{(+-)} & \kappa_{13}^{(+-)}\sqrt{\dfrac{y_3^T}{y_1^T}} & \cdots \\
\kappa_{31}^{(++)}\sqrt{\dfrac{y_1^T}{y_3^T}} & -j\beta_3^{(+)} + \kappa_{33}^{(++)} & \cdots & \kappa_{31}^{(+-)}\sqrt{\dfrac{y_1^T}{y_3^T}} & \kappa_{33}^{(+-)} & \cdots \\
\vdots & \vdots & & \vdots & \vdots & \\
\kappa_{11}^{(-+)} & \kappa_{13}^{(-+)}\sqrt{\dfrac{y_3^T}{y_1^T}} & \cdots & j\beta_1^{(-)} + \kappa_{11}^{(--)} & \kappa_{13}^{(--)}\sqrt{\dfrac{y_3^T}{y_1^T}} & \cdots \\
\kappa_{31}^{(-+)}\sqrt{\dfrac{y_1^T}{y_3^T}} & \kappa_{33}^{(-+)} & \cdots & \kappa_{31}^{(--)}\sqrt{\dfrac{y_1^T}{y_3^T}} & j\beta_3^{(-)} + \kappa_{33}^{(--)} & \cdots \\
\vdots & \vdots & & \vdots & \vdots &
\end{bmatrix}
\begin{bmatrix} a_1^{(+)} \\ a_3^{(+)} \\ \vdots \\ a_1^{(-)} \\ a_3^{(-)} \\ \vdots \end{bmatrix}
\tag{3.1.30}
$$

Under the present circumstances this system holds for the short section between $z = z_1$ and $z = z_1 + \Delta z$ within which the horn in Fig. 3.3 exhibits a nonuniformity. This system may, however, also be generalised to apply to any length of the nonuniform horn, and in such a general situation, represents the coupled wave equations of the nonuniform structure.

Coupled wave equations as we derived them previously in the form of eqns. 2.3.5 and 2.3.9 referred to cylindrical co-ordinates and were based on cylindrical waveguide modes. Our present derivation of coupled wave equations refers to natural co-ordinates and is based on horn modes which are local normal modes in these natural co-ordinates.

The mode amplitudes a_m have been chosen, somewhat arbitrarily, with respect to unit field vectors. To give direct physical meaning to their magnitude they will now be normalised with respect to power. The complex net power carried by the forward and backward travelling components of a mode n obtains from integrating the ρ-component of their complex Poynting vectors over the cross-section.

$$P_n = -b \int_{-1}^{1} [e_{yn}^{(+)}(h_{\eta n}^{(+)})^* + e_{yn}^{(-)}(h_{\eta n}^{(-)})^*]\, u_\eta d\eta$$

$$= ab \frac{\phi}{\sin\phi} [(Y_n^{(+)})^* a_n^{(+)}(a_n^{(+)})^* - (Y_n^{(-)})^* a_n^{(-)}(a_n^{(-)})^*] \quad (3.1.31)$$

If the a_n in this expression are replaced by new amplitude coefficients according to

$$A_n^{(p)} = \left(ab \frac{\phi}{\sin\phi} \frac{Y_n^T}{2} \right)^{1/2} a_n^{(p)} \quad (3.1.32)$$

the power in a mode n of the sector horn relates to the amplitudes of its forward and backward travelling components as

$$P_n = A_n^{(+)}(A_n^{(+)})^* - A_n^{(-)}(A_n^{(-)})^*$$
$$+ j \left[-\frac{\text{Im}\,(Y_n^{(+)})}{\text{Re}\,(Y_n^{(+)})} A_n^{(+)}(A_n^{(+)})^* + \frac{\text{Im}\,(Y_n^{(-)})}{\text{Re}\,(Y_n^{(-)})} A_n^{(-)}(A_n^{(-)})^* \right] \quad (3.1.33a)$$

The 'magnitude squared' of the normalised field amplitude A_n represents the corresponding real or active mode power of the sector horn mode.

$$\text{Re}\,(P_n) = A_n^{(+)}(A_n^{(+)})^* - A_n^{(-)} \cdot (A_n^{(-)})^* \quad (3.1.33b)$$

Replacing the a_n in the field expressions of eqns. 3.1.18 and 3.1.20 by the A_n of eqn. 3.1.32 and calculating R_n according to eqn. 3.1.25*b* with $A_n^{(+)} = A_n^{(-)} = 1$ yields

$$R_n = 1 \quad (3.1.34)$$

Introducing these normalised field amplitudes the system of differential eqns. 3.1.25 transforms to

$$\frac{d}{dz}\begin{bmatrix} A_1^{(+)} \\ A_3^{(+)} \\ \vdots \\ A_1^{(-)} \\ A_3^{(-)} \\ \vdots \end{bmatrix} = \begin{bmatrix} -j\alpha_1^{(+)} & \kappa_{13}^{(++)} & \cdots & \kappa_{11}^{(+-)} & \kappa_{13}^{(+-)} & \cdots \\ \kappa_{31}^{(++)} & -j\alpha_3^{(+)} & \cdots & \kappa_{31}^{(+-)} & \kappa_{33}^{(+-)} & \cdots \\ \vdots & \vdots & & \vdots & \vdots & \\ \kappa_{11}^{(-+)} & \kappa_{13}^{(-+)} & \cdots & j\alpha_1^{(-)} & \kappa_{13}^{(--)} & \cdots \\ \kappa_{31}^{(-+)} & \kappa_{33}^{(-+)} & \cdots & \kappa_{31}^{(--)} & j\alpha_3^{(-)} & \cdots \\ \vdots & \vdots & & \vdots & \vdots & \end{bmatrix} \begin{bmatrix} A_1^{(+)} \\ A_3^{(+)} \\ \vdots \\ A_1^{(-)} \\ A_3^{(-)} \\ \vdots \end{bmatrix} \tag{3.1.35}$$

The diagonal terms in the coupling matrix obtain with

$$a_n^{(p)} = \sqrt{\frac{2}{aY_n^T}}(1-\phi^2/12)A_n^{(p)}$$

as

$$-jp\alpha_n^{(p)} = -jp\beta_n^{(p)} + \frac{a'}{2a}\left(1+\frac{ka}{Y_n^T}\frac{\partial Y_n^T}{\partial ka}\right) + \frac{\phi'\phi}{6} + \frac{\phi'}{2Y_n^T}\cdot\frac{\partial Y_n^T}{\partial\phi} + \kappa_{nn}^{(p,p)} \tag{3.1.36}$$

Since the normalisation of the eigenfields of the horn modes leads to eqn. 3.1.33*b*, and since the total coupling structure is assumed to be lossless, the coefficients in the matrix of eqn. 3.1.35 must obey eqn. 2.4.11.

It follows therefore that

$$\alpha_n^{(+)} = \alpha_n^{(-)}$$

and that

$$\text{Im}\,(\alpha_n^{(p)}) = 0$$

is valid. We confirm this result when we substitute β from eqn. 3.1.20*c* and $\kappa_{nn}^{(p,p)}$ from eqn. 3.1.28 into eqn. 3.1.36 and find that

$$\text{Im}\,(\alpha_n^{(p)}) = O(\phi'\phi^2, \phi'^2, \phi^3) \tag{3.1.37a}$$

$$\text{Re}\,(\alpha_n^{(+)}) = \text{Re}\,(\alpha_n^{(-)}) \tag{3.1.37b}$$

Note that according to eqn. 3.1.24 only terms of lower order in ϕ' and ϕ have been taken into account when the coupling coefficients were evaluated to obtain expressions 3.1.28. The detailed derivation of eqns. 3.1.37, requires results which we obtain only in a subsequent part of this Section, it will therefore be omitted here. In particular we need eqn. 3.1.40 to evaluate $\partial y_n^{(p)}/\partial ka$. Taking eqn. 3.1.37 into consideration allows us to simplify the expression 3.1.36 and write instead

$$\alpha_n^{(p)} = \text{Re}\,(\beta_n^{(+)}) - \text{Im}\,(\kappa_{nn}^{(++)}) \tag{3.1.38}$$

The other relations which follow from eqns. 2.4.11 are

$$\kappa_{mn}^{(p,-p)} = (\kappa_{nm}^{(-p,p)})^*$$

and for $m \neq n$

$$\kappa_{mn}^{(p,p)} + (\kappa_{nm}^{(p,p)})^* = 0$$

They can likewise be verified when the subsequent eqn. 3.1.40 of this Section is utilised. Still other relations which derive from eqns. 2.4.10 are

$$\kappa_{nm}^{(p,-p)} = \kappa_{mn}^{(p,-p)}$$

and for $m \neq n$

$$\kappa_{mn}^{(p,p)} + \kappa_{nm}^{(-p,-p)} = 0$$

they follow from the law of reciprocity. Actually eqns. 2.4.11 were derived when the coupled wave equations are based on cylindrical waveguide modes. However, just as the coupled wave equations may likewise be based on horn modes eqns. 2.4.11 hold also for horn modes if only they are normalised according to eqn. 3.1.34.

In the following chapters of this book the coupled wave equations for the horn wave amplitudes eqn. 3.1.35 with the coefficients of eqns. 3.1.28 and 3.1.38 will be used to analyse abrupt and continuous transition sections. An exact analysis will, however, turn out to be much too involved. In particular the evaluation of the wave admittances $Y_n^{(\pm)}$ turns out to be rather difficult. Therefore approximate expressions for the coefficients κ and α will be used. They will predominantly be needed for interaction between those modes which have the characteristics of propagating modes and are above cutoff. As mentioned before, the results of eqns. 3.1.28 and 3.1.38 are approximations including only terms proportional to

$$\phi', \phi'\phi, \phi^2$$

but neglecting all terms of higher order in ϕ and ϕ'.

To show the main characteristics of the wave admittances a graphical representation of $Y_1^{(+)}$ is given in Fig. 3.5. This diagram represents the complex plane of the wave admittance $y_1^{(+)} = Y_1^{(+)}/Y_0$, which is normalised with respect to the wave admittance $Y_0 = \sqrt{\mu/\epsilon}$ of free space. Lines of constant flare angle ϕ and relative wave number ka are plotted into this complex plane.

When in the field distributions for this lowest-order horn mode with $n - 1$ we replace ϕ by ϕ/n and ka by $(ka \sin\phi/\sin(\phi/n))$ we obtain the field distribution of the horn mode of order n. Thus by applying the corresponding scaling factors to the curves in Fig. 3.5 the wave admittance of any higher-order mode may be read from this diagram according to

$$y_n^{(+)}(\phi, ka) = y_1^{(+)}\left(\phi/n, ka\frac{\sin\phi}{\sin(\phi/n)}\right) \tag{3.1.39a}$$

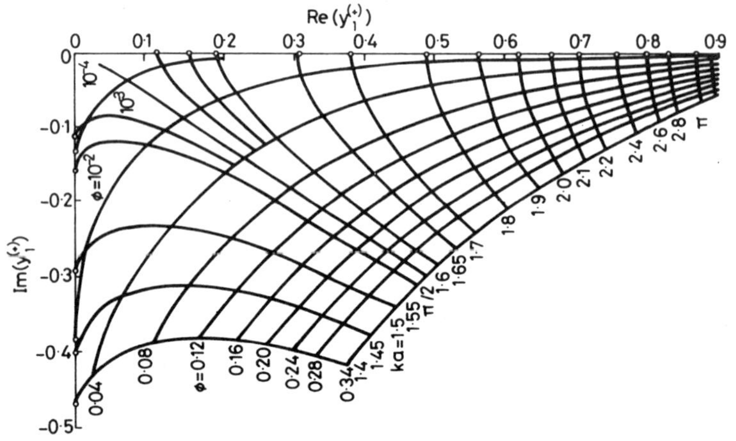

Fig. 3.5 *Complex plane of the normalised wave admittance* $y_1^{(+)}$

For backward travelling modes the wave impedance follows from eqn. 3.1.20*a* as the conjugate complex of the forward wave impedance

$$y_n^{(-)} = (y_n^{(+)})^* \tag{3.1.39b}$$

To derive an approximation for $y_n^{(p)} = Y_n^{(p)}/Y_0$ we start with eqn. 3.1.20*a*. Differentiating this equation with respect to *ka* and using the following recurrence relation for the Hankel function $H_\nu(x)$

$$H_\nu'(x) = \frac{dH_\nu(x)}{dx} = H_{\nu-1}(x) - \frac{\nu}{x}H_\nu(x)$$

$$= -H_{\nu+1}(x) + \frac{\nu}{x}H_\nu(x)$$

leads to

$$-j\sin\phi\frac{\partial y_n^{(+)}}{\partial ka} = \frac{[H_\nu^{(2)}(ka/\sin\phi)]''}{H_\nu^{(2)}(ka/\sin\phi)} - \left\{\frac{[H_\nu^{(2)}(ka/\sin\phi)]'}{H_\nu^{(2)}(ka/\sin\phi)}\right\}^2$$

From the above recurrence relation for $H_\nu(x)$ we derive

$$\frac{H_\nu''(x)}{H_\nu(x)} = -\frac{1}{x}\frac{H_\nu'(x)}{H_\nu(x)} + \left(\frac{\nu^2}{x^2} - 1\right),$$

When we utilise this identity as well as eqns. 3.1.20*a* and 3.1.20*c* we obtain the following expression for the 'wave admittance' squared:

$$(y_n^{(p)})^2 = 1 - \left(\frac{n\pi}{2ka}\frac{\sin\phi}{\phi}\right)^2 - jp\frac{\sin\phi}{ka}\frac{\partial(kay_n^{(p)})}{\partial ka} \qquad (3.1.40)$$

So far this expression is rigorous. If now

$$\phi^2 \ll (kav_n^3)^2 \qquad (3.1.41a)$$

is satisfied with

$$v_n = \left[1 - \left(\frac{n\pi}{2ka}\right)^2\right]^{1/2} \qquad (3.1.41b)$$

$y_n^{(p)}$ can be expanded into a Taylor series with respect to ϕ. Evaluation of eqn. 3.1.40 then leads to

$$y_n^{(p)} = v_n - \frac{jp\phi}{2kav_n^2} + \frac{\phi^2}{2v_n}\left[\frac{5/4 - v_n^2}{(ka)^2 v_n^2} + \frac{1}{3}(1 - v_n^2)\right] + O(\phi^3) \qquad (3.1.41c)$$

When we introduce this result into eqns. 3.1.28 and consider the interaction between the fundamental horn mode and the next higher-order horn mode with $n = 3$, that the change in ϕ couples to it, we obtain the coupling coefficient according to

$$\kappa_{31}^{(p,+)} = \frac{3p}{4\pi^2}\phi'\left[-jka\frac{(v_1 + pv_3)^2}{4} - \frac{\phi}{4}\left\{\frac{v_1}{v_3^2}\left(\frac{7}{2}v_3^2 - 1\right) + p\frac{v_3}{v_1}\left(1 + \frac{3}{2}v_1^2\right)\right\} - O(\phi^2)\right], \qquad (3.1.42a)$$

while the difference in phase constants amounts to

$$\frac{\alpha_1 - \alpha_3}{k} = v_1 - v_3 + \frac{\phi^2}{(ka)^2}\left[\frac{(ka)^2}{2}(v_1 - v_3) + \frac{\frac{1}{8} + \left(\frac{\pi}{2ka}\right)^2}{v_1^2} - \frac{\frac{1}{8} + \left(\frac{3\pi}{2ka}\right)^2}{v_3^2} + \frac{1}{6}\left(\frac{\pi}{2}\right)^2\left(\frac{1}{v_1} - \frac{9}{v_3}\right)\right]$$
$$+ \phi' ka\left(\frac{v_1 - v_3}{3} + \frac{v_1}{\pi^2} - \frac{v_3}{9\pi^2}\right) + O(\phi^3, \phi'\phi) \qquad (3.1.42b)$$

These approximations will be applied in Section 6.4 to analyse waveguide transition sections with very low spurious mode generation.

Any adjoining cylindrical waveguides can be regarded as part of the nonuniform horn with its flare angle ϕ going to zero. For $\phi \to 0$ the forward and backward travelling horn waves become identical with the respective cylindrical waves, as long as these modes are above cutoff.

If, however, the interaction with nonpropagating modes needs to be considered, then the solution of the coupled wave equations 3.1.35 meets with difficulties at those cross-sections, where the flare angle ϕ goes to zero. For $\phi = 0$ and $ka < n\pi/2$ we obtain from eqn. 3.1.39*b* and Fig. 3.5.

$$Y_n^{(+)} = -Y_n^{(-)} \tag{3.1.43}$$

In this case the normal mode fields of the horn modes ⟨travelling⟩ in the positive or negative z-direction are equal. Both waves are identical with the plane wave of the cylindrical waveguide, which, from the narrow side, is evanescent towards the nonuniform section. The field expansion in terms of horn modes is incomplete in this case because the eigenfield of the plane wave which is evanescent towards the other direction cannot be described by the horn mode fields. For $\phi = 0$ all nonpropagating modes have purely reactive wave impedances $Y_n^{(-)}$ and $Y_n^{(+)}$ so that $Y_n^T = 0$. The corresponding coefficients of the coupling matrix will then be infinitely large rendering the whole representation in terms of travelling waves with amplitudes $A_n^{(\pm)}$ unsuitable for the analysis of the sector horn. For $\phi = 0$ therefore only the interaction between propagating modes may be analysed with this representation. To analyse the interaction with evanescent modes in the region of disappearing flare angle necessitates a transformation from the wave amplitudes a_n or A_n to a different set of independent variables which will be studied in the next Section.

3.2 Nonuniform round horn

In this Section we will consider round horns with varying flare angle. To analyse wave propagation we use natural co-ordinates and normal modes of the uniform round horn, just as in the case of the sector horn, but employ a different approach in formulating a system of coupled differential equations for the interaction between waves in nonuniform horns. Instead of introducing effective sources to compensate for the effect of changing flare angle, a field representation in terms of local horn modes will be substituted directly into Maxwell's equations in natural co-ordinates. Coupling between these local horn modes will then also appear as a consequence of the flare angle changing along the horn.

A longitudinal section containing the axis of a round horn with changing flare angles is shown in Fig. 3.6. Natural co-ordinates are locally spherical but otherwise correspond to those of the sector horn. The longitudinal co-ordinate is, however, defined somewhat differently from the definition in Fig. 3.2 for the sector horn. Instead of the axial distance z we use the co-ordinate ζ to describe the reference phase surfaces of the spherical horn mode. At any axial position ζ local spherical co-ordinates r, ϑ, φ are defined by the cone of angle θ tangent to the wall where the sphere through ζ of radius r cuts the horn. With $u = \vartheta/\theta$ as the local polar angle ϑ relative to θ, natural co-ordinates of the round horn are here defined as ζ, u, φ. The matrix of metrical co-efficients of these nonorthogonal co-ordinates follows from Reference 2.5.

$$G_{ij} = \frac{\partial x_1}{\partial y_i}\cdot\frac{\partial x_1}{\partial y_j} + \frac{\partial x_2}{\partial y_i}\cdot\frac{\partial x_2}{\partial y_j} + \frac{\partial x_3}{\partial y_i}\cdot\frac{\partial x_3}{\partial y_j} \tag{3.2.1}$$

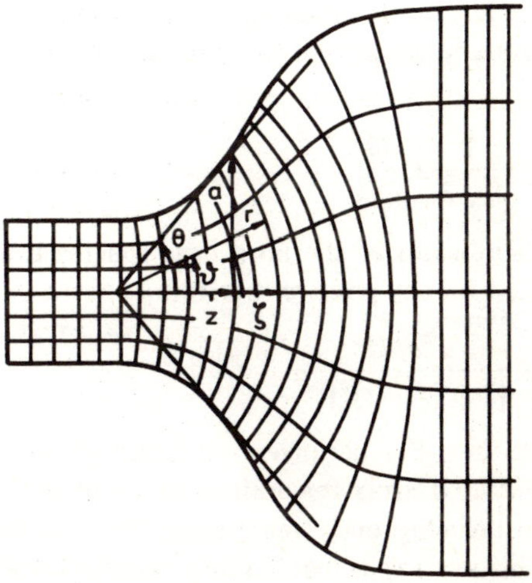

Fig. 3.6 *Natural co-ordinates of the horn*

with Cartesian co-ordinates (x_1, x_2, x_3) and $y_1 = \zeta$, $y_2 = u$, $y_3 = \varphi$. The nonvanishing terms of this matrix are for small flare angles θ

$$G_{\zeta\zeta} = \left[1 + 2\left(\frac{dr}{d\zeta} - 1\right)\sin^2\left(\frac{\theta u}{2}\right)\right]^2 \tag{3.2.2a}$$

$$G_{\zeta u} = (1+\theta^2/6)u(u^2-1)\frac{a^2}{6}\theta\frac{d\theta}{d\zeta} \tag{3.2.2b}$$

$$G_{uu} = r^2\theta^2 \tag{3.2.2c}$$

$$G_{\varphi\varphi} = r^2\sin^2(\theta u) \tag{3.2.2d}$$

From the coefficient determinant

$$G = \begin{vmatrix} G_{\zeta\zeta} & G_{\zeta u} & 0 \\ G_{u\zeta} & G_{uu} & 0 \\ 0 & 0 & G_{\varphi\varphi} \end{vmatrix} = G_{\zeta\zeta}G_{uu}\left(1-\frac{G_{\zeta u}^2}{G_{\zeta\zeta}G_{uu}}\right) \tag{3.2.3a}$$

the angle β between unit vectors $\boldsymbol{n}_\zeta$ in the u-dependent ζ-direction and $\boldsymbol{n}_u$ follows from

$$G = G_{\zeta\zeta}G_{uu}G_{\varphi\varphi}\sin^2\beta \tag{3.2.3b}$$

The deviation α of β from 90° is therefore given by

$$\sin\alpha = \frac{G_{\zeta u}}{\sqrt{G_{uu}G_{\zeta\zeta}}}, \quad \beta = \pi/2-\alpha \tag{3.2.3c}$$

With $G_{\zeta u}$ depending on u according to eqn. 3.2.2b the deviation α is largest in magnitude at $u = 1/\sqrt{3}$ with the corresponding extreme value

$$\alpha = -\frac{1}{9\sqrt{3}}\theta\frac{d\theta}{d\zeta/a} \tag{3.2.4}$$

The natural co-ordinates of the horn have spheres as co-ordinate surfaces ζ = constant for any positive curvature

$$\frac{1}{r_c} = \frac{d^2a/dz^2}{[1+(da/dz)^2]^{3/2}} \tag{3.2.5}$$

of the horn contour. For negative curvature, however, a restriction is imposed by $d\zeta/dz > 0$. Only for positive values of $d\zeta/dz$ are all points inside the horn unambiguously represented by these locally spherical co-ordinates. This restriction requires any contour with negative curvature to have a radius of curvature according to

$$r_c \geqslant \frac{a}{(da/dz)^2}\left\{1+\left(\frac{da}{dz}\right)^2-\left[1+\left(\frac{da}{dz}\right)^2\right]^{1/2}\right\} \tag{3.2.6}$$

Except for this restriction the natural co-ordinates are locally always spherical, and spherical wave functions are adequate to describe any source-free field which satisfies the boundary conditions of the horn locally.

Considering only the excitation of transverse electric fields of circular symmetry the dependence of the corresponding spherical wave functions on the transverse co-ordinate u is given by the Legendre function of the first kind

$$T_p(u) = N_p P_{v_p}(\cos(\theta u)) \tag{3.2.7}$$

N_p is a normalisation factor. In general the order v_p of the Legendre function is nonintegral and follows from the boundary conditions that the tangential electric field of the particular mode vanish at the perfectly-conducting wall of the horn

$$\frac{\partial}{\partial u} P_{v_p}(\cos(\theta u))\bigg|_{u=1} = 0 \tag{3.2.8}$$

The different transverse functions $T_p(u)$ are orthogonal to each other in that they obey

$$\int_0^1 \frac{\partial T_p}{\partial u} \cdot \frac{\partial T_q}{\partial u} \sin(\theta u) du = 0 \tag{3.2.9}$$

for modes of different order ($p \neq q$).

With the normalisation factors given as

$$\frac{1}{N_p^2} = -\frac{\pi v_p(v_p+1)\sin\theta}{(2v_p+1)\theta} \frac{\partial^2}{\partial u \partial v} [P_v(\cos(\theta u))]\bigg|_{\substack{u=1 \\ v=v_p}} P_{v_p}(\cos\theta) \tag{3.2.10}$$

the following orthonormality condition holds for the transverse functions

$$\frac{2\pi}{\theta} \int_0^1 \frac{\partial T_p}{\partial u} \frac{\partial T_q}{\partial u} \sin(\theta u) du = \delta_{pq} \tag{3.2.11}$$

with δ_{pq} as the Kronecker symbol.

Any source-free transverse electric and axially-symmetric field distribution in the nonuniform horn may be expanded in terms of the Legendre functions eqn. 3.2.7. For the transverse field components this expansion is written as

$$E_\varphi = -\sum_p V_p \frac{\partial T_p}{h_u \partial u},$$

$$H_u = \sum_p I_p \frac{\partial T_p}{h_u \partial u} \quad \text{with} \quad h_u = \sqrt{G_{uu}} \tag{3.2.12}$$

The coefficients V_p and I_p in these expansions are still unknown functions of the longitudinal co-ordinate ζ.

Maxwell's equations for the transverse electric and axially-symmetric fields have the following components in the natural co-ordinates:

$$\frac{h_\zeta}{\sqrt{G}} \frac{\partial}{\partial u}(h_\varphi E_\varphi) = -j\omega\mu H_\zeta \tag{3.2.13a}$$

$$\frac{h_u}{\sqrt{G}} \frac{\partial}{\partial \zeta}(h_\varphi E_\varphi) = j\omega\mu H_u \tag{3.2.13b}$$

$$\frac{h_\varphi}{\sqrt{G}} \left\{ \frac{\partial}{\partial \zeta} \left[h_u \left(H_u + \frac{h_{\zeta u}^2}{h_\zeta h_u} H_\zeta \right) \right] - \frac{\partial}{\partial u} \left[h_\zeta \left(H_\zeta + \frac{h_{\zeta u}^2}{h_u h_\zeta} H_u \right) \right] \right\} - j\omega\epsilon E_\varphi \tag{3.2.13c}$$

with

$$h_\zeta = \sqrt{G_{\zeta\zeta}}, \quad h_\varphi = \sqrt{G_{\varphi\varphi}} \quad \text{and} \quad h_{\zeta u} = \sqrt{G_{\zeta u}}$$

Substituting from eqn. 3.2.12 for $E\varphi$ into the ζ-component of eqn. 3.2.13*a* we obtain the expansion in terms of Legendre functions also for the only longitudinal component H_ζ of the field

$$j\omega\mu H_\zeta = -\sum_p V_p \frac{v_p(v_p+1)}{r^2} T_p \tag{3.2.14}$$

To derive generalised transmission-line equations we now substitute for E_φ and H_u from eqn. 3.2.12 into eqn. 3.2.13*b* and multiply this equation by $(h_\zeta/h_u)(\partial T_q/\partial u)$. Integrating it over the local spherical cross-section we obtain by virtue of the orthogonality condition eqn. (3.2.11)

$$\frac{dV_q}{d\zeta} = -j\omega\mu I_q - \sum_p (j\omega\mu\sigma_{pq} I_p + \tau_{pq} V_p) \tag{3.2.15a}$$

where

$$\sigma_{pq} = \iint \frac{h_\zeta - 1}{h_u^2} \frac{\partial T_p}{\partial u} \frac{\partial T_q}{\partial u} dS \tag{3.2.16a}$$

$$\tau_{pq} = \rho_{pq} + \iint \frac{\partial(h_\varphi/h_u)}{\partial \zeta} \frac{1}{h_u} \frac{\partial T_q}{\partial u} \frac{\partial T_p}{\partial u} dS \tag{3.2.16b}$$

with

$$\rho_{pq} = \iint \frac{1}{h_u^2} \frac{\partial T_q}{\partial u} \frac{\partial^2 T_p}{\partial u \partial \zeta} dS \tag{3.2.16c}$$

and the surface integrals extending over the local spherical cross-section.

Next we substitute for E_φ and H_u from eqn. 3.2.12 as well as for H_ζ from eqn. 3.2.14 into eqn. 3.2.13c and multiply this equation by $(h_\zeta/h_u)\cdot(\partial T_q/\partial u)$. Integrating it also over the local spherical cross-section we obtain again by virtue of the orthonormality eqn. 3.2.11

$$\begin{aligned} \frac{dI_q}{d\zeta} = &- \left[j\omega\epsilon + \frac{v_q(v_q+1)}{j\omega\mu r^2} \right] V_q \\ &- \sum_p j\omega\epsilon \left[\left(1 - \frac{v_p(v_p+1)}{(kr)^2} \right) \sigma_{pq} - \frac{v_p(v_p+1)}{(kr)^2} \xi_{pq} \right] V_p \\ &- \sum_p (\rho_{pq} - \alpha_{pq}) I_p - j\omega\epsilon \sum_p \frac{v_p(v_p+1)}{(kr)^2} \eta_{pq} \frac{dV_p}{d\zeta} \end{aligned} \tag{3.2.15b}$$

with

$$\xi_{pq} = \iint \frac{1}{h_u^2} \left\{ \frac{\partial T_q}{\partial u} \frac{\partial h_\zeta}{\partial u} T_p + \frac{r^2}{v_p(v_p+1)} \frac{\partial T_q}{\partial u} \frac{\partial}{\partial \zeta} \left[\frac{v_p(v_p+1)}{r^2} \frac{h_{\zeta u}^2}{h_\zeta} T_p \right] \right\} dS \tag{3.2.16d}$$

$$\eta_{pq} = \iint \frac{h_{\zeta u}^2}{h_\zeta h_u} T_p \frac{\partial T_q}{\partial u} dS \tag{3.2.16e}$$

$$\alpha_{pq} = \iint \frac{1}{h_u^2} \frac{\partial T_q}{\partial u} \frac{\partial}{\partial u} \left[\left(\frac{h_{\zeta u}}{h_u} \right)^2 \frac{\partial T_p}{\partial u} \right] dS \tag{3.2.16f}$$

To assess the results in eqns. 3.2.15 and 3.2.16 and apply them to the analysis of wave propagation in nonuniform horns it appears useful to have approximate solutions for the integrals in eqn. 3.2.16. According to Reference 2 the following approximations hold for the Legendre functions in eqn. 3.2.7 if the flare angle θ is restricted to small values

$$P_{v_p}[\cos(\theta u)] = J_0(x_p u) + O((\theta u)^2) \tag{3.2.17a}$$

$$v_p + \tfrac{1}{2} = (x_p/\theta)[1 + O((\theta/x_p)^2)] \tag{3.2.17b}$$

x_p is the pth zero of the derivative of the Bessel function $J_0(x)$ with respect to the argument x

$$J_0'(x_p) = 0 \tag{3.2.17c}$$

Considering eqn. 3.2.10 for the evaluation of N_p in eqn. 3.2.7 and of the integrals eqn. 3.2.16 we find

$$\begin{aligned}
\sigma_{pp} &= \frac{d\theta}{d\zeta}\left[-\frac{a}{6} + O(\theta^2)\right] \\
p \neq q: \sigma_{pq} &= \frac{d\theta}{d\zeta}\left[-a\frac{4x_p x_q}{(x_p^2 - x_q^2)^2} + O(\theta^2)\right] \\
\xi_{pp} &= O\left(\theta^2 \frac{d\theta}{d\zeta}\right) \\
p \neq q: \xi_{pq} &= \frac{d\theta}{d\zeta}\left[-2a\frac{x_q/x_p}{x_q^2 - x_p^2} + O(\theta^2)\right] \\
\tau_{pp} &= \frac{d\theta}{d\zeta}\left[-\frac{\theta}{9} + O(\theta^2)\right] \\
p \neq q: \tau_{pq} &= O\left(\frac{d\theta}{d\zeta}\,\theta^2\right) \\
\rho_{pq} &= \eta_{pq} = \alpha_{pq} = O\left(\frac{d\theta}{d\zeta}\,\theta^2\right)
\end{aligned} \tag{3.2.17d}$$

If all terms proportional to $\theta^2(d\theta/d\zeta)$ and terms of higher order in θ and $(d\theta/d\zeta)$ are neglected eqns. 3.2.15 reduce to

$$\frac{dV_q}{d\zeta} = -j\omega\mu I_q - \tau_{qq}V_q - j\omega\mu\sum_p \sigma_{pq}I_p \tag{3.2.18a}$$

$$\frac{dI_q}{d\zeta} = -j\omega\epsilon\left[1-\left(\frac{x_p}{ka}\right)^2\right]V_q$$

$$-j\omega\epsilon\sum_p\left[\left(1-\left(\frac{x_p}{ka}\right)^2\right)\sigma_{pq}-\left(\frac{x_p}{ka}\right)^2\xi_{pq}\right]V_p \quad (3.2.18b)$$

Eqns. 3.2.15 or 3.2.18 are the generalised transmission-line equations for the transverse electric and axially-symmetric modes in round horns with the flare angle small but otherwise an arbitrary function of distance except for the restriction according to eqn. 3.2.6. These generalised transmission-line equations represent an infinite set of coupled transmission lines, one for each local normal mode. For constant flare angle θ all the integrals in eqns. 3.2.16 vanish and the normal modes propagate independently of each other.

An exact solution of the infinite system of differential equations is not feasible. An approximation to this solution is most conveniently based on a travelling-wave representation. To this end the voltage and current coefficients V_p and I_p in eqns. 3.2.15 or 3.2.18 are again replaced by the amplitudes of forward and backward travelling waves $A_p^{(+)}$ and $A_p^{(-)}$, respectively

$$V_p(\zeta) = A_p^{(+)}(\zeta)+A_p^{(-)}(\zeta) \quad (3.2.19a)$$

$$I_p(\zeta) = Y_p^{(+)}(kr)A_p^{(+)}(\zeta)-Y_p^{(-)}(kr)A_p^{(-)}(\zeta) \quad (3.2.19b)$$

The coefficients $Y_p^{(+)}$ and $Y_p^{(-)}$ represent wave admittances for forward and backward travelling waves. These wave admittances must be chosen such that when, for the uniform horn with $d\theta/d\zeta = 0$, eqns. 3.2.19 are substituted into eqns. 3.2.15 or 3.2.17, independent wave equations are obtained for any wave amplitude $A_p^{(+)}$ or $A_p^{(-)}$ of the normal modes of the uniform horn. For such uniform round horns the ζ-dependence of the wave amplitudes $A_p^{(\pm)}$ is then given by spherical cylinder functions $R_{v_p}^{(\pm)}(kr)$ according to

$$A_p^{(\pm)}(\zeta) = A_p^{(\pm)}(\zeta_0)\frac{R_{v_p}^{(\pm)}(kr)}{R_{v_p}^{(\pm)}(kr_0)} \quad (3.2.20a)$$

and the wave admittance $Y_p^{(\pm)}(kr)$ at ζ obtains as

$$Y_p^{(\pm)}(kr) = \pm\frac{jY_0}{R_{v_p}^{(\pm)}(kr)}\frac{d[R_{v_p}^{(\pm)}(kr)]}{dkr} \quad (3.2.20b)$$

The radius r of the spherical co-ordinate surface changes linearly with

ζ in this special case of a uniform horn and r_0 designates r at $\zeta = \zeta_0$. For any propagating mode above cutoff the spherical Hankel functions should be chosen according to

$$R_{v_p}^{(\pm)}(kr) = \hat{H}_{v_p}^{\substack{(2)\\(1)}}(kr) = \sqrt{\frac{\pi kr}{2}}\, H_{v_p+1/2}^{\substack{(2)\\(1)}}(kr) \tag{3.2.21}$$

With this choice of functions the modal fields of the horn waves are for $\theta \to 0$, except for a constant factor, identical to those of the respective waves of the cylindrical guide.

The present travelling-wave representation in terms of spherical Hankel functions fails, however, for the flare angle θ approaching zero, that is where the horn tapers into a cylindrical waveguide and when in addition the respective mode is at or below cutoff. At cutoff for $\theta \to 0$ the real parts of the wave admittances $Y_p^{(+)}$ and $Y_p^{(-)}$ go to zero and the coupling coefficients for the interaction with this particular mode become infinitely large. They thus render the travelling-wave representation in terms of spherical Hankel functions unsuitable for analysing the respective wave interaction.

The spherical Hankel function appears likewise unsuitable when the wave admittance at and below cutoff is examined for $\theta \to 0$. With the spherical Hankel functions substituted into eqn. 3.2.21 for $R_{v_p}^{(+)}$ and $R_{vp}^{(\,)}$ the wave admittances turn out to be the complex conjugate of each other

$$Y_p^{(-)} = (Y_p^{(+)})^* \tag{3.2.22}$$

For $\theta \to 0$ and below cutoff these wave admittances are purely imaginary so that one is the negative of the other

$$Y_p^{(-)} = -\,Y_p^{(+)} \tag{3.2.23}$$

This is in direct contrast to the representation in terms of evanescent waves in cylindrical waveguides below cutoff, where the wave admittances are also purely imaginary but equal to each other.

$$Y^{(-)} = Y^{(+)} \tag{3.2.24}$$

To satisfy this latter condition for the wave admittances we select that particular fundamental system of spherical cylinder functions which more directly represents evanescent waves below cutoff, and substitute it for $R_{v_p}^{(+)}$ and $R_{v_p}^{(-)}$ into eqn. 3.2.20. The appropriate fundamental system consists of spherical Bessel and Neumann functions

$$R_{v_p}^{(+)}(kr) = \hat{N}_{v_p}(kr); \quad R_{v_p}^{(-)}(kr) = \hat{J}_{v_p}(kr) \tag{3.2.25}$$

With these functions in eqn. 3.2.20 the wave admittances $Y_p^{(+)}$ and $Y_p^{(-)}$ are equal to each other at and below cutoff for $\theta \to 0$, and the field presentation by horn modes becomes identical to the fields of evanescent cylindrical modes.

For the wave admittances in the travelling-wave representation of eqn. 3.2.19 we will henceforth distinguish two different regions. In case of local horn modes above cutoff, with $kr > v_p$, the wave admittances will be taken as

$$Y_p^{(\pm)}(kr) = \pm jY_0 \frac{\left[\hat{H}_{v_p}^{\overset{(2)}{(1)}}(kr)\right]'}{\hat{H}_{v_p}^{\overset{(2)}{(1)}}(kr)} \tag{3.2.26}$$

where the prime on the spherical Hankel function indicates differentiation with respect to the argument.

In case of local horn modes below cutoff, with $kr < v_p$, the wave admittances will be taken as

$$Y_p^{(+)}(kr) = jY_0 \frac{\hat{N}'_{v_p}(kr)}{\hat{N}_{v_p}(kr)} \quad \text{and} \quad Y_p^{(-)}(kr) = -jY_0 \frac{\hat{J}'_{v_p}(kr)}{\hat{J}_{v_p}(kr)} \tag{3.2.27}$$

For those modes, for which the condition $kr = v_p$ is met at a location within the horn the evanescent wave presentation must be transformed to the propagating wave presentation near this point of transition. If the travelling-wave representation with eqn. 3.2.26 in eqn. 3.2.20 or the evanescent wave representation with eqn. 3.2.27 in eqn. 3.2.20 is substituted for the voltage coefficients V_p and current coefficients I_p into eqn. 3.2.18 a system of differential equations obtains for the forward and backward amplitudes which has the following form:

$$\frac{dA_q^{(+)}}{d\zeta} = \sum_p \left(\kappa_{qp}^{(++)} A_p^{(+)} + \kappa_{qp}^{(+-)} A_p^{(-)}\right), \tag{3.2.28}$$

$$\frac{dA_q^{(-)}}{d\zeta} = \sum_p \left(\kappa_{qp}^{(-+)} A_p^{(+)} + \kappa_{qp}^{(--)} A_p^{(-)}\right) \tag{3.2.28}$$

The coupling coefficients κ_{qp} in these equations are given by

$$\kappa_{pp}^{(\pm\pm)} = \mp jk \frac{Y_p^{(\pm)}}{Y_0}(1+\sigma_{pp}) - \frac{Y_0}{Y_p^T}\frac{\partial Y_p^{(\pm)}/Y_0}{\partial\theta}\frac{d\theta}{d\zeta}$$

$$\pm jk \frac{Y_0}{Y_p^T}\left[\left(\frac{Y_p^{(\pm)}}{Y_0}\right)^2 - (k_p)^2\right]\sigma'_{pp} \tag{3.2.29}$$

$$\kappa_{pp}^{(\pm\mp)} = \pm jk \frac{Y_0}{Y_p^T}\left[\left(\frac{Y_p^{(\mp)}}{Y_0}\right)^2 - (k_p)^2\right]\sigma'_{pp} + \frac{\partial Y_p^{(\mp)}/Y_0}{\partial\theta}\frac{d\theta}{d\zeta}\frac{Y_0}{Y_p^T}$$

$$\kappa_{pq}^{(\pm\pm)} = \pm jk \frac{Y_0}{Y_p^T}\left\{\xi_{qp}\left(\frac{x_q}{ka}\right)^2 - \left[\frac{Y_p^{(\mp)}Y_q^{(\pm)}}{Y_0^2} + (k_q)^2\right]\sigma_{pq}\right\}$$

$$\kappa_{pq}^{(\pm\mp)} = \pm jk \frac{Y_0}{Y_p^T}\left\{\xi_{qp}\left(\frac{x_q}{ka}\right)^2 + \left[\frac{Y_p^{(\mp)}Y_q^{(\mp)}}{Y_0^2} - (k_q)^2\right]\sigma_{qp}\right\}$$

with

$$k_p^2 = 1 - \left(\frac{x_p}{ka}\right)^2; \quad \sigma'_{pp} = \sigma_{pp} + 1 - \frac{da}{d\zeta}\frac{1}{\theta}$$

and

$$Y_p^T = Y_p^{(+)} + Y_p^{(-)}$$

Note that in the expressions 3.2.29 the terms proportional to $\theta^i(d\theta/d\zeta)^k$ with both $i > 1$ and $k > 1$ are neglected.

To obtain eqns. 3.2.29 for the coupling coefficients the following relation

$$\frac{\partial(Y_p^{(\pm)}/Y_0)}{\partial ka} = \pm\frac{j}{\theta}\left[\left(\frac{Y_p^{(\pm)}}{Y_0}\right)^2 - (k_p)^2\right] \tag{3.2.30}$$

has been utilised. This relation can be derived by differentiating eqns. 3.2.26 or 3.2.27 with respect to ka and introducing appropriate formulas for the derivatives of the cylinder functions.

The coupling coefficients in eqn. 3.2.29 can be further evaluated if the respective wave impedances are known. Propagating and evanescent wave admittances for waves above cutoff and waves below cutoff are given by eqns. 3.2.26 and 3.2.27, respectively. Numerical results for the propogating and evanescent wave impedances of the H_{01}-mode with their subscript $p = 1$ are shown in Fig. 3.7 and 3.8. For the propagating waves, eqn. 3.2.22 relates the wave impedance of the forward travelling wave to the wave impedance of the corresponding backward travelling wave.

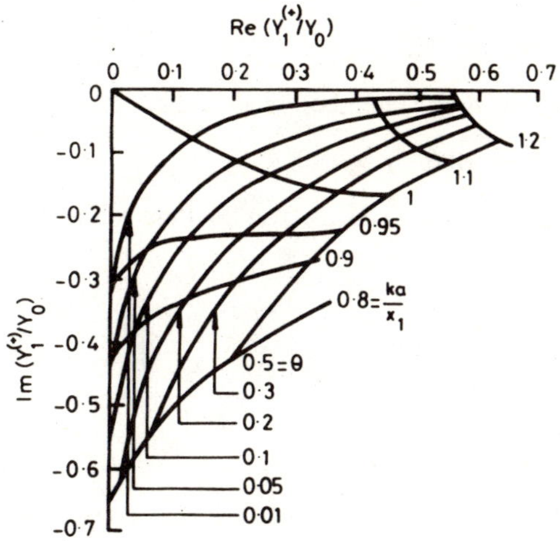

Fig. 3.7 *Propagating forward wave impedance of the H_{01} horn mode*

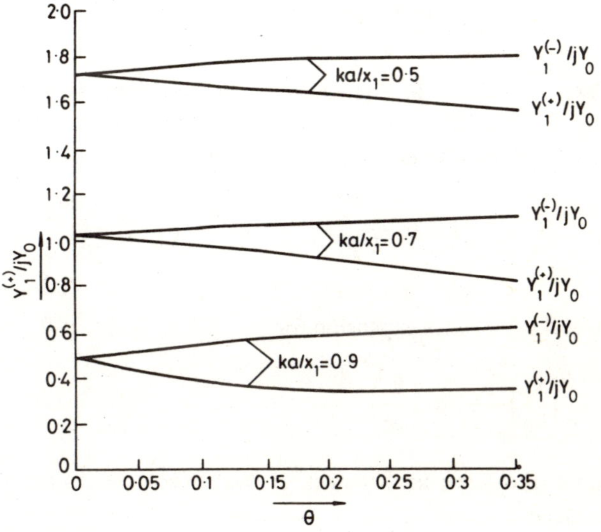

Fig. 3.8 *Evanescent wave impedance of the H_{01} horn mode*

The wave impedance for H_{0p}-waves of higher order $p > 1$ may also be read from the diagrams in Fig. 3.7 and 3.8 when suitable scaling factors are applied. To obtain for small flare angles H_{0p}-fields and impedances from the corresponding H_{01}-quantities the flare angle θ must be scaled according to

$$\theta \to \theta x_1/x_p$$

and the normalised wave number ka according to

$$ka \to ka \sin(\theta x_1/x_p)/\sin\theta$$

The wave admittance of H_{0p}-waves is thus related to the H_{01}-wave admittance by

$$Y_p^{(\pm)}(\theta, ka) = Y_1^{(\pm)}\left(\frac{x_1}{x_p}\theta, \frac{\left(\sin\dfrac{x_1}{x_p}\theta\right)}{\sin\theta}\cdot ka\right) \tag{3.2.31}$$

In the case of

$$\frac{x_p}{ka} > 1$$

and

$$k_p = \left\{1 - \left(\frac{x_p}{ka}\right)^2\right\}^{1/2} \gg \left(\frac{\theta}{ka}\right)^{1/3}$$

the Bessel functions in eqn. 3.2.26 can be expanded into a Taylor series with respect to θ, and the wave admittance for propagating modes is then likewise approximated by such a Taylor series

$$\frac{Y_p^{(\pm)}}{Y_0} = k_p \mp j\theta\,\frac{x_p^2}{2(ka)^3 k_p^2} + \theta^2\,\frac{x_p(1 + 4\,(x_p/ka)^2)}{8\,(ka)^3 k_p^5} + O(\theta^3) \tag{3.2.32}$$

A similar Taylor series expansion for

$$\left(\frac{x_p}{ka}\right)^2 - 1 \gg \left(\frac{\theta}{ka}\right)^{2/3}$$

yields for the wave admittances of evanescent modes the following approximations:

$$\frac{Y_p^{(\pm)}}{Y_0} = -j\left\{\left[\left(\frac{x_p}{ka}\right)^2 - 1\right]^{1/2} \mp \frac{x_p^2}{2(ka)^3\left[\left(\frac{x_p}{ka}\right)^2 - 1\right]}\,\theta + O(\theta^2)\right\}$$

(3.2.33)

References

1 POKROVSKII, V. L., ULINICH, F. R., and SAVVINYKH, S. K.: 'The theory of tapered waveguides', *Radiotekh. & Elektron.*, 1959, **4**, pp. 61–171

2 UNGER, H. G.: 'Wave propagation in horns and through horn junctions', *AEÜ*, 1965, **19**, pp. 459–468

3 BAHAR, E.: Wave scattering in nonuniform waveguides with large flare angles and near cutoff, *IEEE Trans.*, 1968, **MTT-16**, pp. 503–510

4 UNGER, H. G., SAAD, L., and SPORLEDER, F.: 'Electromagnetic waveguide taper and horn analysis in natural coordinates'. Symposia Mathematica, Volume XVIII, Publicazione dell' Istituto Nazionale di Alta Matematica

5 SAAD, L.: 'Eine Theorie zur Berechnung von Hohlleiterübergängen mit Kugelwellen'. Dr.-Ing. Thesis, Technische Universität Braunschweig, 1975

6 STRATTON, J. A.: 'Electromagnetic theory' (McGraw-Hill, 1941)

Chapter 4

Abrupt waveguide transition

4.1 Abrupt transition between cylindrical waveguides

When two cylindrical waveguides of different cross-section are joined together the abrupt transition causes considerable reflection and mode conversion. As an example of such a butt joint we consider the direct connection between two rectangular waveguides with the same height but with different widths $2a_1$ and $2a_2$. In the co-ordinate system of Fig. 4.1 the arrangement is symmetrical with respect to the plane $x = 0$. The abrupt transition in width from $2a_1$ to $2a_2$ is located at $z = 0$.

To apply coupled wave equations we start the analysis by assuming at first a continuous transition of length L as shown in Fig. 4.2. For this arrangement the generalised transmission-line eqns. 2.6.13 appear as

$$\frac{d}{dz}\mathbf{V} = \frac{a'}{a}\mathbf{c}\mathbf{V} - \boldsymbol{\gamma}\mathbf{Z}\mathbf{I} \tag{4.1.1a}$$

$$\frac{d}{dz}\mathbf{I} = -\frac{a'}{a}\mathbf{c}^T\mathbf{I} - \frac{\boldsymbol{\gamma}}{\mathbf{Z}}\mathbf{V} \tag{4.1.1b}$$

$\mathbf{V}$ and $\mathbf{I}$ denote vectors of mode voltages V_n and mode currents I_n of the H_{n0}-modes with $n = 1, 3, 5$, etc. When the fundamental H_{10}-mode is incident we need to consider only these odd-order modes. Any other modes of the rectangular guide are not excited at the symmetrical transition. $\boldsymbol{\gamma}\mathbf{Z}$ and $\boldsymbol{\gamma}/\mathbf{Z}$ are diagonal matrices with their diagonal and only elements given by the propagation constants γ_n and the wave-impedances Z_n of the nth mode according to

$$(\gamma Z)_{nn} = \gamma_n Z_n = j\omega\mu \tag{4.1.2a}$$

$$(\gamma/Z)_{nn} = \gamma_n/Z_n = \gamma_n^2/j\omega\mu \tag{4.1.2b}$$

$a = a(z)$ specifies the contour of the transition with

$$a' = da/dz \tag{4.1.3}$$

representing its slope.

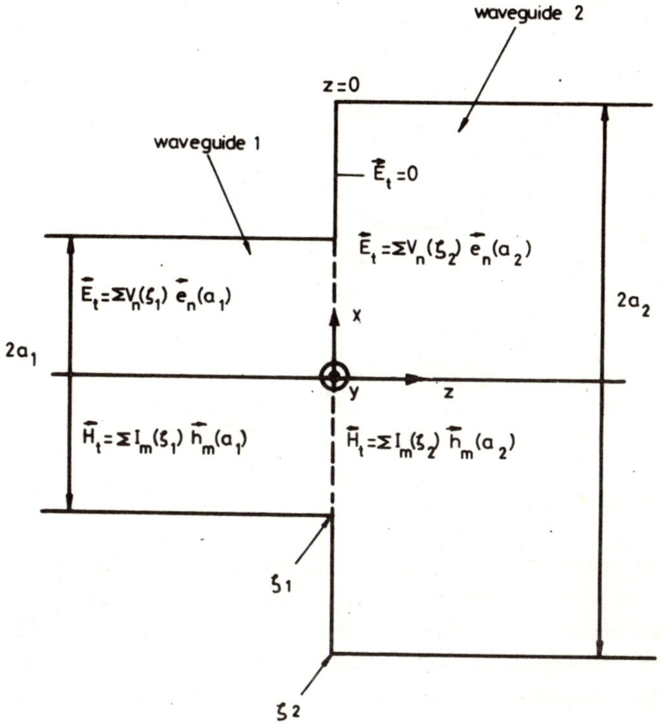

Fig. 4.1 *Abrupt rectangular waveguide junction*

From eqn. 2.6.17 the elements of the matrix **c** are given by

$$c^T_{nm} = c_{mn} = \begin{cases} 0 & \text{for} \quad m = n \\ 2\dfrac{nm}{n^2 - m^2} & \text{for} \quad m \neq n \end{cases} \tag{4.1.4}$$

To apply eqns. 4.1.1 to the abrupt transition we need to go to the limit $L \to 0$. Any singularities in this limit are avoided if, in the region $0 \leqslant z \leqslant L$, we substitute as follows

$$\zeta = \ln(a), \quad \zeta_1 = \ln(a_1), \quad \zeta_2 = \ln(a_2), \quad \frac{d}{dz} = \frac{a'}{a} \cdot \frac{d}{d\zeta} \tag{4.1.5}$$

With these substitutions the system equation 4.1.1 transforms for $L \to 0$ and $a' \to \infty$ to

$$\frac{d}{d\zeta} V = \mathbf{c} V \tag{4.1.6a}$$

$$\frac{d}{d\zeta} I = -\mathbf{c}^T I = \mathbf{c} I \tag{4.1.6b}$$

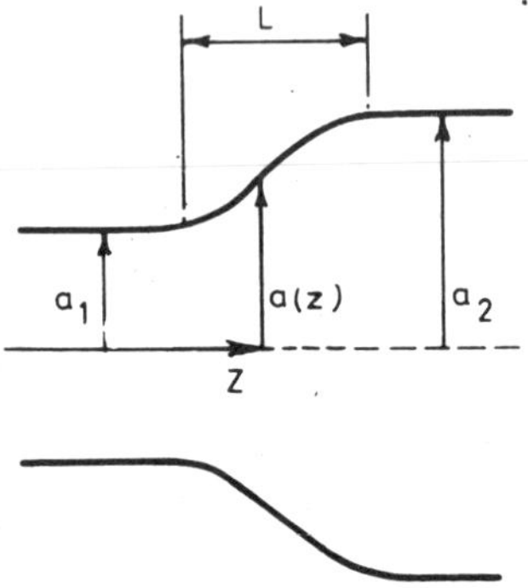

Fig. 4.2 *Continuous transition section*

In the present situation these generalised transmission-line equations couple only mode voltages or mode currents with each other and are hence considerably simpler than the coupled wave equations of eqns. 2.4.1 which exhibit coupling between all wave amplitudes. Therefore the two independent systems eqn. 4.1.6 of differential equations are preferred here to analyse the transition of Fig. 4.1. Usually the analysis of mode coupling in a waveguide structure, such as the transition of Fig. 4.1, aims to determine the scattering matrix. This matrix relates the amplitudes of the waves incident onto the coupling structure to the amplitudes of the waves coming out of it.

As a first step in this analysis eqns. 4.1.6 must be integrated to find the relation between any given vectors $V(\zeta_1)$ and $I(\zeta_1)$ at $\zeta = \zeta_1$ and the resulting vector $V(\zeta_2)$ and $I(\zeta_2)$ at $\zeta = \zeta_2$. As the solution of the system of linear differential eqns. 4.1.6*a* this relation may be written in its most general form as

$$V(\zeta_2) = \mathbf{T} V(\zeta_1) \tag{4.1.7a}$$

Since the two systems of differential eqns. 4.1.6*a* and 4.1.6*b* have identical coupling coefficients the same transformation matrix **T** as for the mode voltages must also be valid for the mode currents:

$$I(\zeta_2) = \mathbf{T}I(\zeta_1) \tag{4.1.7b}$$

For the following calculations it is, however, more useful to invert this solution and write instead

$$I(\zeta_1) = \hat{\mathbf{T}}I(\zeta_2) \tag{4.1.7c}$$

where, according to

$$\hat{\mathbf{T}} = \mathbf{T}^{-1}, \tag{4.1.7d}$$

$\hat{\mathbf{T}}$ denotes the inverse of $\mathbf{T}$.

To determine the matrix $\mathbf{T}$ from eqn. 4.1.6*a* the integration is started at $\zeta = \zeta_1$ with the vector $V(\zeta_1)$ consisting only of

$$V_i(\zeta_1) = \delta_{in} \tag{4.1.8}$$

where δ_{in} is the Kronecker symbol. Under these initial conditions the vector V at $\zeta = \zeta_2$ is equal to the nth column of $\mathbf{T}$.

Under general circumstances the generalised transmission-line equations can be integrated only numerically. Since for such a numerical solution only a finite number of modes can be taken into account the solution will only have a limited degree of accuracy.

For many applications an approximate solution of the coupled wave equations is entirely sufficient. Usually such a solution may be obtained by the following iterative procedure. One starts with the initial values at $\zeta = \zeta_1$ as a first-order approximation

$$V_n^{(1)}(\zeta) = V_n(\zeta_1)$$

$$i \neq n \quad V_i^{(1)}(\zeta) = 0$$

These initial values by themselves are solutions of eqn. 4.1.6*a* for $c_{mn} = 0$. They are hence only slightly modified as long as the coupling remains weak.

For a second-order approximation the initial values are substituted on the right-hand sides of eqn. 4.1.6*a* and these equations are then integrated from ζ_1 to ζ resulting in

$$V_i^{(2)}(\zeta) = \int_{\zeta_1}^{\zeta} c_{in} V_n(\zeta')d\zeta' \doteq c_{in}(\zeta - \zeta_1)V_n(\zeta_1) \tag{4.1.9a}$$

for $i \neq n$ and

$$V_n^{(2)}(\zeta) = V_n(\zeta_1) \tag{4.1.9b}$$

for $i = n$.

The next iteration leads to a still more accurate solution. By substituting $V_i(\zeta) = V_i^{(2)}(\zeta)$ on the right-hand side of eqn. 4.1.6*a*

$$\frac{d}{d\zeta} V_m(\zeta) = \sum_i c_{mi} V_i^{(2)}(\zeta)$$

and again integrating from $\zeta = \zeta_1$, to $\zeta = \zeta_2$ we obtain

$$V_m^{(3)}(\zeta_2) = \left[\delta_{mn} + c_{mn} \ln (a_2/a_1) + \sum_i c_{mi} c_{in} \tfrac{1}{2} \ln^2 (a_2/a_1)\right] V_n(\zeta_1)$$

In this third-order approximation the elements of **T** are given by

$$T_{mn} = \delta_{mn} + c_{mn} \ln (a_2/a_1) + \sum_i c_{mi} c_{in} \tfrac{1}{2} \ln^2 (a_2/a_1) \tag{4.1.10}$$

The iterative solution has the form of a Taylor series in terms of powers of $\ln (a_2/a_1)$. It can be extended to include higher-order terms by continuing the iteration. Such a method of approximately solving the generalised transmission-line equations or the coupled wave equations by iteration can always be applied, in some cases, however, only with certain modifications. There are cases in which the series in eqn. 4.1.10 converges rather slowly. These cases require a larger number of iterations. Also the number of modes which need to be considered to obtain a certain degree of accuracy may be very high.

For the example of an abrupt transition between two rectangular waveguides in Fig. 4.1 Maxwell's equations can be solved rigorously by a field-matching method. By comparing this rigorous field solution with the results in eqn. 4.1.10 we will be able to examine the degree of accuracy of these results. Furthermore it can be shown that eqns. 4.1.6 are in full agreement with the results obtained by the field-matching method. Thus the coupled wave equations 4.1.6 describe the wave propagation through the nonuniform structure in Fig. 4.1 likewise rigorously. To find the exact solution for the waveguide transition in Fig. 4.1 by field-matching methods, we expand the transverse field in the plane $z = 0$ in terms of the eigenfields $\boldsymbol{e}_i$ and $\boldsymbol{h}_i$ of the normal modes of the two connecting waveguides. By matching these field expansions for the transverse electric field $\boldsymbol{E}_t$ on both sides of the junction for $|x| \leqslant a_1$ to each other we obtain

$$\boldsymbol{E}_t = \sum_{n=1,3,5\ldots} V_n(\zeta_1) \boldsymbol{e}_n(a_1) = \sum_{n=1,3,5\ldots} V_n(\zeta_2) \boldsymbol{e}_n(a_2) \tag{4.1.11a}$$

For $a_1 < |x| < a_2$ the perfectly-conducting wall requires the transverse electric field to vanish, hence

$$E_t = 0 = \sum_{n=1,3,5...} V_n(\zeta_2) e_n(a_2) \tag{4.1.11b}$$

Matching the field expansions for the transverse magnetic field H_t in the region $|x| < a_1$ to each other, leads to

$$H_t = \sum_{m=1,3,5...} I_m(\zeta_1) h_m(a_1) = \sum_{m=1,3,5...} I_m(\zeta_2) h_m(a_2) \tag{4.1.12}$$

In the region $a_1 \leqslant |x| \leqslant a_2$ we are not able to specify any boundary conditions for H_t.

The vector functions $e_n(a_1)$, $e_n(a_2)$, $h_m(a_1)$, $h_m(a_2)$ in these field expansions denote the normalized transverse eigenfields according to eqn. 2.6.4 for the waveguides with broad side walls $2a_1$ and $2a_2$, respectively. When only the fundamental (H_{10}-) mode is incident we need to consider only H_{n0}-modes, their transverse eigenfields follow from eqn. 2.6.4 as well as Table 2.1, and are given by

$$e_n(a) = n_y \frac{1}{\sqrt{ab}} \cos\left(\frac{n\pi}{2a} x\right) (-1)^{(n-1)/2} \tag{4.1.13a}$$

$$h_n(a) = n_z \times e_n(a) \tag{4.1.13b}$$

To relate the mode voltages $V_n(\zeta_1)$ in waveguide 1 to the mode voltages $V_n(\zeta_2)$ in waveguide 2 we multiply eqns. 4.1.11a by $e_m(a_2)$ and integrate over the total cross-section S_2 of waveguide 2 at $z = 0$. Taking the orthonormality condition eqns. 2.6.4d into account, we find

$$\iint E_t \cdot e_m(a_2) dS_2 = b \sum_n V_n(\zeta_1) \int_{-a_1}^{a_1} e_n(a_1) \cdot e_m(a_2) dx$$
$$= V_m(\zeta_2) \tag{4.1.14}$$

We compare this equation with eqn. 4.1.7a and note that any element of the matrix **T** is given by

$$T_{mn} = T_{mn}(a_1, a_2) = b \int_{-a_1}^{a_1} e_n(a_1) \cdot e_m(a_2) dx$$
$$= -\frac{2(-1)^{(m+n)/2}}{\sqrt{a_1 a_2}} \int_0^{a_1} \cos\left(\frac{n\pi}{2a_1} x\right) \cos\left(\frac{m\pi}{2a_2} x\right) dx$$
$$= \frac{n(-1)^{(m-1)/2}}{\pi/4} \frac{(a_2/a_1)^{3/2}}{(na_2/a_1)^2 - m^2} \cdot \cos\left(\frac{m\pi}{2} \cdot \frac{a_1}{a_2}\right) \tag{4.1.15}$$

To compare this expression for T_{mn} with eqn. 4.1.10 we expand eqn. 4.1.15 into a Taylor series in powers of $(a_2/a_1 - 1)$. If on the other hand we evaluate the sums $\Sigma c_{mi} c_{in}$ of eqn. 4.1.10 numerically and compare the results we find that eqn. 4.1.15 and eqn. 4.1.10 agree with each other at least for small values of $(a_2/a_1 - 1)$.

To evaluate the boundary condition eqn. 4.1.12 for the magnetic fields at the junction we evaluate

$$\iint \boldsymbol{H}_t \cdot \boldsymbol{h}_n(a_1) dS_1$$

where $\iint dS_1$ denotes integration over the cross-section of waveguide 1 at $z = 0$. For $\boldsymbol{H}_t$ we introduce the transverse magnetic field of eqn. 4.1.12 and obtain from eqn. 4.1.12

$$b \int_{-a_1}^{a_1} \boldsymbol{H}_t \cdot \boldsymbol{h}_n(a_1) dx =$$

$$b \sum_m I_m(\zeta_2) \int_{-a_1}^{a_1} \boldsymbol{h}_n(a_1) \cdot \boldsymbol{h}_m(a_2) dx = I_n(\zeta_1) \qquad (4.1.16)$$

Comparing this result with eqn. 4.1.7c the elements $\hat{T}_{mn}$ of $\hat{\boldsymbol{I}}$ appear as

$$\hat{T}_{nm} = b \int_{-a_1}^{a_1} \boldsymbol{h}_n(a_1) \cdot \boldsymbol{h}_m(a_2) dx \qquad (4.1.17)$$

According to eqn. 4.1.13b this expression is identical with

$$\hat{T}_{nm} = b \int_{-a_1}^{a_1} \boldsymbol{e}_n(a_1) \cdot \boldsymbol{e}_m(a_2) dx = T_{mn} \qquad (4.1.18a)$$

$$\hat{\mathsf{T}} = \mathsf{T}^{-1} = \mathsf{T}^T \qquad (4.1.18b)$$

To show that the results obtained from the field-matching method are in full agreement with the solution of the generalised transmission-line equations, even for large values of $(a_2/a_1 - 1)$, we shall now derive these generalised transmission-line equations from eqns. 4.1.14 and 4.1.16. For this purpose we write eqn. 4.1.14 in the following form:

$$V_m(\zeta_2) = b \int_{-a_1}^{a_1} \boldsymbol{E}_t \cdot \boldsymbol{e}_m(a_2) dx$$

$$= \frac{2}{\sqrt{a_2}} (-1)^{(m-1)/2} \int_0^{a_1} \frac{E_y}{\sqrt{b}} \cos\left(\frac{m\pi}{2a_2} x\right) dx \qquad (4.1.19)$$

where E_y represents the electric field in the cross-section $z = 0$ of waveguide 1. If we differentiate eqn. 4.1.19 with respect to $\zeta_2 = \ln(a_2) = \zeta$ and consider that E_y = constant we obtain

$$\frac{dV_m}{d\zeta} = \frac{dV_m}{da_2} a_2$$

$$= -\frac{V_m}{2} + 2\sqrt{a_2}\,(-1)^{(m-1)/2} \int_0^{a_1} \frac{E_y}{\sqrt{b}} \sin\left(\frac{m\pi}{2a_2}x\right) \frac{m\pi x}{2a_2^2}\, dx \tag{4.1.20}$$

We now evaluate the following Fourier series:

$$\sin\left(\frac{m\pi}{2a_2}x\right) \frac{m\pi x}{2a_2^2} = \sum_{n=1,3,5\ldots} A_n \cos\left(\frac{n\pi}{2a_2}x\right)$$

with its fundamental period ranging over $-a_2 < x < a_2$, and introduce it, as well as $V_n(\zeta_2) = V_n(\zeta) = V_n$ from eqn. 4.1.19, on the right-hand side of eqn. 4.1.20. We can then write this equation in the following form:

$$\frac{dV_m}{d\zeta} = \sum c_{mn} V_n$$

where the coefficients c_{mn} are the same as in eqn. 4.1.4. Thus eqn. 4.1.6*a* is verified by the field-matching method for any value of $a_2/a_1 > 1$.

Differentiating eqn. 4.1.16 with respect to

$$\zeta = \zeta_1 = \ln(a_1)$$

leads to the differential eqns. 4.1.6*b* and thus verifies the remaining part of the generalised transmission-line equations.

From these results we may draw the following conclusions:

(i) The description of field propagation by the infinite system of generalised transmission-line equations or coupled wave equations is a rigorous procedure.

(ii) In general, however, only approximate solutions of these systems of differential equations can be found. In addition the number of modes which need to be considered for a certain degree of accuracy may be rather high. This will be shown in more detail in Section 4.3.

To complement the present results we shall now give a general method by which the scattering matrix of waveguide transitions may be calculated. For this purpose we introduce the diagonal matrices $\hat{\mathbf{Z}}$ and $\mathbf{Z}$ the diagonal and only elements of which are defined by

$$\hat{Z}_{nn} = \sqrt{Z_n(a_1)}; \quad Z_{nn} = \sqrt{Z_n(a_2)} \tag{4.1.21}$$

$Z_n(a_1)$ and $Z_n(a_2)$ represent the wave impedance of the nth mode in waveguide 1 and 2, respectively. They are given by eqn. 2.6.6.

With $\hat{A}$ and $\hat{B}$ designating the amplitude vectors of the waves travelling in waveguide 1 in the forward and backward z-direction, respectively, the following relations follow from eqns. 2.6.10:

$$\begin{aligned} V(\zeta_1) &= \hat{\mathbf{Z}}\hat{A} + \hat{\mathbf{Z}}\hat{B} \\ I(\zeta_1) &= \hat{\mathbf{Z}}^{-1}\hat{A} - \hat{\mathbf{Z}}^{-1}\hat{B} \end{aligned} \tag{4.1.22a}$$

Corresponding relations apply for the vectors A and B of wave amplitudes in waveguide 2

$$\begin{aligned} V(\zeta_2) &= \mathbf{Z}A + \mathbf{Z}B \\ I(\zeta_2) &= \mathbf{Z}^{-1}A - \mathbf{Z}^{-1}B \end{aligned} \tag{4.1.22b}$$

$\mathbf{Z}$ and $\hat{\mathbf{Z}}$ are diagonal matrices and can easily be inverted to calculate $\mathbf{Z}^{-1}$ and $\hat{\mathbf{Z}}^{-1}$. When we substitute for the mode voltage and mode current vectors in eqn. 4.1.7 their representation in terms of travelling wave amplitudes from eqn. 4.1.22 we arrive at

$$\begin{aligned} \mathbf{T}(\hat{\mathbf{Z}}\hat{A} + \hat{\mathbf{Z}}\hat{B}) &= \mathbf{Z}A + \mathbf{Z}B \\ \mathbf{T}(\hat{\mathbf{Z}}^{-1}\hat{A} - \hat{\mathbf{Z}}^{-1}\hat{B}) &= \mathbf{Z}^{-1}A - \mathbf{Z}^{-1}B \end{aligned} \tag{4.1.23}$$

We solve these equations for the amplitude vectors $\hat{B}$ and A of waves, which travel away from the transition, to obtain the following scattering matrix representation in terms of the amplitude vectors $\hat{A}$ and B of waves, which are incident onto the junction

$$\begin{aligned} \hat{B} = {} & (\mathbf{Z}^{-1}\mathbf{T}\hat{\mathbf{Z}} + \mathbf{Z}\mathbf{T}\hat{\mathbf{Z}}^{-1})^{-1}(\mathbf{Z}\mathbf{T}\hat{\mathbf{Z}}^{-1} - \mathbf{Z}^{-1}\mathbf{T}\hat{\mathbf{Z}})\hat{A} \\ & + 2(\mathbf{Z}^{-1}\mathbf{T}\hat{\mathbf{Z}} + \mathbf{Z}\mathbf{T}\hat{\mathbf{Z}}^{-1})^{-1}B \end{aligned} \tag{4.1.24a}$$

$$\begin{aligned} A = {} & 2(\hat{\mathbf{Z}}^{-1}\mathbf{T}^{-1}\mathbf{Z} + \hat{\mathbf{Z}}\mathbf{T}^{-1}\mathbf{Z}^{-1})^{-1}\hat{A} \\ & + (\hat{\mathbf{Z}}^{-1}\mathbf{T}^{-1}\mathbf{Z} + \hat{\mathbf{Z}}\mathbf{T}^{-1}\mathbf{Z}^{-1})^{-1}(\hat{\mathbf{Z}}\mathbf{T}^{-1}\mathbf{Z}^{-1} - \hat{\mathbf{Z}}^{-1}\mathbf{T}^{-1}\mathbf{Z})B \end{aligned} \tag{4.1.24b}$$

To further evaluate this scattering representation we note that for small values of $(a_2/a_1 - 1)$ and according to eqn. 4.1.10 the matrix

$$\mathbf{T} = \mathbf{1} + \mathbf{t}$$

differs from the unitary matrix only by elements t_{ij} which are all small in absolute value

$$|t_{ij}| \ll 1$$

We therefore introduce the diagonal matrix

$$\mathbf{Y} = \mathbf{Z}^{-1}\hat{\mathbf{Z}} + \mathbf{Z}\hat{\mathbf{Z}}^{-1}$$

with which we can write

$$(\mathbf{Z}^{-1}\mathbf{T}\hat{\mathbf{Z}} + \mathbf{Z}\mathbf{T}\hat{\mathbf{Z}}^{-1})^{-1} = [\mathbf{Y} + (\mathbf{Z}^{-1}\mathbf{t}\hat{\mathbf{Z}} + \mathbf{Z}\mathbf{t}\hat{\mathbf{Z}}^{-1})]^{-1},$$

and obtain the following approximation for this expression

$$(\mathbf{Z}^{-1}\mathbf{T}\hat{\mathbf{Z}} + \mathbf{Z}\mathbf{T}\hat{\mathbf{Z}}^{-1})^{-1} \approx$$

$$\mathbf{Y}^{-1} - \mathbf{Y}^{-1}(\mathbf{Z}^{-1}\mathbf{t}\hat{\mathbf{Z}} + \mathbf{Z}\mathbf{t}\hat{\mathbf{Z}}^{-1})\cdot\mathbf{Y}^{-1} = \mathbf{X}$$

The elements of $\mathbf{X}$ can be evaluated quite easily, if the first-order approximation for $\mathbf{t}$ according to eqn. 4.1.10 is used

$$t_{mn} = c_{mn} \ln (a_2/a_1)$$

$$X_{nn} = Y_{nn}^{-1} = \left(\frac{\hat{Z}_{nn}}{Z_{nn}} + \frac{Z_{nn}}{\hat{Z}_{nn}}\right)^{-1}$$

and for $m \neq n$

$$X_{mn} = -Y_{mm}^{-1}(Z_{mm}^{-1}\hat{Z}_{nn} + Z_{mm}\hat{Z}_{nn}^{-1})t_{mn}Y_{nn}^{-1}$$

With these results the input reflection coefficient for the fundamental mode in waveguide 1 is approximately given by

$$\frac{\hat{B}_1}{\hat{A}_1} = \frac{\{Z_1(a_2)/Z_1(a_1)\}^{1/2} - \{Z_1(a_1)/Z_1(a_2)\}^{1/2}}{\{Z_1(a_1)/Z_1(a_2)\}^{1/2} + \{Z_1(a_2)/Z_1(a_1)\}^{1/2}} + O(\ln^2(a_2/a_1))$$

and the excitation of the H_{30}-mode in waveguide 2 by the fundamental mode in waveguide 1 obtains as

$$\frac{A_3}{\hat{A}_1} = 2X_{13} = -2c_{13}\ln(a_2/a_1)$$

$$\frac{\{Z_3(a_1)/Z_1(a_2)\}^{1/2} + \{Z_1(a_2)/Z_3(a_1)\}^{1/2}}{\left(\left\{\frac{Z_1(a_1)}{Z_1(a_2)}\right\}^{1/2} + \left\{\frac{Z_1(a_2)}{Z_1(a_1)}\right\}^{1/2}\right)\left(\left\{\frac{Z_3(a_1)}{Z_3(a_2)}\right\}^{1/2} + \left\{\frac{Z_3(a_2)}{Z_3(a_1)}\right\}^{1/2}\right)}$$

4.2 Abrupt waveguide-to-horn transition

When a waveguide horn with finite flare angles at its input end is directly joined to a cylindrical waveguide of the corresponding cross-section, an abrupt waveguide-to-horn transition is formed at which the flare angles are discontinuous. As one example of considerable practical significance we consider first the junction of a rectangular waveguide and a uniform sector horn as shown in Fig. 4.3. In this junction, which is bound by a plane surface S_1 on the side of the rectangular waveguide and by a circular cylinder S_2 on the side of the sector horn, wave propagation is described by the coupled wave equations of Section 3.1. However, these equations in the form of eqns. 3.1.35, 3.1.28 and 3.1.36 were derived by requiring the expr. 3.1.10 to be satisfied. Eqn. 3.1.10*b* in particular requires

$$\phi' = \frac{d\phi}{dz} \ll 1/a \tag{4.2.1}$$

which is by no means the case for the abrupt transition. We then have

$$\phi' \to \infty$$

at the junction, and this leads to somewhat different values for the differential vector $\boldsymbol{u}_z$ and the fictitious currents $\boldsymbol{i}_n$ and $\boldsymbol{m}_n$ in Section 3.1. These differences vanish, however, when, subsequent to the application of the reciprocity theorem, the integration is performed. Thus the coupled wave eqns. 3.1.35 with the coupling coefficients eqns. 3.1.28 and 3.1.36 are valid in this case, too.

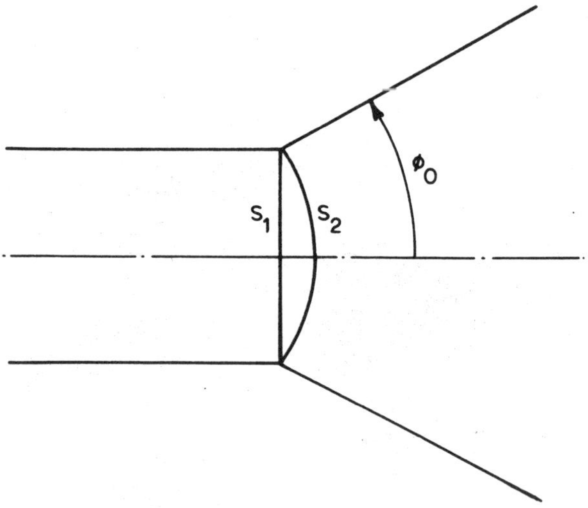

Fig. 4.3 *Waveguide-to-horn junction*

Since in the junction region of Fig. 4.3, between S_1 and S_2 the value of z is constant and $\phi' \to \infty$, the differentiation in eqn. 3.1.35 must be modified according to

$$\frac{d}{dz} = \phi' \frac{d}{d\phi} \tag{4.2.2}$$

Dividing eqn. 3.1.35 on both sides by ϕ' leads to

$$\frac{d}{d\phi} A_m^{(p)} = \sum_n C_{mn}^{(p,+)} A_n^{(+)} + \sum_n C_{mn}^{(p,-)} A_n^{(-)}$$

with

$$m,n = 1,3,5\ldots, \qquad p,q = \pm 1 \tag{4.2.3a}$$

where the coupling coefficients

$$C_{mm}^{(p,p)} = \lim_{\phi' \to \infty} (-jp\alpha_m^{(p)}/\phi') = \frac{j}{\phi'} \mathrm{Im}(\kappa_{mm}^{(p,p)}) \tag{4.2.3b}$$

$$C_{mn}^{(p,q)} = \frac{1}{\phi'} \kappa_{mn}^{(p,q)} \tag{4.2.3c}$$

are given in terms of the elements α and κ of the coupling matrix of eqns. 3.1.36 and 3.1.28.

Under typical operating conditions only the fundamental H_{10}-mode of the rectangular waveguide is incident. If in eqn. 4.2.3*a* only the amplitudes $A_1^{(\pm)}$ of the fundamental mode are taken into account, and if in the coupling coefficients all terms proportional to ϕ are neglected, the coupled differential equations reduce to

$$\frac{d}{d\phi}\begin{bmatrix} A_1^{(+)} \\ A_1^{(-)} \end{bmatrix} = \begin{bmatrix} C_{11}^{(++)} & C_{11}^{(+-)} \\ C_{11}^{(-+)} & C_{11}^{(--)} \end{bmatrix} \begin{bmatrix} A_1^{(+)} \\ A_1^{(-)} \end{bmatrix} \tag{4.2.4a}$$

with

$$C_{11}^{(++)} = -C_{11}^{(--)} = -ja\left(\frac{1}{3} + \frac{1}{\pi^2}\right) k\left\{1 - \left(\frac{\pi}{2ka}\right)^2\right\}^{1/2} - C_{11}^{(-+)} \tag{4.2.4b}$$

$$C_{11}^{(-+)} = -C_{11}^{(+-)} = -\frac{j}{2} \frac{\pi^2/(2ka)^3}{[1-(\pi/(2ka))^2]^{3/2}} \tag{4.2.4c}$$

For the wave admittances $Y^{(\pm)}$ in the elements α and κ of the coupling matrix the expression in eqn. 3.1.41*c* has been substituted.

The most important characteristic of the junction in fundamental operation is its input reflection coefficient. To find an approximate solution for this reflection coefficient at the end of the rectangular waveguide we let a forward travelling wave with amplitude $A_1^{(+)}(0)$ be incident at $\phi = 0$ from the rectangular waveguide. Assuming that

$$|A_1^{(-)}| \ll |A_1^{(+)}|$$

everywhere inside the junction we obtain from the first differential equation of the system (eqn. 4.2.4a) the following solution for this forward travelling wave

$$A_1^{(+)} = A_1^{(+)}(\phi) = A_1^{(+)}(0) \exp\left(\int_0^{\phi} C_{11}^{(++)} d\hat{\phi}\right) \qquad (4.2.5a)$$

Substituting this result into the second eqn. of expr. 4.2.4a and introducing the condition

$$A_1^{(-)}(\phi_0) = 0$$

at $\phi = \phi_0$ leads to

$$A_1^{(-)} = A_1^{(-)}(\phi) = \int_{\phi_0}^{\phi} C_{11}^{(-+)} A_1^{(+)}(\hat{\phi}) \exp\left(\int_{\hat{\phi}}^{\phi} C_{11}^{(--)} d\hat{\hat{\phi}}\right) d\hat{\phi} \qquad (4.2.5b)$$

which can easily be verified by calculating $dA_1^{(-)}/d\phi$ and comparing it with eqn. 4.2.4a.

The input reflection coefficient is derived according to

$$r_{H_{10}} = \frac{A_1^{(-)}(0)}{A_1^{(+)}(0)} = \int_{\phi_0}^{0} C_{11}^{(-+)} \exp\left[\int_0^{\phi} (C_{11}^{(++)} - C_{11}^{(--)}) d\hat{\phi}\right] d\phi$$

$$= -\int_0^{\phi_0} C_{11}^{(-+)} \exp\left[(C_{11}^{(++)} - C_{11}^{(--)})\phi\right] d\phi$$

$$= -C_{11}^{(-+)}\phi_0 + O(\phi_0^2) \qquad (4.2.5c)$$

$$|r_{H_{10}}| = \phi_0 \frac{\pi^2}{2(2ka\{1 - (\pi/(2ka))^2\}^{1/2})^3} \qquad (4.2.6)$$

This is the same result as found in Reference 1 by using a field-matching method. The reflection coefficient as calculated from eqn. 4.2.6 is plotted in Fig. 4.4 for $\phi_0 = \pi/36$ and compared with experimental results. The agreement is obviously quite good, and remains so as long as the flare angle $2\phi_0$ of the horn is small.

To examine the results as they are obtained from the coupled wave equations for abrupt waveguide-to-horn transitions with relatively large flare angles, the round waveguide-to-horn junction of Fig. 4.5 is considered as an example. The wave propagation in the junction region between the plane cross-section at $\zeta = \zeta_0$ and the spherical cross-section at $\zeta = \zeta_1$ is calculated by applying the coupled wave eqns. 3.2.28. When the variable ζ is replaced by the angle θ eqns. 3.2.28 transform to

$$\frac{dA_m^{(p)}}{d\theta} = \sum_n \left(C_{mn}^{(p,+)} A_n^{(+)} + C_{mn}^{(p,-)} A_n^{(-)}\right) \qquad (4.2.7a)$$

with

$$m, n = 1, 2, 3 \ldots, \quad p, q = \pm 1$$

and

$$C_{mn}^{(p,q)} = \kappa_{mn}^{(p,q)} \frac{d\zeta}{d\theta} \qquad (4.2.7b)$$

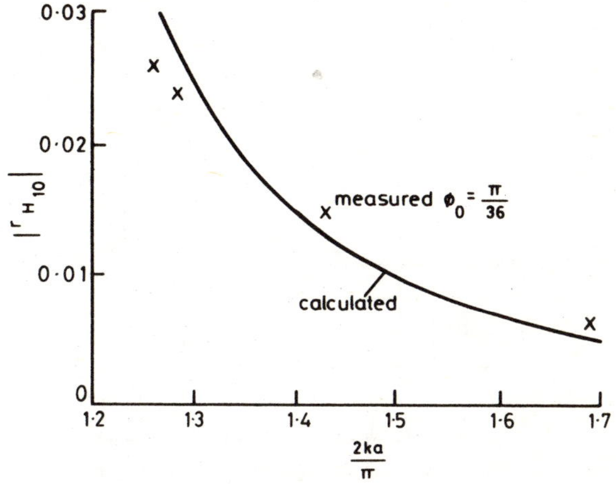

Fig. 4.4 *Measured and calculated reflexion coefficient of rectangular waveguide to sector horn junction*

Between $\zeta = \zeta_0$ and $\zeta = \zeta_1$ the variable ζ relates to the angle θ according to:

$$\zeta - \zeta_0 = \frac{a}{\sin\theta}(1 - \cos\theta) = \frac{a}{2}(\theta + O(\theta^3))$$

$$\frac{d\zeta}{d\theta} = \frac{a}{2}(1 + O(\theta^2)) \qquad (4.2.7c)$$

With the coupling coefficients from eqns. 3.2.29, 3.2.30, 3.2.33 the system of differential eqns. 4.2.7 is completely defined.

To calculate the input reflection coefficient for the lowest order H_{on}-mode an approximate solution similar to eqn. 4.2.5 was derived for the system of coupled wave equations 4.2.7. Eqn. 4.2.5 neglects all higher-order modes and all coupling terms of second and higher order in ϕ. In the case of the round waveguide-to-horn transition the lowest order approximation for the solution of eqn. 4.2.7 has the following form

$$|r_{H_{01}}| = \frac{\theta_0}{4x_1 [(ka/x_1)^2 - 1]^{3/2}} \qquad (4.2.8)$$

$x_1 = 3{\cdot}832$ represents the lowest-order zero of the derivative of the Bessel function of order zero.

Numerical results from an evaluation of this approximation for a junction with $\theta_0 = \pi/10$, are plotted in Fig. 4.6 as curve 1.

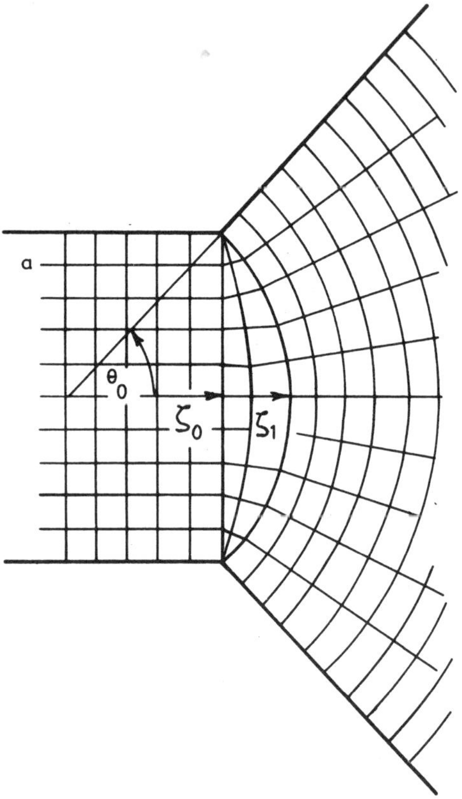

Fig. 4.5 *Round waveguide-to-horn junction*

For a flare angle as large as $2\theta_0 = \pi/5$ we have to expect that higher-order coupling terms contribute significantly to the overall reflection at the junction. To examine these contributions the system of coupled wave eqns. 4.2.7 with the coefficients κ (eqn. 3.2.29) has been integrated numerically. Curve 2 in Fig. 4.6 shows the result of this integration, when still, only the fundamental mode is taken into account, but when the more accurate expressions for its wave admittance $Y_1^{(\pm)}$ according to eqn. 3.2.32 are used, and when in addition higher-order coupling terms for the interaction between its forward and backward travelling components are taken into account. Curve 3 gives the results from the solution of the coupled wave equations which takes into account the interaction between all forward and backward travelling components of both the H_{01}-wave and the H_{02}-wave. The wave admittance $Y_2^{(\pm)}$ for this solution was calculated using eqn. 3.2.33. If in

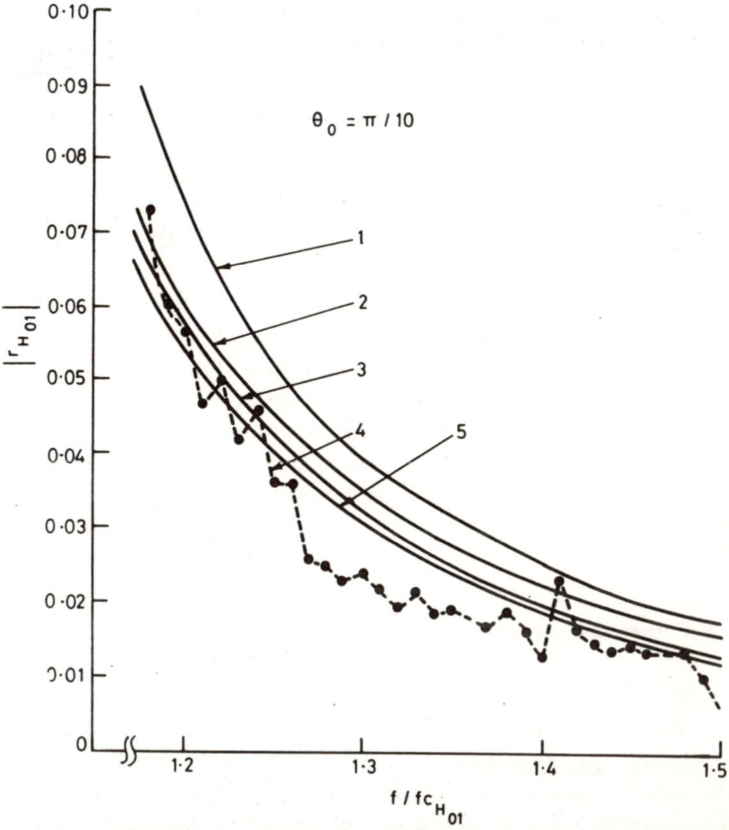

Fig. 4.6 *Reflection coefficient of H_{01} wave at circular waveguide to round horn junction*

Curves 1, 2, 3: Approximate solutions of the coupled wave equations
Curve 1 : Only the H_{01}-mode and only the lowest-order constant coupling terms have been taken into account
Curve 2 : Only the H_{01}-mode has been taken into account but higher-order coupling terms have been included
Curve 3 : Interaction with the H_{02}-mode has been included
Curve 4 : Experimental results
Curve 5 : Theoretical results using a field matching method
f frequency
$f_{cH_{01}}$ H_{01} cutoff frequency

addition H_{03}-interaction is incorporated into the solution of the coupled wave equation, no significant improvement of its accuracy is obtained. Curve 4 in Fig. 4.6 shows measured values of the reflection coefficient.

The theoretical results of curve 5 have been obtained by applying a field-matching process at the surface $\zeta = \zeta_1$ of Fig. 4.5. The particular field representation which was chosen for this process is in principle complete, but only up to five modes were considered in the field expansion. Even though the number of field expansion terms has thus been rather limited, the field-matching method yields nevertheless quite accurate results. Curve 5 can therefore be regarded as representing within its graphical accuracy the correct absolute values of the reflection coefficient. We notice slight differences between the results of curve 5 and those from the solution of the coupled wave equations.

While these differences are quite small for small flare angles they increase significantly towards larger flare angles. In Section 3.2 we introduced various approximations to set up the coupled wave equation in a form convenient for its solution. For such large flare angles as in the present example these approximations are no longer really justified. In the case of reasonably smooth waveguide tapers and transitions, on the other hand, the flare angle remains always much smaller and changes only gradually so that all these approximations are always well justified, and the coupled wave equations can be expected to yield quite accurate solutions.

4.3 Convergence properties of coupled wave equations

Usually solutions of the infinite system of coupled wave equations are obtained by considering the interaction between only a limited number of waves, and neglecting the other higher-order waves completely. Methods to derive such solutions have been demonstrated in the preceding sections, and these as well as other such methods will be applied in the following chapters, too.

To examine the number of waves, which need to be taken into account, to obtain correct results from integrating the coupled wave equations, we apply a very straightforward method. We integrate the coupled wave equations several times, each time with a larger number of waves than the previous time. We then examine the convergence by comparing the results. Coupled wave equations are applied in situations where the convergence is only satisfactory when a very large number of waves is taken into account. In such situations, although the infinite system of coupled wave equations gives a rigorous description of the propagation electromagnetic fields, a limited number of waves may be unsuitable to represent the field correctly. This situation occurs for example if the fields are expanded in terms of modes that are not too well suited for the particular waveguide structure. Later in this Section we shall discuss such a case.

At present, however, let us discuss a few more general statements concerning the convergence of coupled wave equations. We start from the coupled wave equations written in the form of eqn. 2.4.1b

$$\frac{d}{dz}A = (-\gamma+\kappa)A \tag{4.3.1a}$$

To best suite our present purpose we rearrange the elements of the vector A and the coupling matrix as follows:

$$\frac{d}{dz}\begin{bmatrix} A_1 \\ A_{(-1)} \\ A_2 \\ A_{(-2)} \\ \vdots \end{bmatrix} =$$

$$\begin{bmatrix} -\gamma_1+\kappa_{11} & \kappa_{1(-1)} & \kappa_{12} & \kappa_{1(-2)} & \cdots \\ \kappa_{(-1)1} & -\gamma_{(-1)}+\kappa_{(-1)(-1)} & \kappa_{(-1)2} & \kappa_{(-1)(-2)} & \cdots \\ \kappa_{21} & \kappa_{2(-1)} & -\gamma_2+\kappa_{22} & \kappa_{2(-2)} & \cdots \\ \kappa_{(-2)1} & \kappa_{(-2)(-1)} & \kappa_{(-2)2} & -\gamma_{(-2)}+\kappa_{(-2)(-2)} & \cdots \\ \vdots & \vdots & \vdots & \vdots & \end{bmatrix} \begin{bmatrix} A_1 \\ A_{(-1)} \\ A_2 \\ A_{(-2)} \\ \vdots \end{bmatrix} \tag{4.3.1b}$$

In nearly all practical cases we are interested only in the over-all transmission and reflection of the modes of lowest order, which in eqn. 4.3.1b have the subscript number of smallest absolute value. The amplitudes of these modes are combined to form the limited number of components of the vector $\hat{A}$. The amplitudes of all other, higher-order modes are gathered in the vector $\hat{\hat{A}}$, which has hence infinitely many components. The coupling matrix in eqn. 4.3.1b is split up accordingly, so that this equation may be written in the following schematic notation

$$\frac{d}{dz}\left[\begin{array}{c} \hat{A} \\ \hline \hat{\hat{A}} \end{array}\right] = \left[\begin{array}{c|c} -\hat{\gamma}+\hat{\kappa} & \kappa' \\ \hline \kappa'' & -\hat{\hat{\gamma}}+\hat{\hat{\kappa}} \end{array}\right]\left[\begin{array}{c} \hat{A} \\ \hline \hat{\hat{A}} \end{array}\right] \tag{4.3.2}$$

The two vectors $\hat{A}$ and $\hat{\hat{A}}$ are coupled by the coupling matrix according to

$$\frac{d}{dz}\hat{A} = (-\hat{\boldsymbol{\gamma}}+\hat{\boldsymbol{\kappa}})\hat{A} + \boldsymbol{\kappa}'\hat{\hat{A}} \tag{4.3.3a}$$

$$\frac{d}{dz}\hat{\hat{A}} = (-\hat{\hat{\boldsymbol{\gamma}}}+\hat{\hat{\boldsymbol{\kappa}}})\hat{\hat{A}} + \boldsymbol{\kappa}''\hat{A} \tag{4.3.3b}$$

The vector $\boldsymbol{\kappa}'\hat{\hat{A}}$ in the first of these two equations contains in each of its components the infinitely many higher-order modes which interact with the modes of lower order. To estimate what influence this vector has on the transmission and reflection of the lower-order modes with amplitudes $\hat{A}_i$ we introduce local normal modes with amplitudes $\hat{a}_i$. The eigenfield of the local normal mode with subscript i is assumed to consist of the eigenfield of the original lowest order mode with subscript i and an additional field term composed of the eigenfields of higher-order modes. The amplitude vector of the local normal modes is thus written as

$$\hat{a} = \hat{A} + \mathbf{w}\hat{\hat{A}}, \tag{4.3.4}$$

where $\mathbf{w}$ is a matrix of the same format as $\boldsymbol{\kappa}'$. This matrix can be regarded to take into account the influence of the higher-order modes on the propagation of the modes under consideration. Substituting eqn. 4.3.4 into eqns. 4.3.3 yields

$$\frac{d}{dz}\hat{a} - \left(\frac{d}{dz}\mathbf{w}\right)\hat{\hat{A}} - \mathbf{w}\left(\frac{d}{dz}\hat{\hat{A}}\right) = (-\hat{\boldsymbol{\gamma}}+\hat{\boldsymbol{\kappa}})\hat{a}$$
$$+ (\boldsymbol{\kappa}' + \hat{\boldsymbol{\gamma}}\mathbf{w} - \hat{\boldsymbol{\kappa}}\mathbf{w})\hat{\hat{A}} \tag{4.3.5a}$$

$$\frac{d}{dz}\hat{\hat{A}} = (-\hat{\hat{\boldsymbol{\gamma}}} + \hat{\hat{\boldsymbol{\kappa}}} - \boldsymbol{\kappa}''\mathbf{w})\hat{\hat{A}} + \boldsymbol{\kappa}''\hat{a} \tag{4.3.5b}$$

When both these equations are combined, we obtain

$$\frac{d}{dz}\hat{a} = (-\hat{\boldsymbol{\gamma}} + \hat{\boldsymbol{\kappa}} + \mathbf{w}\boldsymbol{\kappa}'')\hat{a}$$
$$+ \left[\boldsymbol{\kappa}' + (\hat{\boldsymbol{\gamma}} - \hat{\boldsymbol{\kappa}})\mathbf{w} + \frac{d}{dz}\mathbf{w} + \mathbf{w}(-\hat{\hat{\boldsymbol{\gamma}}} + \hat{\hat{\boldsymbol{\kappa}}}) - \mathbf{w}\boldsymbol{\kappa}''\mathbf{w}\right]\hat{\hat{A}} \tag{4.3.6}$$

If we now choose $\mathbf{w}$ such that it obeys

$$\boldsymbol{\kappa}' + (\hat{\boldsymbol{\gamma}} - \hat{\boldsymbol{\kappa}})\mathbf{w} + \frac{d}{dz}\mathbf{w} + \mathbf{w}(-\hat{\hat{\boldsymbol{\gamma}}} + \hat{\hat{\boldsymbol{\kappa}}}) - \mathbf{w}\boldsymbol{\kappa}''\mathbf{w} = 0, \qquad (4.3.7)$$

then we obtain for the coupling of the local normal modes

$$\frac{d}{dz}\hat{a} = (-\hat{\boldsymbol{\gamma}} + \hat{\boldsymbol{\kappa}} + \mathbf{w}\boldsymbol{\kappa}'')\hat{a} \qquad (4.3.8)$$

Although an exact solution of eqn. 4.3.7 for $\mathbf{w}$ is usually impossible, eqn. 4.3.7 may nevertheless serve to estimate the elements of the matrix $\mathbf{w}\boldsymbol{\kappa}''$. This latter matrix in turn determines the influence of the higher-order modes in eqns. 4.3.8.

To demonstrate this procedure we consider as an example the continuous rectangular waveguide transition shown in Fig. 2.9*b* and let the fundamental H_{10}-mode be incident. Only cylindrical H_{no}-modes with $n = 1, 3, 5 \ldots$ are generated under these conditions.

The propagation constants of these modes are

$$\gamma_n = \left\{\left(\frac{n\pi}{2a}\right)^2 - k^2\right\}^{1/2} \text{sign}(n) \qquad (4.3.9a)$$

and the coupling coefficients are given by

$$\kappa_{mn} = \frac{a'}{a}\frac{nm}{n^2 - m^2}\left[\sqrt{\frac{\gamma_{|n|}}{\gamma_{|m|}}} + \text{sign}(n\cdot m)\sqrt{\frac{\gamma_{|m|}}{\gamma_{|n|}}}\right], \qquad (4.3.9b)$$

with $a' = da/dz$. sign (n) is defined according to

$$\text{sign}(n) = \begin{cases} 1 & \text{for} \quad n > 0 \\ -1 & \text{for} \quad n < 0 \end{cases} \qquad (4.3.9c)$$

Expanding the elements of $\mathbf{w}$ into Taylor series in terms of a'/a, $(a'/a)^2$ etc. we find, that eqn. 4.3.7 may be written as

$$\boldsymbol{\kappa}' + \hat{\boldsymbol{\gamma}}\mathbf{w} - \mathbf{w}\hat{\hat{\boldsymbol{\gamma}}} + O\left[\left(\frac{a'}{a}\right)^2, \frac{a''}{a}\right] = 0 \qquad (4.3.10a)$$

From this equation the elements of $\mathbf{w}$ are given in a first-order approximation by

$$w_{mi} = \frac{\kappa_{mi}}{-\gamma_m + \gamma_i} \qquad (4.3.10b)$$

they are thus proportional to a'/a. We use eqn. 4.3.10*b* to evaluate the matrix product $\mathbf{w}\boldsymbol{\kappa}''$, and obtain the following approximation for its elements:

$$(\mathbf{w}\boldsymbol{\kappa}'')_{mn} = \sum_i \frac{\kappa_{mi}\kappa_{in}}{-\gamma_m + \gamma_i} \tag{4.3.10c}$$

In this expression the sum has to be extended over all terms with positive and negative subscripts i which correspond to higher-order modes. We substitute for the coupling coefficients from eqns. 4.3.9, and combine the terms with negative and positive index $-i$ and i, to arrive at

$$(\mathbf{w}\boldsymbol{\kappa}'')_{mn} = \left(\frac{a'}{a}\right)^2 \sum_{i=p,p+2,\ldots} \frac{im}{i^2 - m^2}\,\frac{ni}{n^2 - i^2}$$

$$\frac{2\gamma_i\left[\operatorname{sign}((mn))\sqrt{\dfrac{\gamma_{|n|}}{\gamma_{|m|}}} + \sqrt{\dfrac{\gamma_{|m|}}{\gamma_{|n|}}}\right] + 2\gamma_m\left[\operatorname{sign}(n)\dfrac{\gamma_i}{\sqrt{\gamma_{|m|}\gamma_{|n|}}} + \operatorname{sign}(m)\dfrac{\sqrt{\gamma_{|m|}\gamma_{|n|}}}{\gamma_i}\right]}{\gamma_i^2 - \gamma_m^2} \tag{4.3.11}$$

The sum of this expression must be extended over all terms which correspond to higher-order modes with positive index i. If these modes are far enough below cutoff the combined terms of the series are proportional to i^{-3}. Thus the series converges quite rapidly, and the coupling effect of the modes far enough below cutoff on the modes with low index n can be neglected altogether.

This method of investigating the influence of the infinite number of evanescent mode on the coupling of the modes near or above cutoff may be applied to all coupling structures of Chapters 2 and 3 of this book. We can expect to find for all these structures similar convergence characteristics of the coupled wave equations. Only in the case of the two coupled dielectric guides of Section 2.8 may we encounter certain convergence problems. They have been pointed out in the literature.[2]

To learn more about these convergence problems we discuss, as an instructive example, a rectangular waveguide with a dielectric slab of isotropic and homogeneous permittivity ϵ' and uniform width c adjacent to one of the broad side walls of the waveguide, as shown in Fig. 4.7. The permittivity of the homogeneous, isotropic dielectric in the rest of the cross-section is ϵ.

A suitably thin slab modifies the propagation characteristics of the fundamental H_{10}-mode only slightly. A perturbation calculation[3] starts from the propagation constant γ_0 for the fundamental mode in the waveguide without slab and with the uniform permittivity throughout, and arrives at the following first-order perturbational expression

$$j\beta = \gamma_0 + \frac{\omega^2\mu\epsilon}{2j\gamma_0}\,\frac{c}{b}\,\frac{\epsilon' - \epsilon}{\epsilon'} = \gamma_0 - \kappa'_{00} \tag{4.3.12a}$$

The approximation

$$\kappa'_{00} = j\,\frac{\omega^2\mu\epsilon}{2\gamma_0}\,\frac{c}{b}\,\frac{\epsilon'-\epsilon}{\epsilon'}$$

for the change in propagation constant from this perturbation analysis is adequate as long as

$$\frac{\epsilon'-\epsilon}{\epsilon}\,\frac{c}{b} \ll 1 \tag{4.3.12b}$$

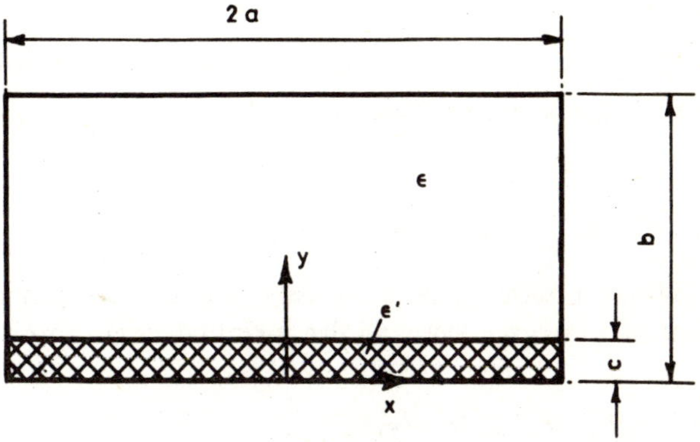

Fig. 4.7 *Rectangular waveguide with a dielectric slab*

We shall now try to obtain this result by solving the appropriate coupled wave equations. To this end the field in the waveguide with the dielectric slab insert is expanded in terms of the eigenfields of the normal modes in a waveguide with equal cross-sectional size but homogeneous dielectric material of permittivity ϵ.

The set of normal modes which the slab couples to the H_{10}-modes consists only of H_{1n}-modes and E_{1n}-modes, the normalised E-fields of which near $y = 0$ follow from Section 2.6. The normalised E-field of a H_{1n}-mode is given by

$$E = n_y E_n^H = n_y \sqrt{\frac{j\omega\mu}{\gamma_{|n|}}\frac{\sqrt{\epsilon_n}}{k_{cn}}}\,\frac{\pi}{2a}\,\frac{1}{\sqrt{ab}}\cos\left(\frac{\pi}{2a}x\right)\cos\left(\frac{n\pi}{b}y\right) \tag{4.3.13a}$$

with $n = 0, 1, 2, \ldots$, while the normalised E-field of an E_{1n}-mode follows from

$$E = n_y \dot{E}_n^E = n_y \sqrt{\frac{\gamma_{|n|}}{j\omega\epsilon}} \frac{\sqrt{2}}{k_{cn}} \frac{n\pi}{b} \frac{1}{\sqrt{ab}} \cos\left(\frac{\pi}{2a}x\right) \cos\left(\frac{n\pi}{b}y\right) \tag{4.3.13b}$$

with $n = 1, 2, 3 \ldots$

$$k_{cn} = \left\{\left(\frac{\pi}{2a}\right)^2 + \left(\frac{n\pi}{b}\right)^2\right\}^{1/2} \tag{4.3.13c}$$

is the cutoff wave number of the mode with subscript n, and γ_n is its propagation constant

$$\gamma_{|n|} = \sqrt{k_{cn}^2 - k^2} = \sqrt{k_{cn}^2 - \omega^2\mu\epsilon} \tag{4.3.13d}$$

Note that ϵ_n in eqn. 4.3.13a represents again Neumann's number

$$\epsilon_n = \begin{cases} 1 & \text{for} \quad n = 0 \\ 2 & \text{for} \quad n \neq 0 \end{cases}$$

These individual normal modes can only propagate independently of each other if within the dielectric slab we introduce currents of magnitude

$$-J_n^{H,E} = -j\omega(\epsilon' - \epsilon)E_n^{H,E}A_n$$

The corresponding currents of opposite sign act as sources, which generate spurious modes. They couple the normal modes of the waveguide without the dielectric insert to each other. The coupling coefficients for this interaction follow from eqn. 2.2.7. In case of coupling from the ith to the mth mode in the same direction, and to the mth mode travelling in opposite direction they are, respectively, given by

$$\kappa_{\substack{mi \\ -mi}} = \mp j\omega(\epsilon' - \epsilon)\frac{1}{2}\int_0^c dy \left[\int_{-a}^{a} dx\,(E_i E_m)\right] \tag{4.3.14}$$

Negative subscripts denote waves in the negative z-direction while positive subscripts belong to waves in positive z-direction.

From eqn. 4.3.14 the coupling coefficients for coupling to the forward travelling H_{10}-mode with its subscript $m = 0$ is given by

$$\kappa_{oi} = \kappa_{io} = -\kappa_{-io}$$

$$= j\omega(\epsilon' - \epsilon)\frac{c}{2b}\frac{k^2}{\epsilon}\left(\frac{\pi}{2a}\right)^2 \frac{\sqrt{\epsilon_i}}{k_{co}k_{ci}} \frac{1}{\sqrt{\gamma_o\gamma_i}} \frac{\sin\left(\frac{i\pi}{b}c\right)}{\frac{i\pi}{b}c} \tag{4.3.15a}$$

if the ith mode denotes a H-mode with positive subscript i, and by

$$\kappa_{oi} = \kappa_{io} = -\kappa_{-io}$$

$$= -j\omega(\epsilon' - \epsilon)\frac{c}{2b}\frac{\pi}{2a}\sqrt{\frac{\mu}{\epsilon}}\frac{\sqrt{2}}{k_{co}k_{ci}}\sqrt{\frac{\gamma_i}{\gamma_0}}\frac{i\pi}{b}\frac{\sin\left(i\frac{\pi}{b}c\right)}{\frac{i\pi}{b}c} \tag{4.3.15b}$$

if the ith mode denotes an E-mode likewise with positive subscript i.

The subscript $i = 0$ belongs to the H_{10}-mode in the forward z-direction. The corresponding coupling coefficient represents the lowest-order approximation to the change in propagation constant as caused by the slab. From eqn. 4.3.15*a* this coefficient follows as

$$\kappa_{00} = -\frac{k^2}{2j\gamma_0}\frac{c}{b}\frac{\epsilon' - \epsilon}{\epsilon} \tag{4.3.16}$$

It differs from eqn. 4.3.12*a* by the factor ϵ/ϵ'.

A more accurate expression for this change in propagation constant is obtained from eqn. 4.3.8, more specifically from the first diagonal element of its coupling matrix

$$\kappa'_{00} = \kappa_{00} + \sum_{i=1,2,3\ldots}(w_{oi}\kappa_{io} + w_{o(-i)}\kappa_{(-i)o}) \tag{4.3.17a}$$

For the w_{oi} in this expression the first-order approximation according to eqn. 4.3.10*a* is given by

$$w_{oi} = \frac{\kappa_{oi}}{\gamma_i - \gamma_0} \tag{4.3.17b}$$

A more accurate analysis shows that the w_{oi} may be represented as a Taylor series in terms of powers of $(\epsilon' - \epsilon)/\epsilon$.

Eqn. 4.3.17*b* gives only the first term of these Taylor-series expansions, which is proportional to $(\epsilon' - \epsilon)/\epsilon$. Eqns. 4.3.17*a* and 4.3.17*b* must be evaluated for all positive and negative subscripts i. The corresponding terms account for spurious modes which propagate in the positive and negative z-direction, respectively. Note that for each value of i a H_{io}- as well as an E_{io}-mode must be considered. In the series of eqn. 4.3.17*a*, however, only the E-modes make significant contributions. If therefore we neglect all contributions from spurious H_{io}-modes we need to include in the sum of

$$\kappa'_{00} = \kappa_{00} + \sum_{i=1,2,3\ldots} \left(\frac{\kappa_{oi}\kappa_{io}}{\gamma_i - \gamma_0} + \frac{\kappa_{oi}\kappa_{io}}{\gamma_i + \gamma_0} \right) \tag{4.3.17c}$$

only those terms that correspond to E_{io}-modes with positive subscript.

Thus substituting from eqns. 4.3.14 and 4.3.15*b*, we obtain

$$\kappa'_{00} = \kappa_{00} \left[1 - \frac{\epsilon' - \epsilon}{\epsilon} \frac{2c}{b} \sum_{i=1,2,3\ldots} \frac{i\pi}{b} \frac{\gamma_i^2}{k_{ci}^2} \frac{1}{\gamma_i^2 - \gamma_0^2} \left(\frac{\sin\left(\frac{i\pi}{b}c\right)}{\frac{i\pi}{b}c} \right)^2 \right] \tag{4.3.18}$$

The series in eqn. 4.3.18 converges very slowly, even for $c \ll b$. For values of i sufficiently large we may introduce the approximation

$$\frac{\gamma_i^2}{k_{ci}^2(\gamma^2 - \gamma_0^2)} \approx \frac{1}{\left(\frac{i\pi}{b}\right)^2}$$

From this approximation we see that many modes with subscripts i contribute to the series. For c sufficiently small compared with b the increments by which the $\sin(x)/x$ function changes its argument from term to term of the series is small enough to replace the summation over i by an integration. We then obtain from eqn. 4.3.18:

$$\kappa'_{00} = \kappa_{00} \left[1 - \frac{\epsilon' - \epsilon}{\epsilon} \frac{2}{\pi} \int_0^\infty \left(\frac{\sin x}{x} \right)^2 dx \right]$$

$$= \kappa_{00} \left[1 - \frac{\epsilon' - \epsilon}{\epsilon} \right]$$

Since this result is valid only for $(\epsilon' - \epsilon)/\epsilon \ll 1$, we can also write it in the following form

$$\kappa'_{00} = \kappa_{00} \frac{\epsilon}{\epsilon'} \tag{4.3.19}$$

which, together with κ_{00} from eqn. 4.3.16, agrees with eqn. 4.3.12*a*.

The summation of the series in eqn. 4.3.18 for $c \ll b$ by way of an integral demonstrates quite clearly that the convergence properties of the coupled wave equations are poor in the present example, and that a solution which takes into account only a small number of higher-order modes will soon lead to very inaccurate overall coupling results.

This example demonstrates also that one needs to be very careful in choosing the appropriate system of normal modes for a nonuniform waveguide section. This is particularly important if, in the subsequent analysis, only an approximate solution of the coupled wave equations becomes available, that accounts only for the interaction of a limited number of modes. As stated earlier, we should always try to work with normal modes which individually, and without any interaction, approximate the actual field in the particular structure for the specified form of excitation as good as possible. These modes will interact only weakly with each other, and a small number of these is only needed at a time to describe the actual field quite accurately.

In the following chapters we shall assume that, for whatever application we consider coupled wave equations, their convergence properties will always be satisfactory enough so that only a limited number of normal modes need to be taken into account to obtain sufficiently accurate overall coupling results. If we are interested, for example, in the transmission of one particular normal mode with index m in a continuous nonuniform guide, then we can see from eqn. 4.3.10c that the propagation of this wanted mode will be most strongly influenced by those spurious modes with index i for which the magnitude of

$$\frac{\kappa_{im}\kappa_{mi}}{\gamma_m - \gamma_i} \qquad (4.3.20a)$$

is at a maximum.

This allows always to estimate readily the contribution of certain spurious modes to the overall wave propagation.

Note, that in the series of eqn. 4.3.10c forward and backward waves with positive and negative subscripts i, respectively, have to be considered.

To estimate the influence of a spurious mode i near cutoff, where absolute values of the respective products $\kappa_{mi}\kappa_{im}$ are quite high, but its propagation constant γ_i is nearly zero, it always facilitates the analysis when both the forward and the backward wave of this spurious mode are considered together. While the absolute values of the expression 4.3.20a for an individual wave i travelling in one direction will be infinite at cutoff, the combined terms for the forward and backward travelling waves in the form of

$$\frac{\kappa_{mi}\kappa_{im}}{\gamma_m - \gamma_i} + \frac{\kappa_{m(-i)}\kappa_{(-i)m}}{\gamma_m - \gamma_{(-i)}} \qquad (4.3.20b)$$

will have a finite value there. This can easily be verified by way of example for the rectangular waveguide transition as shown in Fig. 2.9*b*, when eqns. 4.3.9 are applied.

References

1 PIEFKE, G.: 'Reflexion beim Übergang vom Rechteck-Hohlleiter zum Sektor-Horn,' *AEU*, 1957, **11**, pp. 123–135

2 KUESTER, E. F., and CHANG, D. C.: 'Nondegenerate surface-wave mode coupling between dielectric waveguides,' *IEEE Trans.* 1975, **MTT-23**, pp. 877–882

3 HARRINGTON, R. F.: 'Time-harmonic electromagnetic fields,' (McGraw-Hill, 1961)

Chapter 5

Transitions and couplers with one parasitic wave

A simple example of a nonuniform waveguide in which a parasitic wave is generated is the nonuniform TEM transmission line between two uniform lines with different wave impedances. The TEM wave propagating in one particular direction, is to be transmitted through the nonuniform section with as little reflection as possible. The other wave travelling in the opposite direction is then regarded as the parasitic wave, the excitation of which by the primary wave has to be avoided. This example is particularly simple because the waves on the ideally lossless TEM line do not exhibit any dispersion. Thus very simple expressions can be obtained for the over-all transmission characteristics of the nonuniform transmission line as a function of frequency.

Once these over-all transmission results have been derived for the nonuniform TEM line, they can also be used to analyse and design a more general class of waveguide transitions. This will be shown in the following chapters.

5.1 TEM matching section

In this Section a nonuniform transmission line is considered in which the propagating fields are assumed to be purely TEM. As an example we take the tapered TEM line section of Fig. 5.1 which is inserted between the uniform transmission lines of characteristic impedance Z_1 and Z_2, respectively.

Although the analysis and design of such TEM matching sections are not the primary subject of this book, to determine the transmission characteristics of the line section in Fig. 5.1 will, nevertheless, demonstrate the main principles which are to be applied to calculate suitable waveguide taper contours.

The nonuniform transmission-line model in Fig. 5.1 may represent any tapered TEM line. The coaxial line of Fig. 2.9*a* with a homogeneous lossless material of permeability μ and permittivity ϵ between the conductors is only one particular example. For this example, however, the field configuration is given by well known functions of the cross-sectional co-ordinates, and a normal mode representation according to Section 2.6 can be easily applied to analyse the over-all transmission properties of the nonuniform TEM line. If for the coaxial line as an example, the diameters of the tapered conductors change only gradually, and if the distance between the conducting boundaries is significantly smaller than the wavelength of the electromagnetic waves to be transmitted, we can expect the fields inside the nonuniform section to be purely TEM. Thus only the TEM-mode will be considered with its voltage $u(z)$ across the line and its current $i(z)$ along the line.

TEM-voltage and current obey the well known transmission-line

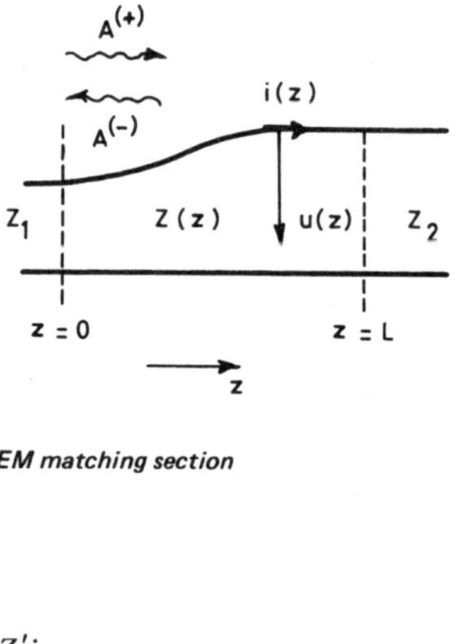

Fig. 5.1 *Lossless TEM matching section*

equations

$$\frac{du}{dz} = -Z'i \tag{5.1.1a}$$

$$\frac{di}{dz} = -Y'u \tag{5.1.1b}$$

where Z' is the series impedance per unit length of the line while Y' is the shunt admittance per unit length of the line.

For the nonuniform TEM-line its quantities Z' and Y' change with the axial distance z. The propagation constant γ and the characteristic impedance Z of the transmission line relate to these primary line parameters as follows:

$$\gamma = j\beta = \sqrt{Z'Y'} \tag{5.1.2a}$$

$$Z = \sqrt{Z'/Y'} \tag{5.1.2b}$$

Note that the characteristic impedance Z of TEM lines represents the voltage/current ratio of a wave on the transmission line and not the ratio of electric to magnetic field strength as it does for waveguide modes. Since the line is assumed to be lossless, Z and β are purely real. For the coaxial line their values are given by

$$Z = \frac{1}{2\pi}\sqrt{\frac{\mu}{\epsilon}}\ln\left(\frac{R}{r}\right) \tag{5.1.3a}$$

$$\beta = 2\pi f\sqrt{\mu\epsilon} \tag{5.1.3b}$$

To solve the line equations 5.1.1, we introduce the wave amplitudes $A^{(+)}$ and $A^{(-)}$ of the TEM waves which propagate in the positive and negative z-direction, respectively,

$$A^{(\pm)} = \frac{1}{2}\left(\frac{u}{\sqrt{Z}} \pm i\sqrt{Z}\right) \tag{5.1.4a}$$

With these relations between line voltage and line current on one side, and the forward and backward wave amplitudes on the other, the wave amplitudes are normalised with respect to power. The real power which they transmit follows from

$$\text{Re}\,(P) = A^{(+)}(A^{(+)})^* - A^{(-)}(A^{(-)})^*$$

When we replace u and i in the transmission-line eqns. 5.1.1 by $A^{(+)}$ and $A^{(-)}$ according to eqn. 5.1.4*a* we obtain

$$\frac{d}{dz}A^{(+)} = -j\beta A^{(+)} + \kappa A^{(-)} \tag{5.1.5a}$$

$$\frac{d}{dz}A^{(-)} = \kappa A^{(+)} + j\beta A^{(-)} \tag{5.1.5b}$$

where

$$\kappa = -\frac{1}{2Z}\frac{dZ}{dz} \tag{5.1.5c}$$

The value of the coupling coefficient κ differs from zero only within the nonuniform section of Fig. 5.1 and varies with z depending on the contour of this section. It is, however, independent of frequency f. The value of the phase constant β on the other hand is a function of frequency but it is independent of z.

When we introduce eqns. 5.1.3 into eqns. 5.1.5 we arrive at the same results as we have obtained them using the approach with eigen-fields of the normal-modes. These results are given by eqns. 2.3.5, 2.6.15.

Eqns. 5.1.5 will presently be applied to analyse gradually tapered TEM matching sections. As a preliminary, however, it is instructive to apply these equations to uniform sections with abrupt transitions for matching purposes. We consider first the abrupt transition between two uniform lines of wave impedances Z_1 and Z_2 (Fig. 5.2). The scattering matrix of this junction relates the forward and backward wave amplitudes $A_1^{(\pm)}$ in line 1 and $A_2^{(\pm)}$ in line 2 to each other. It is obtained by applying Kirchhoff's laws for the voltage and currents at the junction

$$\begin{bmatrix} A_1^{(-)} \\ A_2^{(+)} \end{bmatrix} = \begin{bmatrix} r^{(21)} & t^{(21)} \\ t^{(21)} & -r^{(21)} \end{bmatrix} \begin{bmatrix} A_1^{(+)} \\ A_2^{(-)} \end{bmatrix} \tag{5.1.6a}$$

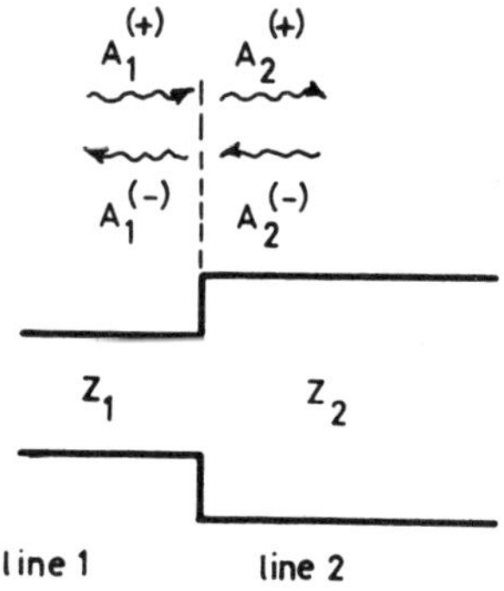

Fig. 5.2 *Abrupt TEM junction*

The input reflection coefficient in line 1 is given by

$$r^{(21)} = \frac{Z_2 - Z_1}{Z_2 + Z_1} \tag{5.1.6b}$$

While the transmission factor obtains as

$$t^{(21)} = \frac{2\sqrt{Z_1 Z_2}}{Z_2 + Z_1} \tag{5.1.6c}$$

Note that for lossless lines both factors are real. Solving eqn. 5.1.6*a* for $A_2^{(+)}$ and $A_2^{(-)}$ leads to the transmission matrix $\mathbf{T}(Z_2, Z_1)$ of the abrupt transition according to

$$\begin{bmatrix} A_2^{(+)} \\ A_2^{(-)} \end{bmatrix} = \mathsf{T}(Z_2, Z_1) \begin{bmatrix} A_1^{(+)} \\ A_1^{(-)} \end{bmatrix} \tag{5.1.7a}$$

where

$$\mathsf{T}(Z_2, Z_1) = \frac{1}{t^{(21)}} \begin{bmatrix} 1 & -r^{(21)} \\ -r^{(21)} & 1 \end{bmatrix} \tag{5.1.7b}$$

That eqn. 5.1.7 is indeed a solution of eqn. 5.1.5 can easily be verified by transforming eqn. 5.1.5

$$\frac{d}{dz} \begin{bmatrix} A_2^{(+)} \\ A_2^{(-)} \end{bmatrix} = \begin{bmatrix} -j\beta & -\dfrac{1}{2Z_2}\dfrac{dZ_2}{dz} \\ -\dfrac{1}{2Z_2}\dfrac{dZ_2}{dz} & j\beta \end{bmatrix} \begin{bmatrix} A_2^{(+)} \\ A_2^{(-)} \end{bmatrix}$$

from the independent variable z to the characteristic impedance Z_2 as the new independent variable. With

$$\frac{d}{dz} = \frac{dZ_2}{dz}\frac{d}{dZ_2}$$

and dividing either side by dZ_2/dz, we obtain in the limit of $dZ_2/dz \to \infty$ the set of equations

$$\frac{d}{dZ_2} \begin{bmatrix} A_2^{(+)} \\ A_2^{(-)} \end{bmatrix} = -\frac{1}{2Z_2} \begin{bmatrix} 0 & 1 \\ 1 & 0 \end{bmatrix} \begin{bmatrix} A_2^{(+)} \\ A_2^{(-)} \end{bmatrix}$$

$$= -\frac{1}{2Z_2} \begin{bmatrix} 0 & 1 \\ 1 & 0 \end{bmatrix} \mathsf{T}(Z_2, Z_1) \begin{bmatrix} A_1^{(+)} \\ A_1^{(-)} \end{bmatrix}$$

If on the other hand eqn. 5.1.7a is differentiated with respect to Z_2 the following equations result:

$$\frac{d}{dZ_2} \begin{bmatrix} A_2^{(+)} \\ A_2^{(-)} \end{bmatrix} = \left[\frac{d}{dZ_2}\mathsf{T}(Z_2, Z_1)\right] \begin{bmatrix} A_1^{(+)} \\ A_1^{(-)} \end{bmatrix}$$

These equations are identical to the previous equations, which in turn were derived from eqn. 5.1.6.

Eqn. 5.1.7 can be applied to analyse the well known quarter-wave transformer in terms of the wave amplitudes as they are defined by Fig. 5.3. The transmission matrix $\mathsf{T}'(Z_2)$ of the homogeneous line section with the characteristic impedance Z_2 is given by

$$T'(Z_2) = \begin{bmatrix} \exp(-j\phi_2) & 0 \\ 0 & \exp(j\phi_2) \end{bmatrix} \quad \text{with} \quad \phi_2 = \beta L_2 \tag{5.1.8b}$$

This matrix relates the wave amplitudes at the output of the section to its input amplitudes by

$$\begin{bmatrix} A_2^{(+)} \\ A_2^{(-)} \end{bmatrix} = \mathsf{T}'(Z_2) \begin{bmatrix} \hat{A}_2^{(+)} \\ \hat{A}_2^{(-)} \end{bmatrix} \tag{5.1.8a}$$

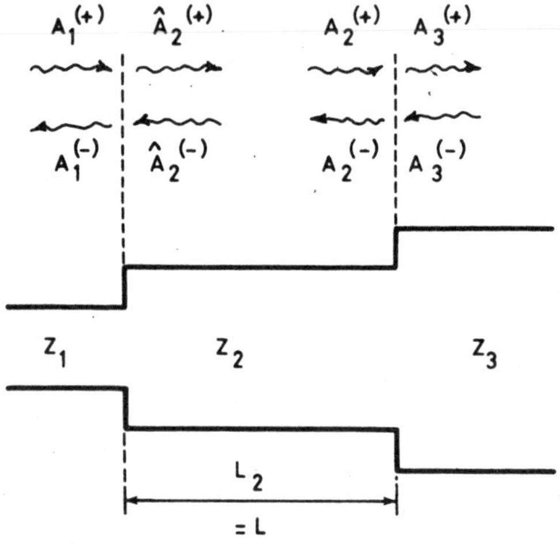

Fig. 5.3 λ/4-matching section

The total transmission matrix $\hat{\mathsf{T}}$ of the nonuniform line of Fig. 5.3 including the two abrupt transitions on both sides is then given by

$$\hat{\mathsf{T}} = \begin{bmatrix} \hat{\mathsf{T}}^{(++)} & \hat{\mathsf{T}}^{(+-)} \\ \hat{\mathsf{T}}^{(-+)} & \hat{\mathsf{T}}^{(--)} \end{bmatrix} = \mathsf{T}(Z_3, Z_2)\mathsf{T}'(Z_2)\mathsf{T}(Z_2, Z_1) \tag{5.1.9b}$$

It provides us with the relation

$$\begin{bmatrix} A_3^{(+)} \\ A_3^{(-)} \end{bmatrix} = \hat{\mathsf{T}} \begin{bmatrix} A_1^{(+)} \\ A_1^{(-)} \end{bmatrix} \tag{5.1.9a}$$

between the overall input and output wave amplitudes.

The input reflection coefficient for the waves incident on line 1 is obtained from the second equation of system 5.1.9*a* when we let $A_3^{(-)} = 0$. As the ratio of $A_1^{(-)}$ to $A_1^{(+)}$ under this particular output condition it is given by

$$R = \frac{A_1^{(-)}}{A_1^{(+)}} = -\frac{\hat{T}^{(-+)}}{\hat{T}^{(--)}}. \tag{5.1.10}$$

When we evaluate the triple product of matrices of eqn. 5.1.9*b* we obtain for the input reflection coefficient

$$R \exp(j\phi_2) = \frac{r^{(32)} \exp(-j\phi_2) + r^{(21)} \exp(j\phi_2)}{1 + r^{(32)} r^{(21)} \exp(-j\phi_2)} \tag{5.1.11}$$

From this general formula the impedance can be matched only if the real quantities $r^{(32)}$ and $r^{(21)}$ are chosen to be equal:

$$r^{(32)} = \frac{Z_3 - Z_2}{Z_3 + Z_2} = r^{(21)} = \frac{Z_2 - Z_1}{Z_2 + Z_1} \tag{5.1.12a}$$

which leads to the well known condition

$$(Z_2)^2 = Z_1 Z_3 \tag{5.1.12b}$$

for the quarter-wave transformer.

If the impedances differ only so little from each other that they obey

$$\left(\frac{\sqrt{Z_3} - \sqrt{Z_1}}{\sqrt{Z_3} + \sqrt{Z_1}}\right)^2 \ll 1 \tag{5.1.13a}$$

then we may disregard the term in the denominator of eqn. 5.1.11 that contains the product of the two small quantities $r^{(32)}$ and $r^{(21)}$. Under this condition the quarter-wave transformer with $r^{(32)} = r^{(21)}$ has the following input reflection as a function of frequency:

$$\frac{R}{R_0} \exp(j\phi_2) = \cos(x\pi) = F(x) \tag{5.1.13b}$$

In this expression R_0,

$$R_0 = r^{(32)} + r^{(21)} \tag{5.1.13c}$$

represents the input reflection at zero frequency ($f \to 0$) while x is the normalised frequency given by

$$x = \beta_2 L/\pi = 2fL/c \tag{5.1.13d}$$

with $c = 1/\sqrt{\mu\epsilon}$ as the intrinsic phase velocity in the dielectric material between the conductors.

A similar procedure can be applied to calculate the transmission characteristics of multisection transmission-line transformers, which consist of a series connections of uniform lines with abrupt transitions at the junctions. For the two-section line transformer of Fig. 5.4 the over-all transmission matrix obtains as

$$\begin{bmatrix} A_4^{(+)} \\ A_4^{(-)} \end{bmatrix} = \hat{\mathsf{T}} \begin{bmatrix} A_1^{(+)} \\ A_1^{(-)} \end{bmatrix} \tag{5.1.14a}$$

$$\hat{\mathsf{T}} = \mathsf{T}(Z_4, Z_3)\,\mathsf{T}'(Z_3)\,\mathsf{T}(Z_3, Z_2)\,\mathsf{T}'(Z_2)\,\mathsf{T}(Z_2, Z_1) \tag{5.1.14b}$$

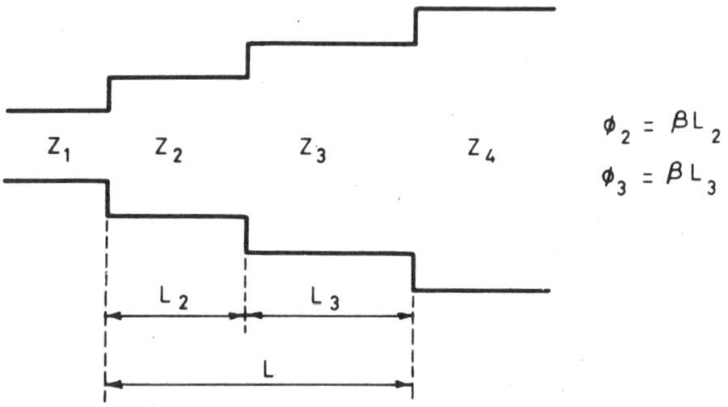

Fig. 5.4 *Two-line matching section*

and the input reflection coefficient R follows from eqn. 5.1.10:

$$R \exp[j(\phi_2 + \phi_3)] =$$

$$\frac{r^{(21)} \exp[j(\phi_3 + \phi_2)] + r^{(32)}(1 - r^{(21)} r^{(43)}) \exp[j(\phi_3 - \phi_2)] + r^{(43)} \exp[-j(\phi_3 + \phi_2)]}{1 + r^{(32)}[r^{(21)} \exp(-j2\phi_2) + r^{(43)} \exp(-j2\phi_3)] + r^{(21)} r^{(43)} \exp[-j2(\phi_2 + \phi_3)]} \tag{5.1.15}$$

In this expression the reflection coefficients r of the individual transitions are real quantities and independent of frequency while ϕ_2 and ϕ_3 vary proportional to frequency. To obtain as many zeros of the absolute value of the numerator as is possible, the values of r and ϕ have to be chosen such that the numerator is always real:

$$r^{(21)} = r^{(43)} \tag{5.1.16a}$$

$$\phi_2 = \phi_3 \tag{5.1.16b}$$

If then again, we neglect all terms proportional to products $r^{(ij)} r^{(mn)}$, which are small quantities, we obtain

$$\frac{R}{R_0} \exp\left[j(\phi_2 + \phi_3)\right] = y + (1-y)\cos(\pi x) = F(x) \tag{5.1.17a}$$

with

$$y = r^{(32)}/R_0$$

where, in correspondence with eqn. 5.1.13, the parameters R_0 and x are defined as

$$R_0 = r^{(21)} + r^{(32)} + r^{(43)} \tag{5.1.17c}$$

and

$$x = \beta L/\pi, \quad L = L_2 + L_3 \tag{5.1.17d}$$

Multisection transmission line transformers are often used to match two lines of different wave impedances to each other for specified frequency bands. According to eqns. 5.1.13 and 5.1.17 they show bandpass characteristics with the transmission characteristic being periodic in frequency. Note that the transmission characteristics as given above have been obtained by neglecting all higher-order modes. In transmission lines for TEM-waves such higher-order modes are usually far below cutoff and strongly evanescent. In the abrupt junction of Fig. 5.2 these evanescent modes store reactive energy and will, in particular at high frequencies, act as shunt reactances. For such high frequencies they must be considered in the design procedure.

Let us now return to continuous matching sections with only a gradual change in cross-section. Any higher-order-mode effects are still much weaker in such continuous matching sections than at abrupt transitions and will hence not be considered. The main feature of the continuous matching sections as compared to the multisection transformer with abrupt transitions is that highpass characteristics can be obtained for its transmission.

The principles which we develop to obtain good highpass characteristics for continuous TEM matching sections will in the subsequent chapter also be applied to design waveguide tapers with low spurious mode generation.

We start again with the coupled wave equations and derive a solution for the case where at $z = 0$ the matching section of Fig. 5.1 is excited by a forward travelling TEM-wave of unit amplitude

$$A^{(+)}(z = 0) = 1 \tag{5.1.18a}$$

and a zero backward amplitude

$$A^{(-)}(z = 0) = 0 \tag{5.1.18b}$$

The distribution function $\kappa(z)$ for the coupling function is assumed to be known.

An analytical solution of eqn. 5.1.5 cannot be found for an arbitrary coupling function $\kappa(z)$. Thus let us, at the moment, use an approximation for which we assume, that in eqn. 5.1.5*a* the inequality

$$|\beta A^{(+)}| \gg |\kappa A^{(-)}| \qquad (5.1.19)$$

is satisfied. Under this assumption we obtain as a zero-order approximation the solution

$$A^{(+)}(z) = \exp(-j\beta z) \qquad (5.1.20a)$$

and

$$A^{(-)}(L) = \int_0^L \kappa A^{(+)}(z) \exp[j\beta(L-z)]\,dz = \hat{T}^{(-+)}$$

$$= \int_0^L \kappa \exp[j\beta(L-2z)]\,dz \qquad (5.1.20b)$$

The expressions 5.1.20*a* for $z = L$ and 5.1.20*b* can be regarded as the elements $\hat{T}^{(++)}$ and $\hat{T}^{(-+)}$ of the transmission matrix for the nonuniform section between $z = 0$ and $z = L$. This transmission matrix relates output to input amplitudes of the nonuniform section by

$$\begin{bmatrix} A^{(+)}(z = L) \\ A^{(-)}(z = L) \end{bmatrix} = \begin{bmatrix} \hat{T}^{(+-)} & \hat{T}^{(+-)} \\ \hat{T}^{(-+)} & \hat{T}^{(--)} \end{bmatrix} \begin{bmatrix} A^{(+)}(z = 0) \\ A^{(-)}(z = 0) \end{bmatrix} \qquad (5.1.20c)$$

The element $T^{(--)}$ of the transmission matrix is obtained if the nonuniform section is at $z = 0$ excited with

$$A^{(+)}(z = 0), \quad A^{(-)}(z = 0) = 1$$

Just as in eqn. 5.1.20*a* we obtain under these excitation conditions

$$\hat{T}^{(--)} = \exp(j\beta L) \qquad (5.1.20d)$$

According to eqn. 5.1.10 the input reflection coefficient R as seen from line 1 is the ratio of the elements $\hat{T}^{(-+)}$ and $\hat{T}^{(--)}$ of the transmission matrix. It is hence given by

$$R = -\frac{\hat{T}^{(-+)}}{\hat{T}^{(--)}} = -\int_0^L \kappa \exp(-j2\beta z)dz \qquad (5.1.21)$$

Let us now ask under what conditions the nonuniform section between the uniform transmission lines with characteristic impedances Z_1 and Z_2, respectively, will have low input reflection.

In the limit of zero frequency, for

$$\beta L = 0 \qquad (5.1.22a)$$

the exact solution of eqns. 5.1.5 is according to eqn. 5.1.6

$$R = \frac{Z_2 - Z_1}{Z_2 + Z_1} \tag{5.1.22b}$$

On the other hand, if we still maintain eqn. 5.1.22*a* but evaluate eqn. 5.1.21 we obtain with eqn. 5.1.5*c* for the input reflection

$$R = -\int_0^L \kappa dz = \tfrac{1}{2} \ln (Z_2/Z_1) \tag{5.1.22c}$$

This approximation differs from the exact result in eqn. 5.1.22*b* because it was obtained under the assumption of eqn. 5.1.19. For $\beta L \ll 1$ the inequality 5.1.19 is in general only satisfied if, at the same time, the absolute value of R remains small.

From the results of eqn. 5.1.22 it now becomes evident that if for good performance of the matching section we require

$$|R| \ll \frac{Z_2 - Z_1}{Z_2 + Z_1},$$

the value of βL has to be sufficiently large. This means, that for a finite length L, and with $\beta = 2\pi f\sqrt{\mu\epsilon}$, low input reflection over a wide band can only be obtained for frequencies f above a certain minimum limit frequency.

To utilise eqn. 5.5.21 in designing the nonuniform line section for good highpass matching characteristics we define a number of design parameters. We first transform to

$$\xi = 2\beta_0(z - L/2) \tag{5.1.23a}$$

as the new variable of integration in eqn. 5.1.21. The value β_0 of the phase constant of the TEM wave refers to a characteristic frequency f_0, which has the meaning of a cutoff or limit frequency of the pass band of the matching section with its highpass character. The passband is then defined by

$$\eta = f/f_0 > 1 \tag{5.1.23b}$$

and we will aim to design the matching section for satisfactory performance at frequencies of $f > f_0$. We abbreviate by

$$\theta = 2\beta_0 L/2 \tag{5.1.23c}$$

the limit of integration for the new variable in eqn. 5.1.21.

We furthermore use the following notation for the distribution of coupling as a function of the new variable

$$CK(\xi) = -\frac{\kappa}{2\beta_0} = \frac{1}{2\beta_0 2Z}\frac{dZ}{dz} = \frac{d\ln(Z)}{2d\xi} \tag{5.1.23d}$$

The coefficient C in this distribution function is a constant quantity, the value of which will be specified later.

When we introduce these design parameters into eqn. 5.1.21 the integral for the input reflection assumes the following form:

$$R(\eta) = \int_{\theta-}^{\theta} CK(\xi) \exp(-j\eta\xi) d\xi \qquad (5.1.24)$$

Suppose now that $CK(\xi)$ were given as a function of ξ. We could then, from eqn. 5.1.23d, determine the distribution function $Z(z)$ for the characteristic impedance of the matching section from

$$Z(\xi)/Z(-\theta) = Z(\xi)/Z_1 = \exp\left[2\int_{-\theta}^{\xi} CK(\xi')d\xi'\right] \qquad (5.1.25)$$

where the relation between z and ξ is given by eqn. 5.1.23a.

Thus the task of designing an efficient matching section reduces to the problem of finding a suitable function $CK(\xi)$ for $-\theta < \xi < \theta$.

This function $CK(\xi)$ has to be chosen to satisfy the following requirements:

(a) according to eqn. 5.1.24 a suitable characteristic with low enough magnitude must be obtained for the input reflection coefficient R as a function of $\eta \geqslant 1$
(b) according to eqn. 5.1.23c the required value of θ should be as small as possible to obtain, for a specified value of β_0, a minimum length L of the matching section and
(c) the boundary conditions

$$Z = Z_1 \quad \text{at} \quad z = 0; \quad \xi = -\theta$$

and

$$Z = Z_2 \quad \text{at} \quad z = L; \quad \xi = +\theta$$

must be satisfield.

According to eqn. 5.1.25 they require

$$2C\int_{-\theta}^{\theta} K(\xi)d\xi = \ln(Z_2/Z_1) \qquad (5.1.26a)$$

To meet these requirements, independent of the particular ratio Z_2/Z_1 of the characteristic impedances, at both ends of the matching section, we let the coefficient C in eqn. 5.1.23d be

$$2C = \ln(Z_2/Z_1) \qquad (5.1.26b)$$

With this choice of C we obtain instead of eqn. 5.1.26a

$$\int_{\theta-}^{\theta} K(\xi)d\xi = 1 \qquad (5.1.27)$$

and can now search for suitable distribution functions independent of the particular impedance ratio.

Of the various requirements for the distribution function which we specified, the first calls only for a suitable reflection coefficient as a function of frequency but otherwise is quite arbitrary. For the synthesis of distribution functions we need to be somewhat more specific and therefore require, that for frequencies above the limit frequency f_0 the magnitude of R is always smaller than a given maximum value R_{max}.

$$|R| \leqslant R_{max} \quad \text{for} \quad \eta \geqslant 1 \qquad (5.1.28)$$

For this maximum value of R we will always have a specification that is bound by

$$R_{max} < |(Z_2 - Z_1)/(Z_2 + Z_1)| \qquad (5.1.29)$$

If R_{max} were not specified to be smaller than this upper limit we could connect the lines directly to each other with an abrupt transition, and not exceed the required maximum reflection.

5.2 Optimum coupling distribution for short coupling sections

The problem is to synthesise a continuous distribution function for $K(\xi)$ that will fulfil the requirement for the maximum reflection, to stay below a given maximum value R_{max}, above a given frequency f_0, in a nonuniform matching section, that is as short as possible.

To solve this problem we define

$$F(\eta) = \int_{-0}^{\theta} K(\xi) \exp(-j\eta\xi) d\xi \qquad (5.2.1a)$$

and

$$D = R_{max}/C \qquad (5.2.1b)$$

and find that, according to eqn. 5.1.24, the requirement (eqn. 5.1.28) is identical to

$$|F(\eta)| \leqslant D \quad \text{for} \quad \eta \geqslant 1 \qquad (5.2.1c)$$

The quantity D is called the discrimination of the coupling distribution function $K(\xi)$, and conveniently expressed in decibels by taking $-20 \log_{10}(D)$.

The coupling distribution function $K(\xi)$ is different from zero only in the nonuniform section, where the transmission line changes its characteristic impedance; therefore

$$K(\xi) \neq 0 \quad \text{only for} \quad |\xi| \leqslant \theta \tag{5.2.1d}$$

while

$$K(\xi) = 0 \qquad \text{for} \quad |\xi| > \theta$$

The section of finite $K(\xi)$ should be kept as short as possible. With $F(\eta)$ according to eqn. 5.2.1a, eqn. 5.1.27 can now be written as

$$F(0) = 1 \tag{5.2.1e}$$

The problem is to find a coupling distribution function $K(\xi)$ which for a given value of D satisfies all eqns. 5.2.1 with a minimum value of θ. We shall find, that this problem has a well defined optimum solution, which is already known from the literature. For practical application of matching sections it may, however, be convenient to choose different coupling distribution functions. Therefore in this chapter a more general solution of the synthesis problem is outlined, which includes the optimum solution.

At first let us prove that for the optimum solution, $K(\xi)$ must be an even function of ξ, and hence must satisfy the condition

$$K(\xi) = K(-\xi) \tag{5.2.2}$$

If $K(\xi)$ were a general distribution it would contain an even function, and in addition an odd function of ξ. The odd component would not contribute to eqn. 5.2.1e for $\eta = 0$. It would, however, contribute to eqn. 5.2.1a for $\eta \neq 0$ and since this contribution would always be imaginary while the contribution of the even part is real the magnitude of $F(\eta)$ is always increased by an additional odd term of $K(\xi)$. This does not necessarily mean that D will increase due to a properly chosen odd term, because the increase of $|F(\eta)|$ may occur for those values of η where $|F(\eta)|$ is smaller than D. But for any TEM matching section an additional odd term in the function $K(\xi)$ does not improve the design, it cannot reduce D or θ. Therefore for an optimum solution, $K(\xi)$ must be an even function of ξ.

For a general solution of $K(\xi)$ we try the Fourier series representation

$$K(\xi) = \sum_{i=1,3,5\ldots}^{N} a_i \cos\left(i \frac{\pi}{2} \frac{\xi}{\theta}\right) \tag{5.2.3a}$$

where N is an odd integer. This set introduced into eqn. 5.2.1a leads to

$$F(x) = \cos(\pi x) \sum_{i=1,3,5\ldots}^{N} a_i x_0 (-1)^{(i-1)/2} \frac{i}{(i/2)^2 - x^2} \tag{5.2.3c}$$

where

$$x = \eta\theta/\pi; \quad x_0 = \theta/\pi \tag{5.2.3b}$$

We consider first the simplest solution of eqn. 5.2.3*a*, for which we let $N = 1$. With only one term in the Fourier-series expansion, eqn. 5.2.1*e* requires

$$a_1 x_0 = 0{\cdot}25 \tag{5.2.4a}$$

The resulting function $F(x)$ is plotted in Fig. 5.5, curve *a*.

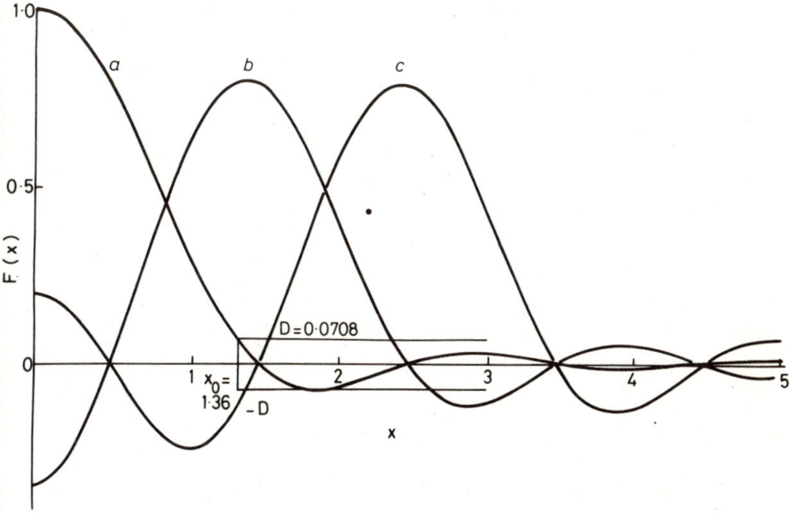

Fig. 5.5 *Function F(x) according to eqn. 5.2.3c*
Curve a $a_i x_0 = 0{\cdot}25{\cdot}\delta_{i1}$
Curve b $a_i x_0 = 0{\cdot}25{\cdot}\delta_{i3}$
Curve c $a_i x_0 = 0{\cdot}25{\cdot}\delta_{i5}$
$\delta_{ik} = \begin{cases} 1 & \text{for } i = k \\ 0 & \text{for } i \neq k \end{cases}$

From this diagram we can read the discrimination which we obtain for the coupling distribution in form of a cosine function. It depends obviously on which minimum value we let the length parameter x have for $\eta > 1$. If we let

$$x > x_0 = 1{\cdot}36 \tag{5.2.4c}$$

then D will be

$$D = 0{\cdot}0708 \tag{5.2.4b}$$

and this maximum reflection occurs not only at $x = x_0$ but also at $x = 1{\cdot}9$.

Note, that according to eqn. 5.2.1c the frequency parameters η must be equal to unity when x is at its minimum value and when the absolute value of $F(x)$ is equal to or less than D. Thus, from eqn. 5.2.1c, we obtain with eqn. 5.2.3b

$$|F(x)| \leqslant D \quad \text{for} \quad x \geqslant x_0 = \theta/\pi \tag{5.2.5}$$

In Fig. 5.5 we have also plotted the reflection characteristics $F(x)$ for coupling distributions $K(\xi)$ according to eqn. 5.2.3a that consist only of the second term with $i = 3$ and only of the third term with $i = 5$ of the Fourier series. The quantities $a_3 x_0$ and $a_5 x_0$ have in each of these cases again been chosen equal to 0·25.

By inspecting the curves in Fig. 5.5 we visualise a simple method to improve our solution with $N = 1$. If, in eqn. 5.2.3a, we add a term with $i = 5$ and a small positive value of a_5, then we will slightly shift the second zero of $F(x)$ near $x = 2{\cdot}5$ to smaller values of x, while all other zeros remain at the same location. At the same time we reduce the magnitude of $F(x)$ for values of x that are smaller than this particular zero point; on the other hand, the absolute values of $F(x)$ for $x > 2{\cdot}5$ increase somewhat. Thus, to a first approximation, we reduce the magnitude of the extreme values in a certain frequency range if we reduce the distance between the adjacent zeros and vice versa. In the above case the introduction of an additional term with $i = 5$ and a small and positive coefficient a_5 reduces the absolute value of the first extremum, and hence also reduces D. The minimum value of D is achieved when the absolute values of the first and the second extremum are equal. At the same time we have to increase the value of x_0 slightly to obtain the optimum value of D. This introduction of an additional small term with $i = 5$ gives a significant improvement. Adding further terms we can force the absolute values of even more extrema of $|F(x)|$ to be equal to each other.

In the following an algorithm is described which allows us to determine the coefficients a_i of the Fourier-series expansion in eqn. 5.2.3a such that the reflection characteristic $F(x)$ assumes a finite number n of equal ripples. For this algorithm we specify the values of $N = 2n + 1$ and x_0. According to eqn. 5.2.3c the lowest zero of $F(x)$ which is independent of the value of a_i occurs at $x = x_{n+1} = (N + 2)/2$. The maximum number of zeros between x_0 and x_{n+1} is $(N - 1)/2 = n$.

The maximum number of extrema between x_0 and x_{n+1} is n as well. We now denote by x_k the values of x where these extrema occur, with

$$x_0 < x_1 < \ldots < x_k < \ldots < x_n < x_{n+1} \tag{5.2.6a}$$

At $x = x_k$ the derivative dF/dx must vanish and at $x = 0$ the value of $F(x)$ must be unity. These conditions provide us with a set of linear equations of the form

$$\left[W \right] \begin{bmatrix} a_1 x_0 \\ a_3 x_0 \\ \vdots \\ \end{bmatrix} = \begin{bmatrix} 0 \\ 0 \\ \vdots \\ 1 \end{bmatrix} \qquad (5.2.6b)$$

The elements W_{ki} of the matrix **W** are for all $k \leqslant n+1$ given by

$$W_{ki} = (-1)^{\frac{i-1}{2}} i \left[\frac{2x \cos(\pi x_k)}{[(i/2)^2 - x_k^2]^2} - \frac{\pi \sin(\pi x_k)}{(i/2)^2 - x_k^2} \right] \qquad (5.2.6c)$$

The elements $W_{n+1,i}$ are simply obtained from eqn. 5.2.3*c* with $x = 0$

$$W_{n+1,i} = (-1)^{(i-1)/2} 4/i \qquad (5.2.6d)$$

Eqns. 5.2.6*b* form a complete set of linear equations to calculate the value of the $a_k x_0$. With these amplitudes thus determined the values of $F(x_k)$ for $0 \leqslant k \leqslant n$ can be evaluated according to eqn. 5.2.3*c*.

Usually we shall find that owing to the arbitrary choice of the values of x_k the absolute values of $F(x_k)$ differ strongly one from another. To find a better solution we change the distances between the zeros of $F(x)$ in such a way that this distance is increased if the extreme absolute value of F in this region is too low and vice versa. The same effect can be achieved if the distance between the x-values of the extrema is varied directly. The design is substantially improved, when we calculate

$$\Delta x_k = (x_{k+1} - x_k)/|F(x_k)| \quad \text{for} \quad 0 \leqslant k \leqslant n \qquad (5.2.6e)$$

and determine a new set of x'_k according to

$$x'_0 = x_0$$

$$x'_{k+1} = x'_k + \Delta x_k c \qquad (5.2.6f)$$

Here the coefficient c is chosen such that

$$x_{n+1} - x_0 = C \sum_{k=0}^{n} \Delta x_k \qquad (5.2.6g)$$

With this new set of values x_k we start again from eqn. 5.2.6*a* and improve the coupling distribution frunction iteratively. Fig. 5.6 shows the starting curve of $F(x)$ and the final result for $x_0 = 2{\cdot}5$ and $n = 4$. If the first extremum for $x > x_{n+1}$ has an absolute value which is higher than those for the extrema in between $x_0 < x < x_{n+1}$ then for the given x_0 the value of N must be increased.

The above calculation for $n = 4$ was carried out numerically for different values of x_0. The values of D which were obtained from this numerical evaluation are marked by x in Fig. 5.7, while the computed values of $a_i x_0$ are given in Table 5.2. Some of the corresponding coupling distribution functions $K(\xi)$ are plotted in Fig. 5.8.

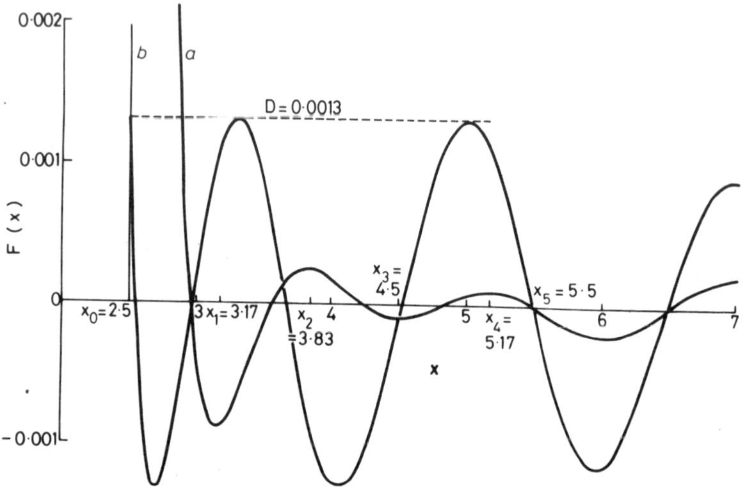

Fig. 5.6 *Determination of coupling distribution functions according to eqns. 5.2.6 with $x_0 = 2{\cdot}5$ and $n = 4$*

a Starting curve F(x) with arbitrary choice of x_i
b Final result for F(x) with n = 4 equal ripples

The contour of the nonuniform matching section which results from this synthesis procedure is not always very smooth. Actually the lower we require the values of D and x_0 to be the more pronounced will be the ripple of $K(\xi)$ when n is increased. On the other hand the higher the value of x_0 is specified the higher the number n can be chosen, while still maintaining a smooth function $K(\xi)$.

To find the optimum coupling distribution for the example of $D = 0{\cdot}0708$ the value of x_0 has been calculated as a function of n. For this purpose the algorithm according to eqns. 5.2.6 was applied to find for a given x_0 the reflection characteristics with n equal ripples. Subsequently x_0 was varied until the specified value of D was obtained. Fig. 5.9 shows the results.

In Fig. 5.10 the associated functions $K(\xi)$ are plotted. For large n these functions approach the so called Dolph-Chebyshev distribution

Table 5.1. *Values of x_0 and $a_i x_0$ of coupling distribution functions according to eqn. 5.2.3a for which the discrimination is given by curve 5 of Fig. 5.7*

$x_0 = \theta/\pi$	$a_1 x_0$	$a_3 x_0$	$a_5 x_0$	$a_7 x_0$	$a_9 x_0$
1·50000000	0·24996398	0·00941538	0·00717514	−0·00755549	0·00594087
1·60000000	0·25376701	0·01749447	0·00434247	−0·00516219	0·00412673
1·70000000	0·25713639	0·02514177	0·00222744	−0·00340349	0·00281253
1·80000000	0·26013587	0·03237364	0·00071346	−0·00213266	0·00187186
1·90000000	0·26280370	0·03918664	−0·00028031	−0·00125095	0·00122284
2·00000000	0·26522193	0·04566012	−0·00087671	−0·00062638	0·00075567
2·10000000	0·26740886	0·05179074	−0·00112038	−0·00021216	0·00043639
2·20000000	0·26939508	0·05759888	−0·00106821	0·00004659	0·00022364
2·30000000	0·27120601	0·06310360	−0·00076810	0·00019243	0·00008670
2·40000000	0·27286296	0·06832277	−0·00026059	0·00025832	0·00000289
2·50000000	0·27438750	0·07328515	0·00042176	0·00026965	−0·00004449
2·60000000	0·27578407	0·07797065	0·00124387	0·00024605	−0·00006727
2·70000000	0·27707656	0·08242991	0·00218622	0·00020231	−0·00007437

$$K(\xi) = \frac{D}{2}\left\{\frac{I_1\,[\theta\sqrt{1-(\xi/\theta)^2}\,]}{\sqrt{1-(\xi/\theta)^2}} + \delta(\xi-\theta) + \delta(\xi+\theta)\right\} \tag{5.2.7a}$$

for which the reflection characteristic is given by

$$F(x) = \frac{\cos\sqrt{(\pi x)^2-\theta^2}}{\cosh\theta} = \frac{\cos\,[\theta\sqrt{\eta^2-1}]}{\cosh\theta} \tag{5.2.7b}$$

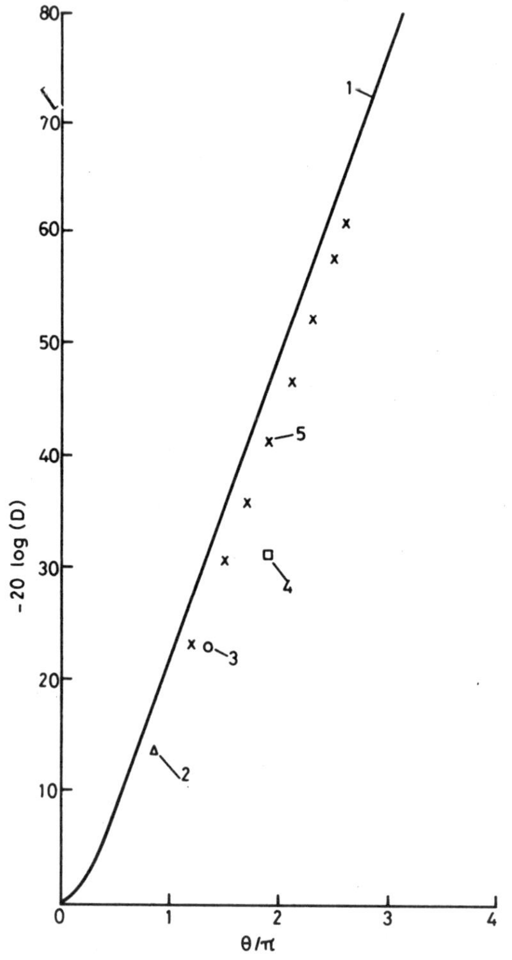

Fig. 5.7 *Discrimination of coupling distribution functions K(ξ)*
Curve 1 Dolph-Chebyshev distribution
2 $K(\xi)$ = constant
3 $K(\xi) \sim \cos(\pi\xi/2\theta)$
4 $K(\xi) \sim \cos^2(\pi\xi/2\theta)$
5 $K(\xi)$ calculated according to eqns. 5.2.6 with $n = 4$

I_1 in the Dolph-Chebŷshev distribution denotes the modified Bessel function of the first order. $\delta(t)$ is Dirac's delta function. For more details on this distribution the reader is referred to References 1 & 2.

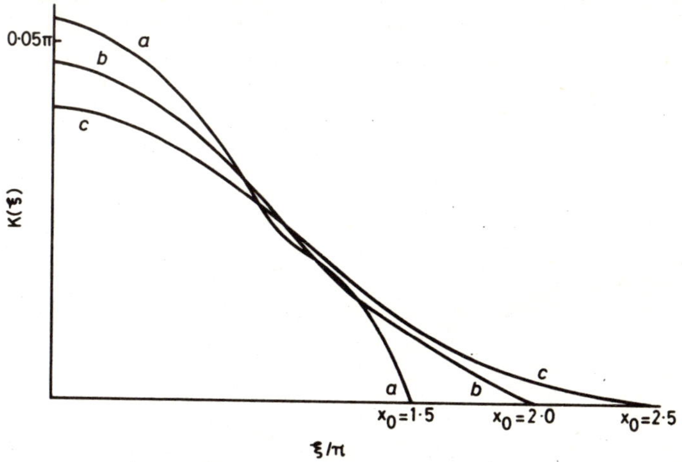

Fig. 5.8 ***Coupling distribution functions calculated according to eqns. 5.2.6 with n = 4 for different values of x_0***

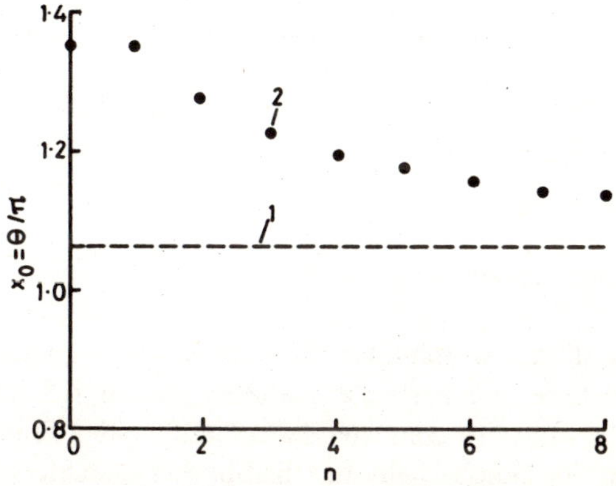

Fig. 5.9 ***Required values of $x_0 = \theta/\pi$ for D = 0·0708***
Curve 1 Dolph-Chebŷshev distribution
Curve 2 Coupling distribution calculated according to eqns. 5.2.6 with n equal ripples

The impulses of the δ-functions at $\xi = \pm\theta$ require steps in the impedance curve at both ends of the matching section. Such impedance steps can only be accomplished when the cross-section of the transmission line changes abruptly. At such abrupt transitions, however, the field is not any longer purely TEM and the reactive energy stored in the evanescent fields of higher-order modes manifests itself as a shunt reactance at the location of the abrupt transition. These reactances are not accounted for in the design procedure and consequently may degrade the performance of the matching section. For this reason the Dolph-Chebyshev distribution may be unsuitable for matching sections as well as other waveguide transitions.

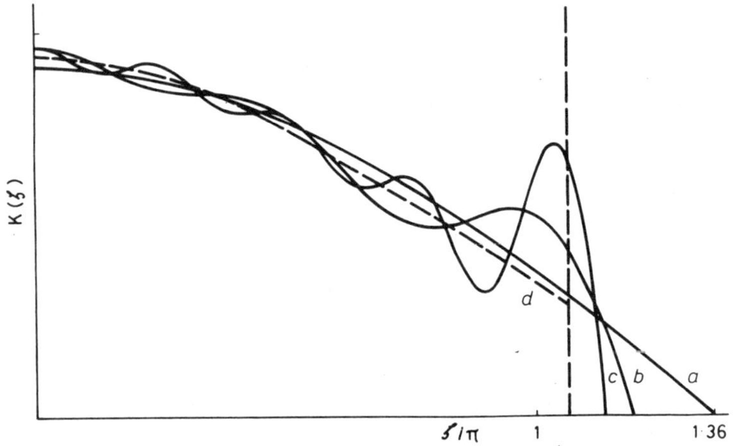

Fig. 5.10 *Distribution functions with D = 0·0708 calculated according to eqns. 5.2.6*
a n = 0 (N = 1)
b n = 4 (N = 9)
c n = 8 (N = 17)
d Dolph-Chebyshev distribution

But even if the Dolph-Chebyshev distribution is not a suitable coupling function for a particular matching section, it is still useful as a model with which to compare the actual coupling function and its performance. The Dolph-Chebyshev distribution guarantees the smallest value of θ for a given value of D. This becomes evident when we examine eqn. 5.2.7*b*. For $\eta \geqslant 1$ the absolute value of all extrema are equal to each other and given by

$$D = 1/\cosh\theta \tag{5.2.7c}$$

This equal-ripple characteristic represents the optimum solution to the problem and the associated coupling distribution function is as short as it possibly can be. In this case all zeros of the reflection characteristic are optimally spaced, and the lowest zero appears therefore at the minimum possible values of x for a given D. The discrimination as a function of the length parameter $x_0 = \theta/\pi$ for the Dolph-Chebyshev distribution according to eqn. 5.2.7b is shown in Fig. 5.7. Also plotted in this figure are corresponding results for other coupling distribution functions which are frequently used to design transmission line or waveguide tapers.

These considerations show that there may be many ways to determine suitable coupling distribution functions (Reference 3). The procedure which is based on eqns. 5.2.6 represents just one alternative. It yields, however, near-optimum results for nonuniform line sections with smooth contours. Because of these smooth contours the fields are very nearly TEM all through the nonuniform line section and the results of the TEM analysis are valid to a high degree of accuracy.

The method which we have presented here for the design of near optimum or even the optimum coupling distribution functions leads to a reflection characteristic with equal ripples. This method can be modified for the ripples not to be equal to each other but differ one from another in a manner which we have at our disposal. In order that the absolute values of $F(x)$, at the points $x = x_k$ where they are at an extremum, assume values according to

$$|F(x_k)| \sim f(x_k)$$

we only need to require instead of eqn. 5.2.6e that the following equation be satisfied:

$$\Delta x_k = (x_{k+1} - x_k) f(x_k)/|F(x_k)| \qquad (5.2.6h)$$

The function $f(x_k)$ must be specified to be real and positive at all points $x = x_k$. With this procedure we can in particular control the magnitude of those ripples which occur at the lower frequencies of the reflection characteristic. This modified design procedure offers also a possibility to obtain smooth functions $K(\xi)$ for a large number of n. For this purpose the magnitude of the controlled ripples of $F(x)$ for large values of x have to be matched to the values of the uncontrolled ripples occuring at even larger x values.

The synthesis procedure for the coupling distribution enables us to find suitable functions $K(\xi)$. Once this function has been determined the nonuniform matching section can be designed. The algorithm which this design follows is summarised in Table 5.2.

An additional remark on this design procedure concerns the deviations which can be tolerated for the coupling distribution function. If we decrease the value of D and at the same time require the length to remain near its minimum value, the coupling distribution must be close to the optimum distribution, and hence very accurate.

For large values of $\eta\theta$ and D, however, the coupling distribution may deviate substantially from the optimum distribution; as long as it remains a smooth function of ξ the reflection characteristic will have only

Table 5.2 *Design of a continuous TEM matching section with highpass matching characteristics*

Calculate $D = 2R_{max}/\ln(Z_2/Z_1)$

Determine according to the formulas of this section a suitable coupling distribution function

$$K(\xi)$$

with a minimum value of θ and the corresponding response function

$$F(\eta) \leqslant D \quad \text{for} \quad \eta > 1$$

Calculate the distribution function for the characteristic impedance

$$Z(\xi)/Z_1 = \exp\left[\ln(Z_2/Z_1)\int_0^{\xi} K(\xi')d\xi'\right]$$

with the axial distance

$$z = (\xi + \theta)/(2\beta_0); \quad \beta_0 = 2\pi f_0\sqrt{\mu\epsilon}$$

The input reflection coefficient is then given by

$$R(f) = \ln(Z_2/Z_1)F(f/f_0)/2$$

small absolute values. Therefore the probability that $F(\eta)$ is degraded by small deviations of the coupling distribution function is small in this case. Degradation, however, is very likely to occur for small values of $\eta\theta$ and D. Such degradation may also occur in cases where eqn. 5.2.1*a* is not accurate enough to represent the reflection characteristic $F(\eta)$ in terms of $K(\xi)$. The actual reflection may then be much higher then we expect it from the optimum design.

One further comment concerns the application of this design procedure to matching sections for waveguides instead of TEM-transmission lines.

For the TEM matching section the desired length L is directly proportional to θ and thus, for a given value of D, we must choose the coupling distribution function with the smallest value of θ to obtain the shortest possible matching section. For this reason only symmetrical coupling distribution functions $K(\xi)$ have been taken into consideration when near-optimum solutions have been determined.

The customary method of designing overmoded waveguide transitions starts likewise from the coupling distribution function. In this case the relation between θ and L is, however, more complicated; the required length depends also on the shape of $K(\xi)$. In this case it may be desirable to choose coupling distribution functions which are not just an even function $K_e(\xi)$, but contain an additional odd term $K_0(\xi)$. We will then have

$$K(\xi) = K_e(\xi) + K_0(\xi)$$

with

$$K_e(\xi) = K_e(-\xi)$$

and

$$K_0(\xi) = -K_0(-\xi)$$

Numerical calculations have shown, however, that for a given value of D a possible odd term will decrease in magnitude, if we choose θ smaller and smaller. For the lower limit of θ, which is obtained for the Dolph-Chebyshev distribution, any odd term must vanish altogether.

In concluding this discussion of the design or continuous matching sections, let us compare its results with those of the stepped impedance transformer of the preceding Section.

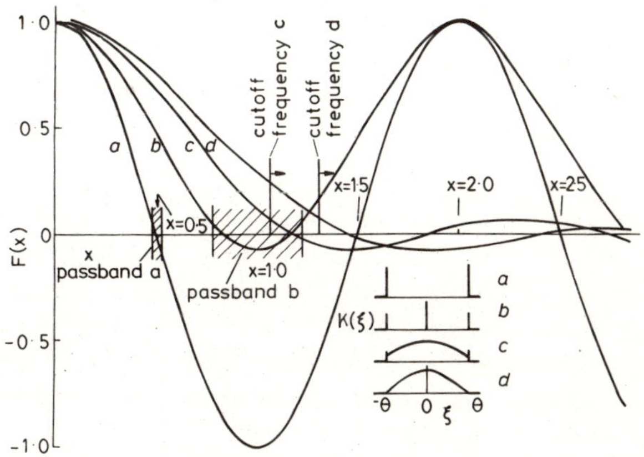

Fig. 5.11 *Comparison of minimum lengths of different matching sections*
Curve a λ/4-matching section
Curve b two-line matching section
Curve c continuous matching section with Dolph-Chebyshev distribution
Curve d continuous matching section with $K(\xi) \sim \cos[\pi\xi/(2\theta)]$

The normalisation of the function $F(x)$ in eqn. 5.1.13*b* for the quarter-wave transformer, and of eqn. 5.1.17*a* for the two section impedance transformer is identical with that of eqn. 5.2.3*c* and eqn. 5.2.7*b* for continuous matching sections. F is proportional to the absolute value of the input reflection coefficient R, while x is proportional to the overall length as well as to frequency.

Curves *a* and *b* of Fig. 5.11 show $F(x)$ for the stepped impedance matching sections according to eqns. 5.1.13*b* and 5.1.17*a*. The value of y in eqn. 5.1.17*a* has been chosen such that the ripple in the passband has as its largest absolute value $|F| = 0{\cdot}0708$. This is the same value which, according to Fig. 5.7, is obtained for D if $K(\xi)$ is chosen proportional to $\cos\{(\pi/2\theta)\xi\}$. The corresponding result is given in curve *d* of Fig. 5.11. Curve *c* shows the function $F(x)$ for the Dolph-Chebyshev distribution with $\theta/\pi = 1{\cdot}06$.

Fig. 5.11 allows us to compare the minimum lengths of the different matching sections and the passbands which they offer.

References

1 COLLIN, R. E.: 'Theory and design of wide-band multisection quarter-wave transformers', *Proc. IRE.*, 1955, **43**, pp. 179–185

2 KLOPFENSTEIN, R. W.: 'A transmission line taper of improved design', *ibid.*, 1956, **44**, pp. 31–35

3 HECKEN, R. P.: 'A near-optimum matching section without discontinuities', *IEEE Trans.*, 1972, **MTT-20**, pp. 734–739

Chapter 6

Gradual waveguide transitions

In Chapters 2 and 3 the coupled wave equations have been set up for gradual waveguide transitions. Starting from these equations we shall now develop design procedures to determine suitable transition contours with low spurious mode generation. The detailed design objectives for such waveguide transitions depend on the particular application; they may differ considerably for different applications.

In the following we will investigate a few typical cases which serve particularly well as examples to show how the coupled wave equations can be used to correctly analyse and design inhomogeneous waveguide sections.

6.1 Highly overmoded waveguide transitions with low mode conversion

The usual approach starts from the coupled wave equations for cylindrical modes, which, for the sake of brevity, are written here in matrix form

$$(d/dz)\mathbf{A} = (-\boldsymbol{\gamma} + \boldsymbol{\kappa})\mathbf{A} \tag{6.1.1}$$

z denotes the longitudinal co-ordinate, while $\mathbf{A}$ is a vector, which has the individual wave amplitudes A_n as its components. All A_n with a positive index n are supposed to denote amplitudes of modes with index n which propagate in the positive z-direction, whereas A_{-n} is the amplitude of the respective mode n which propagates in the negative z-direction. $\boldsymbol{\gamma}$ is a diagonal matrix, with $\gamma_{nn} = \gamma_n = -\gamma_{-n}$ representing the propagation constant of the nth mode. The elements κ_{nm} of the matrix $\boldsymbol{\kappa}$ describe the coupling from the mth mode to the nth mode. For the transition between waveguides of different cross-sections the values of the elements of $\boldsymbol{\gamma}$ and $\boldsymbol{\kappa}$ depend on the contour of the transition and on frequency.

When we make the same assumptions as in Section 2.6 for the form of the inhomogeneous waveguide in the transition section we obtain from eqns. 2.6.3, 2.6.11

$$\gamma_n = j\sqrt{k^2 - k_{cn}^2} = j\beta_n \tag{6.1.2a}$$

$$\kappa_{nn} = \kappa_{(-n)(-n)} = 0 \tag{6.1.2b}$$

$$\kappa_{n(-n)} = \kappa_{(-n)n} = C_{nn} - \frac{d \ln (Z_n)}{2dz} \tag{6.1.2c}$$

for any one mode n, while for $n \neq m$ and $n > 0, m > 0$ we obtain

$$\kappa_{mn} = \kappa_{(-m)(-n)} = (C_{mn}\sqrt{Z_n/Z_m} - C_{nm}\sqrt{Z_m/Z_n})/2 \tag{6.1.2d}$$

$$\kappa_{m(-n)} = \kappa_{(-m)n} = (C_{mn}\sqrt{Z_n/Z_m} + C_{nm}\sqrt{Z_m/Z_n})/2 \tag{6.1.2e}$$

Z_n is the wave impedance for the nth mode, with

$$Z_n = \sqrt{\frac{\beta_n}{\omega\epsilon}} \quad \text{for E-modes and} \tag{6.1.2f}$$

$$Z_n = \sqrt{\frac{\omega\mu}{\beta_n}} \quad \text{for H-modes.} \tag{6.1.2g}$$

k_{cn} denotes the cutoff wave number of the nth mode. k is the intrinsic wave number of the homogeneous dielectric material inside the transition.

$$k = \omega\sqrt{\mu\epsilon} \tag{6.1.2h}$$

The walls of the waveguide transition are assumed to be perfect conductors.

The quantities C_{mn} represent coupling factors. They are, for a given transition contour, determined only by the geometrical change of the cross-sectional sizes as a function of the axial distance z, and hence independent of frequency.

The values of the cutoff wave numbers k_{cn} are also independent of frequency. They are, since a function of z, only determined by the form and size of the respective local cross-section.

The coupled wave equations 6.1.1 for cylindrical modes are usually applied to calculate the overall mode conversion in waveguide transitions. Their advantage is that the coefficients of the system of linear differential equations, which they represent, are readily obtained by applying the results which are given in Section 2.6.

An exact integration of eqn. 6.1.1 with the aim of calculating certain parameters of the scattering matrix of nonuniform waveguides with arbitrary contours is, however, not quite feasible. The number of coupled modes in eqns. 6.1.1 is infinite and it is difficult to decide where to truncate the system and still obtain results of sufficient accuracy. The integration can, moreover, only be performed numerically, and must be repeated for every frequency throughout the operating band.

For the integration along the transition we need to know the initial values for the amplitudes A_n. Before hand, all these initial values are not known to us, and it is difficult to make physically meaningful assumptions for them. If, in particular, nonpropagating modes are taken into consideration the amplitudes of their backward components grow exponentially from any chosen initial value, and we encounter numerical problems. In addition in this case the coefficients of eqns. 6.1.1 become infinite at the cutoff cross-sections.

Most of these difficulties can, however, be overcome, at least in principle, and some of the methods to be applied will be investigated in more detail in the following sections. For waveguide transitions though, one has to be satisfield quite often with approximate solutions of eqns. 6.1.1.

To find simple solutions we take into account that under operating conditions usually only one mode is to be transmitted through the transition. In the following we will call this mode the wanted mode, and designate it by the index n. Only the mode conversion from this wanted mode into other spurious modes is of practical interest. We designate these unwanted spurious modes by indices m.

Under these circumstances the following problem has to be solved: Only the wanted mode with unit amplitude $A_n = 1$ is incident at one port of the transition. Throughout the transition it interacts with spurious modes and converts power to them. The amplitudes A_m of these spurious modes at the output port of the transition must be determined.

This problem may be solved without too much difficulty when the transition is sufficiently gradual and when, in addition, all modes that need to be taken into account are above cutoff throughout the transition.

As an example we consider interaction between the wanted mode n and two spurious modes m and p. Of these two spurious modes let us focus attention on the amplitude A_m of mode m, and consider mode p only to the extent to which it influences A_m. The coupled wave equations for these three modes have the following form:

$$\frac{d}{dz}\begin{bmatrix} A_n \\ A_m \\ A_p \end{bmatrix} = \begin{bmatrix} -j\beta_n & \kappa_{nm} & \kappa_{np} \\ \kappa_{mn} & -j\beta_m & \kappa_{mp} \\ \kappa_{pn} & \kappa_{pm} & -j\beta_p \end{bmatrix} \begin{bmatrix} A_n \\ A_m \\ A_p \end{bmatrix} \tag{6.1.3a}$$

All coefficients κ and β depend on z.

At the input of the transitions for $z = 0$, we have as initial values

$$\begin{bmatrix} A_n \\ A_m \\ A_p \end{bmatrix}_{z=0} = \begin{bmatrix} 1 \\ 0 \\ 0 \end{bmatrix}$$

Dividing the first of eqns. 6.1.3a by A_n it may be written as

$$\frac{d(\ln A_n)}{dz} = -j\beta_n + \underline{\kappa_{nm} A_m/A_n} + \underline{\kappa_{np} A_p/A_n} \tag{6.1.3b}$$

From the first and the second of eqns. 6.1.3a it follows that

$$\frac{d(A_m/A_n)}{dz} = \kappa_{mn}\left[1 - \underline{\frac{\kappa_{nm}}{\kappa_{mn}}\left(\frac{A_m}{A_n}\right)^2} + \underline{\frac{\kappa_{mp}}{\kappa_{mn}}\frac{A_p}{A_n}} - \underline{\frac{\kappa_{np}}{\kappa_{mn}}\frac{A_m}{A_n}\frac{A_p}{A_n}}\right] - j(\beta_m - \beta_n)A_m/A_n \tag{6.1.3c}$$

An equation corresponding to eqn. 6.1.3c but for A_p/A_n instead of A_m/A_n may similarly be derived from the first and the third of eqns. 6.1.3a.

If the underlined terms in eqn. 6.1.3b and 6.1.3c are small enough so that we may neglect them then these differential equations may be integrated to yield as solutions

$$A_n = \exp\left[-j\int_0^z \beta_n dz'\right] \tag{6.1.4a}$$

$$A_m/A_n = \int_0^z \kappa_{mn} \exp\left[-j\int_{z'}^z (\beta_m - \beta_n)dz''\right] dz' \tag{6.1.4b}$$

We thus obtain for the spurious mode amplitude A_m at the output of the transition

$$\hat{A}_m = A_m(z = L)$$

$$= \exp\left(-j \int_0^L \beta_n dz\right) \int_0^L \kappa_{mn} \exp\left[-j \int_z^L (\beta_m - \beta_n) dz'\right] dz \tag{6.1.4c}$$

To estimate the error which we have made by neglecting the underlined terms in eqns. 6.1.3*b* and 6.1.3*c* we integrate eqn. 6.1.4*b* by parts

$$A_m/A_n = -j \frac{\kappa_{mn}}{\beta_m - \beta_n}$$

$$+ j \int_0^z \left\{\frac{d[\kappa_{mn}/(\beta_m - \beta_n)]}{dz}\right\} \exp\left[-j \int_{z'}^z (\beta_m - \beta_n) dz''\right] dz' \tag{6.1.5a}$$

At present we consider only gradual transition and interaction only between modes which are above cutoff throughout the transition. Under these conditions the first term on the right-hand side of eqn. 6.1.5*a* has always a much larger maximum of its absolute value than the second term, at any frequency within the operating band. We therefore have

$$|A_m/A_n|_{max} \approx \left|\frac{\kappa_{mn}}{\beta_m - \beta_n}\right|_{max}$$

This result allows us a first check as to whether the underlined terms in eqns. 6.1.3*b* and 6.1.3*c* are really small enough to be neglected. A more accurate test on how these terms influence the overall spurious mode generation should not only consider the above maximum value, but also the change of these terms as a function of z all along the transition.

The results in eqns. 6.1.4 for the amplitudes of modes above cutoff will be applied to specific examples in the three following sections. In Section 6.3, in particular, the approximations which were made to obtain these results will be more closely examined.

If one of the spurious modes which interact with the wanted mode goes through cutoff within the transition its coupling coefficient κ grows beyond all limits. This is not only the case for a representation in terms of cylindrical waves, but also when the coupled wave equations are written in terms of horn waves. For such spurious modes near, or at, cutoff the underlined terms in eqn. 6.1.3*c* will certainly not remain small compared with unity, and the above approximate solution no longer applies. We will examine this situation in more detail in Section 6.5.

6.2 Circular and rectangular waveguide tapers

The waveguide transitions which we consider in this Section are shown in Fig. 6.1. The geometry of these transitions is uniquely characterised by the function $a(z)$. In the case of the circular waveguide taper the wanted mode n is the H_{01}-mode with $n = 1$. It couples only to H_{om}-modes with $m = 2, 3 \ldots$

The cutoff wave numbers are

$$k_{cn} = \frac{x_n}{a} \qquad (6.2.1a)$$

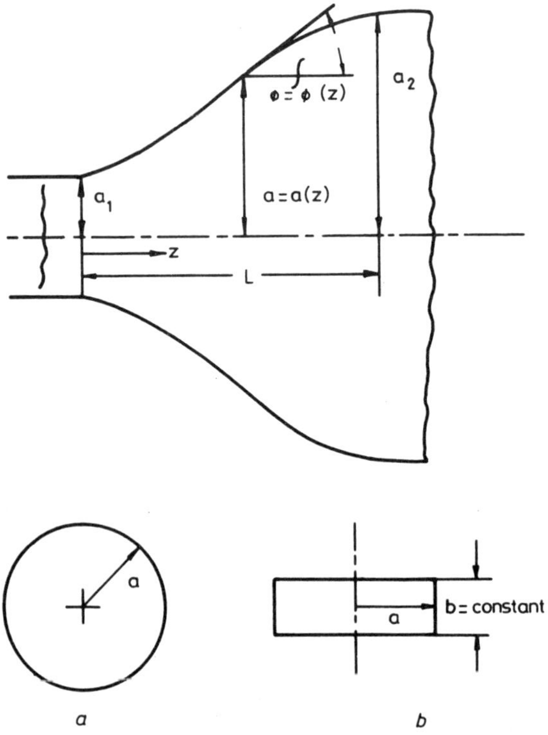

Fig. 6.1 *Waveguide transition*
a Cross-section of a circular waveguide taper
b Cross-section of a rectangular waveguide transition

where x_n is the nth zero of the derivative of the Bessel function $J_0(x)$

$$J_0'(x_n) = 0 \tag{6.2.1b}$$

In the rectangular transition the wanted mode is the fundamental H_{10}-mode. Because we assume the transition to be symmetric with respect to the longitudinal centre section the wanted mode couples only to spurious H_{m0}-modes with $m = 3, 5, 7 \ldots$

In this case we obtain for the cutoff wave numbers

$$k_{cn} = \frac{x_n}{a} \tag{6.2.1c}$$

with

$$x_n = \frac{n\pi}{2} \tag{6.2.1d}$$

With this notation the coupling factors for either of the waveguide transitions of Fig. 6.2 follows from eqn. 2.6.17*b*:

$$C_{mn} = -\frac{1}{a}\frac{da}{dz}\,2\frac{x_m x_n}{x_m^2 - x_n^2} \tag{6.2.2}$$

For a given transition contour $a(z)$ the integrals (expr. 6.1.4) for the amplitudes of any one of the spurious modes are completely determined by eqns. 6.1.2, 6.2.1 and 6.2.2.

The approximate solution according to eqn. 6.1.4 assumes that all spurious modes to be considered remain above cutoff throughout the taper. To facilitate the transition analysis even further it is often assumed, that all these spurious modes are so far above cutoff that

$$ka \gg x_m \tag{6.2.3a}$$

holds along the transition.[1-3] Under this condition we obtain

$$\gamma_m - \gamma_n = -\frac{j}{2ka^2}(x_m^2 - x_n^2) \tag{6.2.3b}$$

$$\kappa_{mn} = C_{mn} = -\frac{1}{a}\frac{da}{dz}\,2\frac{x_m x_n}{x_m^2 - x_n^2} \tag{6.2.3c}$$

$$\kappa_{(-m)n} = 0 \tag{6.2.3d}$$

and eqn. 6.1.4*c* can be written as

$$\hat{A}_m \exp\left(j\int_0^L \beta_n\,dz\right) =$$

$$-2\frac{x_m x_n}{x_m^2 - x_n^2}\int_0^L \frac{1}{a}\frac{da}{dz}\exp\left[j\frac{x_m^2 - x_n^2}{2k}\int_z^L \frac{dz'}{a^2}\right]dz \tag{6.2.4}$$

This integral expression for spurious mode amplitudes allows us to decide which of the spurious modes interacts most critically with the wanted mode and is hence excited with the largest amplitude.

First we note that under present conditions the wanted mode interacts to any significant extent only with spurious modes that travel in the same direction as the wanted mode, and that only these spurious modes need to be considered. The integral for the spurious mode amplitude in eqn. 6.2.4 is multiplied by the factor

$$\frac{x_m x_n}{x_m^2 - x_n^2}$$

This factor becomes largest for that particular spurious mode which with its eigenvalue x_m is closest to the eigenvalue x_n of the wanted mode. In the case of the circular waveguide taper the above factor is largest for the H_{02}-mode with $m = 2$, and with the rectangular waveguide transition for the H_{30}-mode with $m = 3$. Later on we will find that in transitions with smooth contours the average excitation of these particular spurious modes by the wanted mode with $n = 1$ is indeed at maximum. We will therefore call these next-neighbour modes the main spurious modes and assign to them the mode index m. In the following a design procedure for waveguide transitions will be developed which, as its primary objective, controls the amplitude level of the main spurious mode. It aims to keep the main spurious mode as low as possible in amplitude.

For this design procedure we refer again to eqn. 6.2.4. In the limit of very high frequencies the exponential function of the integrand differs little from unity. In this limit therefore the integral has the value $\ln(a_2/a_1)$ and the amplitude of the main spurious mode is given by

$$|\hat{A}_m| = C = 2\frac{x_m x_n}{x_m^2 - x_n^2}\int_0^L \frac{d[\ln(a)]}{dz}\,dz$$

$$= 2\frac{x_m x_n}{x_m^2 - x_n^2}\ln\left(\frac{a_2}{a_1}\right) \qquad (6.2.5)$$

The contributions to the spurious mode amplitude in this expression add all in phase along the transition. To obtain an overall spurious mode amplitude which is significantly lower than this limiting value the maximum absolute value of the exponent in the integral of eqn. 6.2.4

$$\frac{x_m^2 - x_n^2}{2k}\int_0^L \frac{dz}{a^2}$$

must be sufficiently large, so that the contributions to the spurious mode along the transition turn in phase and interfere destructively. For a finite length L of the transition this interference will be destructive only below a certain limit frequency f_0. With such an upper limit frequency as reference we make the following substitutions in eqn. 6.2.4:

$$\eta = f_0/f \tag{6.2.6a}$$

$$k_0 = 2\pi f_0 \sqrt{\mu\epsilon} \tag{6.2.6b}$$

$$\xi + \theta = \frac{1}{j\eta}\int_0^z (\gamma_n - \gamma_m)dz' = \int_0^z \frac{x_m^2 - x_n^2}{2k_0}\frac{dz'}{a^2} \tag{6.2.6c}$$

$$2\theta = \frac{1}{j\eta}\int_0^L (\gamma_n - \gamma_m)dz' = \int_0^L \frac{x_m^2 - x_n^2}{2k_0}\frac{dz'}{a^2} \tag{6.2.6d}$$

$$CK(\xi) = -\frac{j\kappa_{mn}}{\dfrac{1}{\eta}(\gamma_n - \gamma_m)} = k_0 a \frac{da}{dz}\frac{4x_m x_n}{(x_m^2 - x_n^2)^2} \tag{6.2.6e}$$

and obtain

$$\hat{A}_m = \hat{A}_m(\eta) = -\exp\left(-j\int_0^L \beta_n dz\right)\exp(j\eta\theta)\int_{-\theta}^{\theta} CK(\xi)\exp(j\eta\xi)d\xi \tag{6.2.7}$$

Since $K(\xi)$ is a real function, and since both exponential factors which multiply the integral are of unit amplitude and change only the phase of the main spurious mode we can write for the absolute value of its amplitude

$$|\hat{A}_m| = |W_m(\eta)| \tag{6.2.8a}$$

with

$$W_m(\eta) = \int_{-\theta}^{\theta} CK(\xi)\exp(-j\eta\xi)d\xi \tag{6.2.8b}$$

This expression for the spurious mode generation in overmoded waveguide transitions is identical with eqn. 5.1.24, which in Chapter 5 was used as the basic coupling formula to design TEM matching sections. The design of transitions in overmoded waveguide may likewise be based on this formula.

Once we have found a suitable coupling distribution function $K(\xi)$ we still need to determine the corresponding transition contour. Suppose that $CK(\xi)$, θ and k_0 are known; we then replace d/dz in eqn. 6.2.6e by $(d/d\xi)\cdot(d\xi/dz)$ with $d\xi/dz$ from eqn. 6.2.6c and find

$$\ln\left(\frac{a(\xi)}{a_1}\right) = \frac{x_m^2 - x_n^2}{2x_m x_n} C \int_{-\theta}^{\xi} K(\xi')d\xi' \qquad (6.2.9a)$$

Thus the dimension a of the transition is obtained as a function of ξ. By substituting $a(\xi)$ into eqn. 6.2.6c and solving this equation for z we obtain

$$z(\xi) = \frac{2k_0}{x_m^2 - x_n^2} \int_{-\theta}^{\xi} a^2(\xi')d\xi \qquad (6.2.9b)$$

With both these eqns. 6.2.9 the contour $a(z)$ may be evaluated for any given $K(\xi)$, θ and k_0 by using numerical methods.

Note that with the value of C from eqn. 6.2.5 and with eqn. 6.2.9a we have

$$\int_{-\theta}^{\theta} K(\xi)d\xi = 1 \qquad (6.2.10)$$

This relation for the coupling distribution function corresponds to eqn. 5.1.27 in Chapter 5.

Because the coupling eqn. 6.2.8b corresponds directly to eqn. 5.1.24 we can now apply the method of the preceding chapter to design short waveguide transitions with a specified spurious mode level. Table 6.1 gives the algorithm that solves the following problem:

Design the transition of minimum length between the homogeneous waveguides with sizes a_1 and a_2 such that, for frequencies f lower than f_0, the overall mode conversion $|W|$ from the wanted mode to the main spurious mode is less than a specified maximum value W_{max}.

Table 6.1 *Design procedure for waveguide transitions according to Fig. 6.1*

(*a*) Calculate C according to eqn. 6.2.5 and

$$D = W_{max}/C$$

(*b*) Determine a coupling distribution function $K(\xi)$ (e.g. according to the results of Section 5.2) with a minimum value for θ and the corresponding $F(\eta) \leqslant D$ for $\eta = f_0/f > 1$.

(*c*) Calculate the transition contour $a(\xi)$ and $z(\xi)$ by applying eqns. 6.2.9

The expected coupling factor W is then obtained as

$$W(\eta) = CF(\eta)$$

It may be assumed that

$$W_{max} < 2\frac{x_m x_n}{x_m^2 - x_n^2} \ln \frac{a_2}{a_2} \tag{6.2.11}$$

because otherwise the design problem, as long as only the main spurious mode is taken into consideration, can be solved by a butt joint between the two different waveguides.

Just as for the TEM matching section, the design of the waveguide transition has been reduced to the problem of finding a suitable coupling distribution function $K(\xi)$. Before we discuss this problem in more detail, let us briefly return to the question as to whether our previously specified main spurious mode with index m is indeed generated most strongly in a well designed waveguide transition. If we designate any of the higher-order spurious modes by the index p we find from eqn. 6.2.4 with the substitutions eqns. 6.2.5 and 6.2.6 that the wanted mode n excites this higher-order spurious mode with an amplitude according to

$$\begin{aligned} |A_p(\eta)| &= \frac{(x_m^2 - x_n^2)^2}{(x_p^2 - x_n^2)^2} \frac{x_p}{x_m} C \left| \int_{-\theta}^{\theta} K(\xi) \exp\left[j \frac{x_p^2 - x_n^2}{x_m^2 - x_n^2} \eta\xi \right] d\xi \right| \\ &= \left[\frac{x_m^2 - x_n^2}{x_p^2 - x_n^2} \right]^2 \frac{x_p}{x_m} \left| W\left(\frac{x_p^2 - x_n^2}{x_m^2 - x_n^2} \eta \right) \right| \end{aligned} \tag{6.2.12}$$

This result shows that the overall coupling factor for any higher-order spurious mode, which is sufficiently above cutoff throughout the transition, will always be much smaller than for the main spurious mode, if only $|W(\eta)|$ of eqn. 6.2.8*b* decreases with increasing η. The condition that the extrema of $|W(\eta)|$ decrease with increasing η is usually fulfilled for smooth contours. The excitation of higher-order spurious modes, which come near or even go below cutoff within the transition, cannot any longer be calculated from eqn. 6.2.12. But for gradual tapers it is safe to assume that the wanted mode excites them with even smaller amplitudes. We are therefore well justified to design the transition contour with respect to weak excitation of the main spurious mode only.

In designing the waveguide transition we need to choose a suitable coupling distribution function $K(\xi)$. The best possible choice of $K(\xi)$ for given values of D, a_1, a_2 and k_0 should lead to a minimum length L of the transition for a specified maximum value of $|W(\eta)|$ with $\eta > 1$. From eqns. 6.2.9 and 6.2.5 we find for this transition length

$$L = \frac{2k_0}{x_m^2 - x_n^2} a_1^2 \int_{-\theta}^{\theta} (a_2/a_1)^{2\int_{-\theta}^{\xi} K d\xi'} d\xi \tag{6.2.13}$$

This equation does not give as simple a relation between L and θ as we found it for the TEM matching section in Chapter 5. In that case the length L was directly proportional to θ. There the $K(\xi)$ with the smallest θ gave the shortest length.

In the case of the waveguide transition, however, the required length depends not only on θ but also on the distribution of the function $K(\xi)$. The smallest possible value of θ is obtained when the Dolph-Chebÿshev distribution is chosen for $K(\xi)$. For transitions with ratios a_2/a_1 not much larger than unity this Dolph-Chebÿshev distribution will not only give the smallest possible θ-value, but also result in a minimum transition length. Even if a_2/a_1 is substantially larger than unity, and as long as eqn. 6.2.7 applies, the Dolph-Chebÿshev distribution for $K(\xi)$ will still lead to a waveguide transition of nearly minimum length.

Note, that an asymmetric coupling distribution function in eqn. 6.2.13 does not improve the transition. Any asymmetry of $K(\xi)$ will only increase θ from its smallest possible value for a given value of D, just as it was stated in Section 5.2.

From these considerations we can expect that the smooth coupling distribution functions which we have found in Section 5.2 will also result in suitable contours for waveguide transitions when we introduce them in the design procedure as given in Table 6.1. The next Section will show, however, that this design procedure has some limitations. Eqn. 6.2.8*b* gives only an approximation for the excitation of spurious modes. It may not always be accurate enough and the total excitation of spurious modes may be much stronger than that to be expected from eqn. 6.2.8*b*. This deviation is particularly critical when very low values of D and $\eta\theta$ are specified for the design, and when the remarks at the end of Section 5.2 apply. Thus for short waveguide transitions with large dimensional ratio a_2/a_1 and when extremely low spurious mode generation is specified, we must modify the design method.

6.3 Quasidiagonalisation of coupled wave equations

In the preceding Section the overall conversion from the wanted mode to the main spurious mode was calculated from the coupled wave eqns. 6.1.1 by using their approximate solution eqn. 6.1.4*c*.

Eqn. 6.1.4*c* represents the exact solution of

$$\frac{d}{dz}\begin{bmatrix} A_n \\ A_m \end{bmatrix} = \begin{bmatrix} -\gamma_n & 0 \\ \kappa_{mn} & -\gamma_m \end{bmatrix} \begin{bmatrix} A_n \\ A_m \end{bmatrix} \tag{6.3.1}$$

for the initial conditions $A_n = 1$ and $A_m = 0$ at $z = 0$. This solution (eqn. 6.1.4*c*) neglects not only the influence of the main spurious mode on the propagation of the wanted mode, but also the interaction with other forward travelling spurious modes, and with all waves which travel in the negative *z*-direction. Therefore the solution (eqn. 6.1.4*b*) can be applied only in the case of transitions which are gradual enough.

In the present Section we will continue to assume that all modes under consideration are far above cutoff throughout the transition, so that only those modes which travel in the forward *z*-direction interact with each other. With this assumption we obtain more accurate formulas for mode conversion by transforming from the amplitudes in the coupled wave eqns. 6.1.1 to a different set of amplitudes that very nearly diagonalises the coupling matrix of eqns. 6.1.1. For this quasi-diagonalisation of the coupled wave equations, as suggested in Reference 2 and carried out in Reference 4, the new set of amplitudes represents local modes with fields that satisfy the boundary conditions at the walls of the transition individually. We relate these local normal mode amplitudes a_n to the wave amplitudes A_n of eqns. 6.1.1 by the following linear transformation:

$$A = (\mathbf{1} + \mathbf{w})a \tag{6.3.2}$$

With $\mathbf{1}$ as the unit matrix, all differences between the cylindrical mode amplitudes A_n and the amplitudes a_n of the new set of local normal modes are contained in the matrix $\mathbf{w}$. If the transition has only small flare angles the local fields are still nearly cylindrical, and local normal modes, which satisfy the boundary conditions at the transition walls, differ little from cylindrical waves. Under these circumstances the elements of the matrix $\mathbf{w}$ will be small compared with unity. Thus it can be taken for granted that

$$[\mathbf{1} + \mathbf{w}]^{-1} = \mathbf{1} - \mathbf{w} \tag{6.3.3}$$

With the transformation eqn. 6.3.2 the coupled wave eqns. 6.1.1 can be written as

$$\left(\frac{d}{dz}\mathbf{w}\right)a + (\mathbf{1} + \mathbf{w})\left(\frac{d}{dz}a\right) = (-\boldsymbol{\gamma} + \boldsymbol{\kappa})(\mathbf{1} + \mathbf{w})a \tag{6.3.4}$$

Premultiplying this matrix equation by $(\mathbf{1} + \mathbf{w})^{-1}$ we obtain $-(\mathbf{1} + \mathbf{w})^{-1}(\boldsymbol{\gamma} - \boldsymbol{\kappa})(\mathbf{1} + \mathbf{w})$ as the matrix that multiplies $\mathbf{a}$ on its right-hand side. We abbreviate this matrix by

$$\boldsymbol{\gamma} + \boldsymbol{\varepsilon} = (\mathbf{1} + \mathbf{w})^{-1}(\boldsymbol{\gamma} - \boldsymbol{\kappa})(\mathbf{1} + \mathbf{w}) \tag{6.3.5}$$

and, by an appropriate choice of $\mathbf{w}$, make it a diagonal matrix. The coupled wave equations for the new amplitudes of local normal modes will then appear as

$$\frac{d}{dz}a = \left[-(\boldsymbol{\gamma}+\boldsymbol{\varepsilon}) - (\mathbf{1}+\mathbf{w})^{-1}\frac{d}{dz}\mathbf{w}\right]\mathbf{a} \tag{6.3.6}$$

With the approximation according to eqn. 6.3.3 they further simplify to

$$\frac{d}{dz}a = \left[-(\boldsymbol{\gamma}+\boldsymbol{\varepsilon}) - (\mathbf{1}-\mathbf{w})\frac{d}{dz}\mathbf{w}\right]a \tag{6.3.7}$$

The diagonal and only elements of $\boldsymbol{\gamma}+\boldsymbol{\varepsilon}$ in this equation are the propagation constants of local normal modes. If $\mathbf{w}$ would not change along the transition any mode n with amplitude a_n would propagate with $\gamma_{nn}+\epsilon_{nn}=\gamma_n+\epsilon_n$ independently of all the other local normal modes. Any variation of $\mathbf{w}$ will cause these modes to interact, but in gradual tapers this interaction remains rather weak.

The propagation constants $\gamma_n+\epsilon_n$ are the eigenvalues of the coupling matrix $\boldsymbol{\gamma}+\boldsymbol{\kappa}$, and for the local normal mode n follow as solutions of the homogeneous system of linear equations

$$(\boldsymbol{\gamma}+\boldsymbol{\kappa})(\mathbf{1}+\mathbf{w})_n = (\gamma_n+\epsilon_n)(\mathbf{1}+\mathbf{w})_n, \tag{6.3.8}$$

where $(\mathbf{1}+\mathbf{w})_n$ is the nth column vector of the matrix $(\mathbf{1}+\mathbf{w})$. To derive this solution we use the following schematic representation of eqn. 6.3.8

$$\begin{bmatrix} & \vdots & & \vdots & \vdots & \\ \cdots\ \gamma_m - \gamma_n(-\epsilon_n) & & \cdots -\kappa_{mn} \cdots & -\kappa_{ml} & \cdots \\ & \vdots & & \vdots & \vdots & \\ -\kappa_{nm} & & \cdots -\epsilon_n & \cdots & & \\ & & & & & \end{bmatrix} \begin{bmatrix} \vdots \\ w_{mn} \\ \vdots \\ 1+w_{nn} \\ \vdots \\ w_{ln} \\ \vdots \end{bmatrix} = \begin{bmatrix} 0 \\ \vdots \\ 0 \\ \vdots \\ 0 \\ \vdots \end{bmatrix} \tag{6.3.9}$$

Any nontrivial solution of this homogeneous system requires its coefficient determinant to vanish. This condition represents the characteristic equation of the coupling matrix $\boldsymbol{\gamma}-\boldsymbol{\kappa}$. Simple approximate solutions of the characteristic equation are obtained when all cylindrical modes under consideration are far enough from degeneracy, or the transition sufficiently gradual, that

$$|\kappa_{mn}| \ll |\gamma_m - \gamma_n| \tag{6.3.10}$$

Under these circumstances the local normal mode n has a propagation constant that differs from the propagation constant γ_n of the corresponding cylindrical mode approximately by

$$\epsilon_n = \sum_{m \neq n} \frac{\kappa_{mn}\kappa_{nm}}{\gamma_m - \gamma_n} \tag{6.3.11a}$$

Additional terms in this approximation are proportional to κ^i with integer powers $i > 2$.

As one prerequisite for the exprs. 6.3.10 to be satisfied the respective cylindrical modes must be sufficiently far above cutoff. A mode m which is near cutoff contributes to ϵ_n not only with its forward but also with its backward travelling wave. Both waves with index m and $-m$ must be considered simultaneously. As mentioned before in Section 4.3 and 6.1 this contribution is always finite so that

$$|\epsilon_n| \ll |\gamma_m - \gamma_n| \tag{6.3.11b}$$

is still valid. Its evaluation, however, requires more extensive computation. To avoid these additional complications we consider, in what follows, only such modes of index n, m and l which are sufficiently far above cutoff. The summation in eqn. 6.3.11*a* and in subsequent equations, will then have to be taken only over these modes. We also disregard any interaction with waves travelling in the negative z-direction, and have therefore only positive subscripts n, m, l in the following calculations.

The restriction that no degenerate modes occur in eqn. 6.1.1 is satisfied in the case of the circular and rectangular transition of Fig. 6.1 when they are excited by the H_{01}-mode and the fundamental mode, respectively.

With eqns. 6.3.10 and 6.3.11 we obtain from eqn. 6.3.9 for the value of w_{mn} with $m \neq n$

$$w_{mn} = \frac{1}{\gamma_m - \gamma_n}\left(\kappa_{mn} - \sum_{\substack{l \neq n \\ l \neq m}} \frac{\kappa_{ml}\kappa_{ln}}{\gamma_l - \gamma_n}\right) \tag{6.3.12}$$

Any additional terms in this approximation are again proportional to κ^i, with integer powers $i > 2$. The column vectors of the transformation matrix $\mathbf{1}+\mathbf{w}$ are eigenvectors of the symmetrical matrix $-\boldsymbol{\gamma}+\boldsymbol{\kappa}$, and hence orthogonal to each other. If we normalise these vectors such that

$$(\mathbf{1}+\mathbf{w})^T\cdot(\mathbf{1}+\mathbf{w})^* = \mathbf{1}$$

any of its diagonal elements w_{nn} consists of terms proportional to κ^2 and of higher-order terms,

$$w_{nn} \sim \kappa^2 \tag{6.3.13}$$

Therefore, and because of approximations which will be made later on, we need not consider these terms in the following.

We have now determined all propagation and coupling coefficients in the coupled wave eqn. 6.3.7 for local normal modes, and can proceed to solve them. Their general form corresponds to the coupled wave equations for cylindrical modes. Their coupling coefficients for these local normal modes are, however, substantially smaller than for cylindrical modes. This is particularly true for transitions with smooth contours and low overall spurious mode generation.

By applying the same procedure that lead to eqn. 6.1.4b as the solution for cylindrical modes we likewise obtain a solution in terms of local normal modes. Only the wanted mode with index n is incident and while propagating through the transition it excites the main spurious mode with index m most strongly. Since the local normal modes become identical with the cylindrical modes in the uniform waveguides at both ends of the transition the over-all excitation is the same in the cylindrical mode presentation as in the local normal mode presentation. Introducing eqns. 6.3.11–6.3.13 into eqn. 6.3.7 we arrive at a solution for the over-all conversion which corresponds to eqn. 6.1.4b and has the following form:

$$|\hat{A}_m| = |\hat{a}_m| = \left|\int_0^L \left\{\frac{d}{dz}\frac{1}{\gamma_m-\gamma_n}\left(\kappa_{mn}+\sum_{\substack{l\neq m\\ l\neq n}}\frac{\kappa_{ml}\kappa_{ln}}{\gamma_l-\gamma_m}\right)\right.\right.$$
$$\left. - \sum_{\substack{l\neq m\\ l\neq n}}\frac{\kappa_{ml}}{\gamma_m-\gamma_l}\frac{d}{dz}\frac{\kappa_{ln}}{\gamma_l-\gamma_n}\right\}$$
$$\left.\times\exp\left[\int_z^L\left(\gamma_n-\gamma_m+\underline{\sum_{l\neq n}\frac{\kappa_{nl}\kappa_{ln}}{\gamma_l-\gamma_n}}-\underline{\sum_{l\neq m}\frac{\kappa_{ml}\kappa_{lm}}{\gamma_l-\gamma_m}}\right)dz'\right]dz\right| \tag{6.3.14}$$

For cylindrical modes which are far enough above cutoff, eqns. 6.2.3 apply and we may write

$$\frac{\kappa_{ml}}{\gamma_m-\gamma_l}\frac{d}{dz}\left(\frac{\kappa_{ln}}{\gamma_l-\gamma_n}\right)=\frac{1}{2}\frac{d}{dz}\left(\frac{\kappa_{ml}}{\gamma_m-\gamma_l}\frac{\kappa_{ln}}{\gamma_l-\gamma_n}\right)$$

The exponential function with the underlined terms in eqn. 6.3.14 is expanded into a Taylor series. Integration of eqn. 6.3.14 by parts then yields

$$|\hat{A}_m| = \left|\int_0^L \left[\kappa_{mn} + \sum_{\substack{l\neq m\\ l\neq n}} \kappa_{ml}\kappa_{ln}\frac{(\gamma_m+\gamma_n)/2-\gamma_l}{(\gamma_l-\gamma_n)(\gamma_m-\gamma_l)} + \kappa_{mn}\int_z^L\left(\sum_{l\neq n}\frac{\kappa_{nl}\kappa_{ln}}{\gamma_l-\gamma_n} - \sum_{l\neq m}\frac{\kappa_{ml}\kappa_{lm}}{\gamma_l-\gamma_m}\right)dz'\right] \times \exp\left\{\int_z^L(\gamma_n-\gamma_m)dz'\right\}dz\right| \tag{6.3.15}$$

The expression in brackets includes all terms proportional to κ, to κ^2, and to $\kappa\int\kappa^2 dz$. Terms of even higher order in κ are neglected.

With eqns. 6.2.3 the series in eqn. 6.3.15 can be further evaluated. Introducing the coupling distribution function $CK(\xi)$ of eqns. 6.2.5 and 6.2.6 we obtain

$$|\hat{A}_m| = \int_{-\theta}^{\theta}\left(CK - j\alpha_1\frac{C^2K^2}{\eta} + j\alpha_2\frac{C^3K}{\eta}\int_\xi^\theta K^2 d\xi'\right)\exp(j\eta\xi)d\xi \tag{6.3.16}$$

with

$$\alpha_1 = \sum_{\substack{l\neq m\\ l\neq n}}^{l\leqslant p}\frac{(x_m^2-x_n^2)^3[x_l^2-(x_m^2+x_n^2)/2]x_l^2}{(x_l^2-x_m^2)^2(x_l^2-x_n^2)^2x_nx_m} \tag{6.3.17a}$$

and

$$\alpha_2 = -2+\sum_{\substack{l\neq m\\ l\neq n}}^{l\leqslant p}\frac{(x_m^2-x_n^2)^3}{x_m^2x_n^2}x_l^2\left[\frac{x_m^2}{(x_l^2-x_m^2)^3}-\frac{x_n^2}{(x_l^2-x_n^2)^3}\right] \tag{6.3.17b}$$

The series expressions of eqns. 6.3.17*a* and *b* must be summed over all higher-order mode terms which we see necessary to take into account.

For the rectangular transition of Fig. 6.1*b* the quantities α_1 and α_2 have been evaluated with $n=1$, $m=3$ and $l=5, 7, 9 \ldots$. Fig. 6.2 shows them as functions of the index p for the highest-order mode that has been taken into account in this evaluation. Only a few terms need to be considered to get close to the final results

$$\alpha_1 = 20/27 \quad \text{and} \quad \alpha_2 = 128/81 \tag{6.3.17c}$$

The corresponding values for the circular taper of Fig. 6.1*a* are

$$\alpha_1 = 1/1{\cdot}56 \quad \text{and} \quad \alpha_2 = 0 \tag{6.3.17d}$$

Let us now compare the supposedly more accurate results for mode conversion in eqn. 6.3.16 with those in eqn. 6.2.8. The average values of those coupling terms in the integrand of eqn. 6.3.16 which are proportional to K^2 and to $K \int K^2 d\xi$ will most likely be much smaller than the main term, which is proportional to K. This is particularly true for transitions with weak spurious mode excitation. Nevertheless these higher-order terms may contribute significantly to the overall mode conversion. The reason is that in good transitions and for small values of η the spurious mode generation, which the main term causes, is carefully compensated by destructive interference. The higher-order terms depend quite differently on ξ and may modify the spurious mode excitation substantially.

As an example we consider a rectangular waveguide transition which is designed according to Section 6.2. For its coupling distribution function we choose

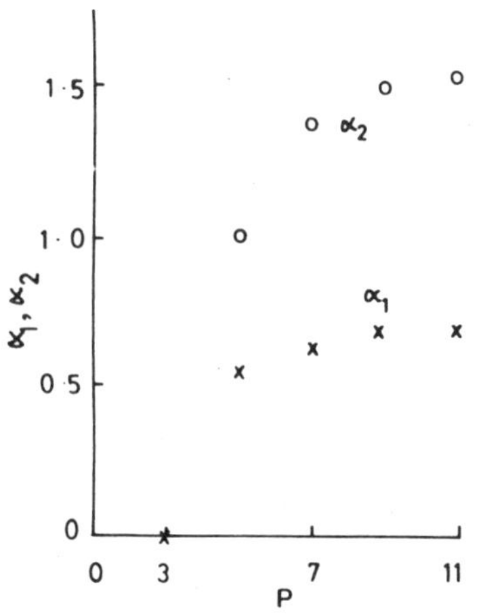

Fig. 6.2 *Convergence properties of the series eqns. 6.3.17 for the rectangular transition ($\alpha_2 = -2$ for $p = 3$)*

$$K = 0{\cdot}1095\cos(\xi/5) + 0{\cdot}0284\cos(3\xi/5) = K_0 \qquad (6.3.18)$$

with

$$2\theta = 5\pi$$

This function of K satisfies eqn. 6.2.10.

Fig. 6.3 shows the output amplitude $\hat{A}_m$ of the main spurious mode from eqn. 6.3.16 as well as the values of

$$W_1 = \left|\int_{-\theta}^{\theta} K \exp(j\eta\xi)d\xi\right| \quad \text{of eqn. 6.2.8,}$$

$$W_2 = \left|\int_{-\theta}^{\theta} \frac{20}{27\eta} K^2 \exp(j\eta\xi)d\xi\right| \qquad (6.3.19a)$$

$$W_3 = \left|\int_{-\theta}^{\theta} \frac{128}{81\eta} K \left(\int_{\xi}^{\theta} K^2 d\xi'\right) \exp(j\eta\xi)d\xi\right| \qquad (6.3.19b)$$

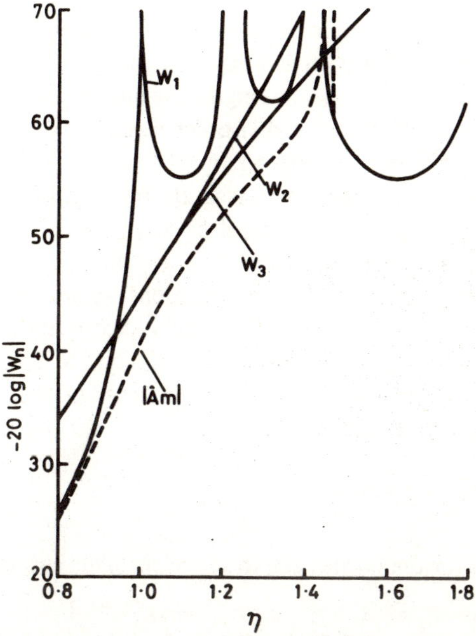

Fig. 6.3 *Mode conversion of a rectangular transition designed according to Table 6.1 with K(ξ) of eqn. 6.3.18 and C = 1 (a₂/a₁ = 3·8)*

W_2 and W_3 represent the absolute values of the higher-order coupling terms in eqn. 6.3.16 for $C = 1$. The difference between the mode conversion according to eqn. 6.3.16 and that according to eqn. 6.2.8 is significant in particular for values of η which are close to unity. The quantity η assumes such values close to unity for operating frequencies

f approaching the highest design frequency f_0. The difference between results from the presentation in terms of local normal modes and that in terms of cylindrical modes becomes even more pronounced for larger values of C. Eqn. 6.3.16 takes only account of terms up to the order C^3. For large values of C terms of still higher-order in C would be needed to determine mode conversion accurately.

Under normal circumstances and for moderate values of C the terms W_2 and W_3 give the strongest degradation from the predicted performance of transitions, when they are designed according to Section 6.2.

From these results we may draw the following conclusions. To calculate the conversion between two modes in a transition accurately, both the influence of the main spurious mode on the propagation of wanted modes and the coupling effects of higher-order modes need to be considered. The effect of higher-order mode coupling on the amount of converted power increases when the ratio a_2/a_1 of cross-sectional sizes at both ends of the transition increases. The effect of the higher-order mode coupling becomes likewise more significant when the coupling distribution function is optimised, and low mode conversion is obtained for a short transition length. To find out in which of these ways the coupling effects of higher-order modes are most critical we can use the results of eqns. 6.3.16 and 6.3.17.

So far we have used the more accurate solution of eqn. 6.3.16 of the coupled wave equations in terms of local normal modes only to analyse given waveguide transitions. This solution offers also the opportunity to design waveguide transitions somewhat better than with cylindrical modes. Such a transition design should of course include the higher-order coupling terms of eqn. 6.3.16. To develop this design method we integrate the first term in the integrand of eqn. 6.3.16 and neglect the small term.

$$\alpha_2 C^3 K \int_{-\theta}^{\theta} K^2 d\xi'$$

We thus obtain for the output amplitude of the main spurious mode

$$|\hat{A}_m| = \left| \frac{1}{\eta} \int_{-\theta}^{\theta} \left(CK' - \alpha_1 C^2 K^2 - \alpha_2 C^3 K \int_{-\theta}^{\xi} K^2 d\xi \right) \exp(j\eta\xi) d\xi \right| \tag{6.3.20}$$

The prime on K denotes differentiation with respect to ξ.

We are now faced with the problem of choosing a distribution for the function in the parenthesis in eqn. 6.3.20 such that the over-all mode conversion remains as low as possible. To this end we let

$$CK' - \alpha_1 C^2K^2 - \alpha_2 CK \int_{-\theta}^{\xi} C^2K^2 d\xi' = C_1 K_0' + C_2 K_0 \qquad (6.3.21)$$

where the prime on K and K_0 again denotes differentiation with respect to ξ.

We first examine the function $CK(\xi)$ which according to eqn. 6.2.9 is uniquely determined by the transition contour.

The boundary conditions for this function are

$$CK(\xi = -\theta) = CK(\xi = +\theta) = 0 \qquad (6.3.22a)$$

and, according to eqn. 6.2.9*a*

$$\int_{-\theta}^{\theta} CK d\xi = C = \frac{2x_m x_n}{x_m^2 - x_n^2} \ln (a_2/a_1) \qquad (6.3.22b)$$

Introducing eqn. 6.3.21 into eqn. 6.3.20 leads to

$$|\hat{A}_m| = \left| (C_1 + jC_2/\eta) \int_{-\theta}^{\theta} K_0 \exp (j\eta\xi) d\xi \right| \qquad (6.3.23a)$$

Usually $|C_2|$ will be small compared with $|C_1|$ and we have $C_1 \approx C$. Thus, eqn. 6.3.23*a* is approximated by

$$|\hat{A}_m| \approx \left| C \int_{-\theta}^{\theta} K_0 \exp (j\eta\xi) d\xi \right| \qquad (6.3.23b)$$

According to this result $K_0(\xi)$ may be interpreted as an effective coupling distribution function. This coupling distribution must again be chosen by following the guide lines which were given in Chapter 5.2.

Once a suitable distribution for $K_0(\xi)$ has been found the function $CK(\xi)$ may be calculated by integrating eqn. 6.3.21. The real constants C_1 and C_2 must be determined such that the boundary conditions of eqns. 6.3.22 are satisfied. With the formulation of eqn. 6.3.21 this is always possible. These equations must, however, be evaluated by numerical methods. One starts the integration at $\xi = -\theta$ with $C_1 = C$ and an arbitrary value of C_2, as well as with $CK(\xi = -\theta) = 0$. As a result of the integration one obtains $CK(\xi = \theta)$ and $\int_{-\theta}^{\theta} CK d\xi$. The values of C_1 and C_2 are then varied to satisfy the correct boundary conditions for $CK(\xi)$. Note, however, that when the initial trial values of C_1 and C_2 are too far off the function $CK(\xi)$ will show singularities.

Except for this particular manner in which K is determined all other steps in the design procedure for the waveguide transition correspond to those of Table 6.1.

For illustration we present the results from this design procedure for a typical transition between circular waveguides as shown in

Fig. 6.1*a*. We specify the diameter ratio as $a_2/a_1 = 4$ and obtain $C = 2{\cdot}16$. For the distribution K_0 we choose the function as given by eqn. 6.3.18.

The following values of C_1 and C_2 satisfy the boundary conditions for $CK(\xi)$:

$$C_1 = 2{\cdot}27; \quad C_2 = 0{\cdot}30.$$

Fig. 6.4 compares $CK(\xi)$ from this design procedure with $CK_0(\xi)$.

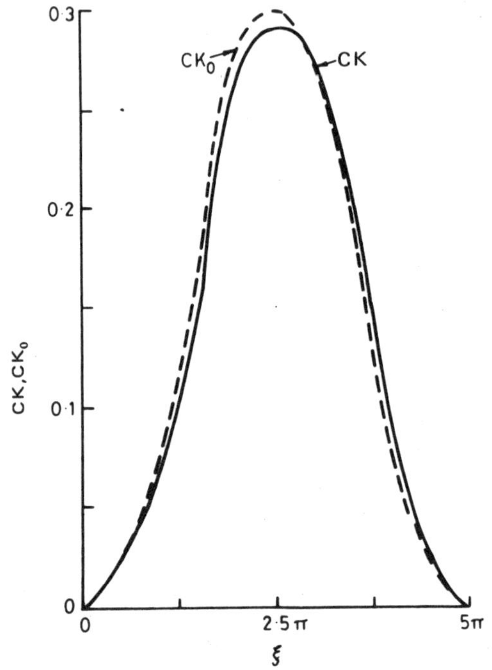

Fig. 6.4 *Coupling distribution function CK of the improved circular taper,* $a_2/a_1 = 4$
CK_0 is the corresponding coupling distribution according to the design principle of Section 6.2.

The contour of the waveguide transition

$$A = a/a_1 = \exp\left[\ln(a_2/a_1)\int_{-\theta}^{\xi} K(\xi')d\xi'\right] \qquad (6.3.24a)$$

as a function of

$$Z = x_m x_n z/(k_0 a_1^2) = \int_{-\theta}^{\xi} A^2 d\xi' / \ln(a_2/a_1) \qquad (6.3.24b)$$

is shown in Fig. 6.5. When we replace CK in eqn. 6.3.24*a* by CK_0 we return to the contour $A'(z)$, as it results from the design which considers only the main spurious mode of the cylindrical mode set, and no other higher-order modes. The difference $A - A'$ is also plotted in Fig. 6.5. In this particular example it amounts maximally to 4%, and is hence quite significant.

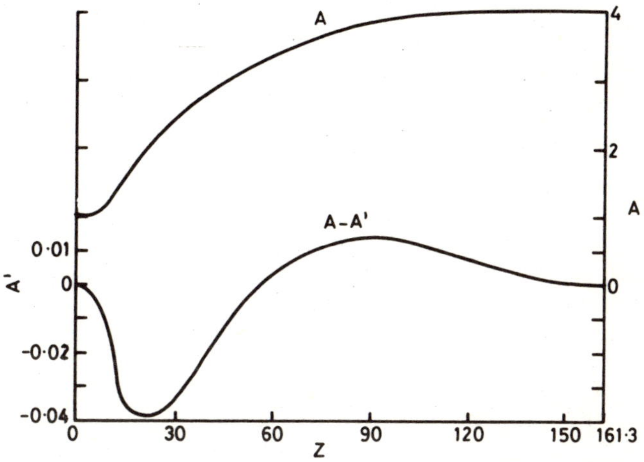

Fig. 6.5 *Contour A(Z) of the improved circular taper, $a_2/a_1 = 4$ A' (Z) is the corresponding contour according to Section 6.2*

The preceding analysis and design of waveguide transitions, which is based on local normal modes, and not on cylindrical modes, needs to cope only with much smaller coupling factors in the coupled wave equations. The over-all mode conversion can therefore be calculated much more accurately. At first sight this appears as a distinct advantage. In the design procedure we need, however, to make a number of approximations and cannot maintain the accuracy to its full extent.

It is therefore not always advisable to work with the representation in terms of local normal modes. The cylindrical mode representation and its simpler design procedure will, under certain conditions, lead to waveguide transitions which are just as close to the optimum, and perform just as well.

To decide on which method of analysis and which design procedure to choose we return to eqns. 6.1.3 and use the following abbreviations:

$$x = A_m/A_n; \quad y = A_p/A_n$$

$$\hat{K} = -\eta \frac{\kappa_{mn}}{\beta_m - \beta_n}; \quad \xi + \theta = -\frac{1}{\eta}\int_0^z (\beta_m - \beta_n)dz \quad (6.3.25a)$$

We assume in this discussion that all modes n, m and p are *far* above cutoff throughout the transition. The ratios

$$\frac{\beta_p - \beta_n}{\beta_m - \beta_n}; \quad \frac{\kappa_{pn}}{\kappa_{mn}}; \quad \frac{\kappa_{nm}}{\kappa_{mn}}; \quad \frac{\kappa_{np}}{\kappa_{pn}}; \quad \frac{\kappa_{mp}}{\kappa_{nm}} \quad (6.3.25b)$$

of differences in phase constants and of coupling coefficients will then be nearly constant along the transition, and nearly independent of frequency. Under this assumption, and in terms of the quantities in eqn. 6.3.25*a*, we approximate eqn. 6.1.3*c* by

$$\frac{dx}{d\xi} = \hat{K}\left[1 - \frac{\kappa_{nm}}{\kappa_{mn}}x^2 + \frac{\kappa_{mp}}{\kappa_{mn}}y\right] + j\eta x \quad (6.3.26a)$$

A term which is proportional to xy has been disregarded here because it is small enough to be of no significance. An equation corresponding to eqn. 6.1.3*c* but with A_m/A_n replaced by A_p/A_n and vice versa may likewise be approximated by

$$\frac{dy}{d\xi} = \frac{\kappa_{pn}}{\kappa_{mn}}\hat{K} + j\eta \frac{\beta_p - \beta_n}{\beta_m - \beta_n} y \quad (6.3.26b)$$

In this equation all terms have been omitted which are either proportional to x or to y^2. All these omitted terms are of secondary significance when we study the influence of the higher-order spurious mode p on the interaction between the wanted mode n and the main spurious mode m.

To solve eqn. 6.3.26 we have previously used the first-order approximation

$$\hat{x}(\eta, \xi) = e^{j\eta\xi}\int_{-\theta}^{\xi} \hat{K}(\xi')e^{-j\eta\xi'}d\xi' \quad (6.3.27a)$$

$$y(\eta, \xi) = \frac{\kappa_{pn}}{\kappa_{mn}}\int_{-\theta}^{\xi} \hat{K}(\xi') \exp\left[j\eta \frac{\beta_p - \beta_n}{\beta_m - \beta_n}(\xi - \xi')\right]$$

$$= \frac{\kappa_{pn}}{\kappa_{mn}}\hat{x}\left(\eta \frac{\beta_p - \beta_n}{\beta_m - \beta_n}, \xi\right) \quad (6.3.27b)$$

This solution neglects the underlined terms in eqn. 6.3.26*a* completely. We now improve this solution by taking account of these terms at least approximately. To this end we substitute for x and y in these terms from eqn. 6.3.27, and obtain the more accurate solution

$$x(\eta,\theta) = \hat{x}(\eta,\theta) + \Delta_1 + \Delta_2 \tag{6.3.28}$$

with

$$\Delta_1 = j\eta \exp(j\eta\theta)\frac{\kappa_{nm}}{\kappa_{mn}}\frac{2}{3}\int_{-\theta}^{\theta}[\hat{x}(\eta,\xi)]^3 \exp(-j\eta\xi)d\xi \tag{6.3.28a}$$

and

$$\Delta_2 = -j\eta\left(\frac{\beta_p-\beta_n}{\beta_m-\beta_n}-\frac{1}{2}\right)\frac{\kappa_{pn}\kappa_{mp}}{\kappa_{mn}^2}\exp(j\eta\theta)\times$$

$$\int_{-\theta}^{\theta}\left[\hat{x}\left(\frac{\beta_p-\beta_n}{\beta_m-\beta_n}\eta,\xi\right)\right]^2 \exp\left[j\eta\xi\left(2\frac{\beta_p-\beta_n}{\beta_m-\beta_n}-1\right)\right]d\xi \tag{6.3.28b}$$

The first additional term Δ_1 accounts for the influence which the main spurious mode has on the propagation of the wanted mode while the second additional term accounts for the effect which the higher-order spurious mode has on the interaction between the wanted mode and the main spurious mode.

To estimate the order of magnitude of these additional terms it suffices to substitute an approximate expression for $\hat{K}$ when we evaluate $\hat{x}(\xi)$ from eqn. 6.3.27*a*. In the case of the representation in terms of cylindrical modes we have from eqn. 6.2.6

$$\hat{K} = -CK(\xi)$$

while for the representation in terms of local normal modes $\hat{K}$ follows from eqns. 6.3.7 and 6.3.12 as

$$\hat{K} \approx -\eta\frac{dw_{mn}}{dz}\frac{1}{(\beta_m-\beta_n)} = \frac{dw_{mn}}{d\xi} \approx \frac{j}{\eta}C\frac{dK(\xi)}{d\xi}$$

As a typical example, and to obtain representative results for Δ_1 and Δ_2, we take the coupling distribution functions $K(\xi)$ from Section 5.2 which we evaluated there for $n = 4$ and whose coefficients appear in Table 5.2. We furthermore let $C = 1$ and $|\kappa_{mn}/\kappa_{nm}| = 1$. Referring now to the symmetrical transition between rectangular waveguides according to Fig. 6.1*b*, and to its fundamental (H_{10})-mode as the wanted mode we have the H_{30}-mode as the main spurious mode and the H_{50}-mode as the spurious mode of next higher-order. For these particular modes we have

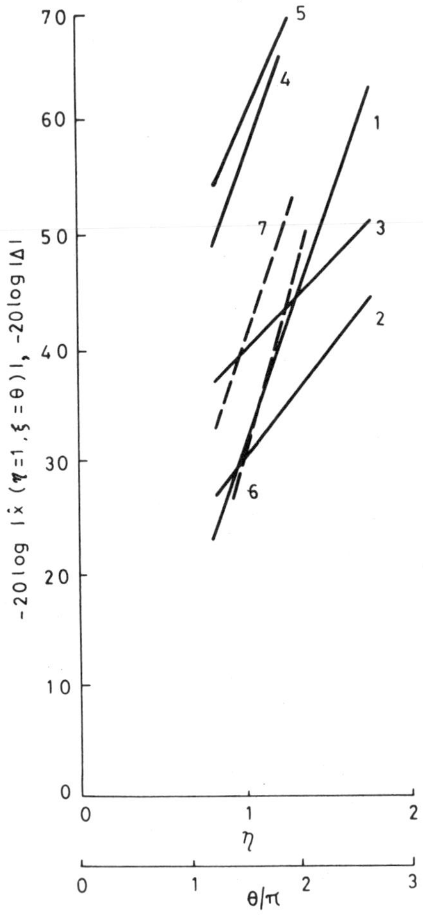

Fig. 6.6 *Influence of higher-order coupling terms according to eqns. 6.3.28 for rectangular waveguide transitions with C = 1*
$K(\xi)$ calculated according to eqns. 5.2.6 with $n = 4$
1 First-order coupling term of eqn. 6.3.27*a* according to Fig. 5.7, curves 5
2 $\Delta_1(\theta/\pi)$ according to eqn. 6.3.28a for cylindrical modes, $\eta = 1$
3 $\Delta_2(\theta/\pi)$ according to eqn. 6.3.28b for cylindrical modes, $\eta = 1$
4 $\Delta_1(\theta/\pi)$ according to eqn. 6.3.28a for local normal modes, $\eta = 1$
5 $\Delta_2(\theta/\pi)$ according to eqn. 6.3.28b for local normal modes, $\eta = 1$
6 $\Delta_1(\eta)$ according to eqn. 6.3.28a for cylindrical modes, $\theta/\pi = 1{\cdot}5$
7 $\Delta_2(\eta)$ according to eqn. 6.3.28b for cylindrical modes, $\theta/\pi = 1{\cdot}5$

$$\frac{\beta_p - \beta_n}{\beta_m - \beta_n} = 3$$

and in case of the representation in terms of cylindrical modes

$$\frac{\kappa_{pn}\kappa_{mp}}{\kappa_{mn}\kappa_{nm}} = 0{\cdot}823,$$

while the representation in terms of local normal modes leads to

$$\frac{\kappa_{pn}\kappa_{mp}}{\kappa_{mn}\kappa_{nm}} = 0{\cdot}244$$

The absolute values of Δ_1 and Δ_2, as we evaluate them for this representative example, are plotted in curves 2 . . . 5 of Fig. 6.6 as functions of the length parameter θ/π for $\eta = 1$. In addition the values of Δ_1 and Δ_2 are given in curves 6, 7 for $\theta/\pi = 1{\cdot}5$ as a function of η for cylindrical modes.

Although these results pertain only to a specific example they are nevertheless representative enough for well designed transitions to allow the following general conclusions.

The coupling integral of eqn. 6.1.4*c* approximates the solution of the coupled wave equations for optimum or near optimum waveguide transitions well enough only if C as well as θ are sufficiently small. The influence of higher-order mode coupling diminishes quite rapidly when η becomes larger.

6.4 Waveguide taper design with coupled horn waves

The local normal modes, as we defined them in the preceding section, are eigenvectors of the coupling matrix $-\boldsymbol{\gamma} + \boldsymbol{\kappa}$ of the coupled wavc equations, which in turn are based on a representation in terms of cylindrical modes. A transformation from these cylindrical modes to the local normal modes very nearly diagonalises the coupling matrix and leads to coupled wave equations with much smaller coupling terms. If the original coupling matrix would be constant anywhere along the taper then this transformation would completely diagonalise it, and the local normal modes would be truly normal modes, which propagate independently of each other through this section of constant coupling.

Although wc introduced the local normal modes just as a mathematical entity they nevertheless correspond quite closely to the horn modes of waveguide horns and horn transitions. These horn modes also propagate independently of each other through uniform horns, and interact only when the horn changes its flare angle, or becomes otherwise nonuniform.

We should therefore expect similar results when we base the analysis and design of circular or rectangular waveguide transitions not on the local normal modes but on their respective horn modes. The horn modes are normal modes of the uniform horn. But also in nonuniform horns the fields of individual horn modes satisfy boundary conditions much better than the eigenfields of individual cylindrical modes. As a consequence, the coupling factors in the system of coupled wave equations are much lower for the horn modes than for the cylindrical modes. This is especially true for those modes which are above cutoff. In this case the horn modes are quite similar to the local normal modes of the preceding section.

Let us first consider waveguide transitions according to Fig. 6.1 in which the wanted and the main spurious modes are well above cutoff all along the transition. In Section 3.1 we have normalised the horn wave amplitudes of the rectangular horn with respect to power. For vanishing flare angle $\phi \to 0$ these horn modes become therefore identical to the normal modes of the adjoining cylindrical waveguide as long as these modes are above cutoff. The horn modes of the round horn in Section 3.2 have not yet been normalised in this fashion and are not directly identical to cylindrical modes for vanishing flare angle.

In the following we shall therefore refer only to the rectangular waveguide transition; the results for this transition in Section 3.1 can be more readily applied here.

The coupled wave eqns. 3.1.35 are approximately solved by eqn. 6.1.4b, and the overall conversion to the main spurious mode appears as

$$|\hat{A}_m| = \left| \int_0^L \kappa_{mn}^{(++)} \exp\left[-j \int_z^L (\alpha_m^{(+)} - \alpha_n^{(+)}) dz' \right] dz \right| \qquad (6.4.1)$$

with $\kappa_{mn}^{(++)}$ according to eqn. 3.1.28a and $\alpha_i^{(+)} = \alpha_i$ according to eqn. 3.1.36.

Reconversion effects and the interaction with other than the main spurious mode cause additional terms such as in eqn. 6.3.15. These additional terms are proportional to $(\phi')^2$. In eqn. 6.4.1 they must be added to $\kappa_{mn}^{(++)}$ as well as to $\alpha_m^{(+)} - \alpha_n^{(+)}$.

When the coupling coefficient $\kappa_{ik}^{(++)}$ was derived in Chapter 3.1 only terms proportional to ϕ' and $\phi'\phi$ were taken into account. Therefore the additional coupling terms from eqn. 6.3.15 could not improve the accuracy of eqn. 6.4.1 since they are proportional to (ϕ').[2] They may hence be neglected in the expressions of $\kappa_{mn}^{(++)}$ and $\alpha_m^{(+)} - \alpha_n^{(+)}$. Any terms of still higher order in ϕ' and ϕ may likewise be disregarded.

For a rectangular waveguide transition according to Fig. 6.1*b* eqn. 6.4.1 can now readily be evaluated with $\kappa_{31}^{(++)}$ and $\alpha_3 - \alpha_1$ from eqns. 3.1.42. If the main spurious mode is far above cutoff, so that

$$2ka \gg 3\pi \qquad (6.4.2a)$$

the contour parameters v_1 and v_3 in eqn. 3.1.41*b* are just equal to unity:

$$v_1 = v_3 = 1 \qquad (6.4.2b)$$

We introduce ξ and K of eqns. 6.2.6 into eqn. 6.4.1 and integrate by parts. This leads to the same expression for the over-all mode conversion as it was obtained in eqn. 6.3.16 by transforming from the cylindrical modes to the local normal modes, and thus nearly diagonalising the coupled wave equations for cylindrical modes. More details on this evaluation of eqn. 6.4.1 are given in Reference 4.

As a conclusion from this evaluation we note that, when both the wanted mode and the spurious mode are far above cutoff, the horn mode analysis is not any more accurate than the analysis with a representation in terms of cylindrical modes. To obtain more accurate results from the horn mode analysis it must take higher-order coupling terms into account.

There are, however, also situations in which the simple horn mode analysis is quite superior to the analysis in terms of cylindrical modes and even superior to the analysis in terms of local normal modes. For one such case we note that eqn. 6.3.16 was derived from the local normal mode presentation under the assumption (expr. 6.2.3*a*), that

$$ka \gg x_m \qquad (6.4.3a)$$

On the other hand, in the horn mode presentation eqn. 6.4.1 holds together with eqns. 3.1.42*a* and *b* when eqn. 3.1.41*a* is satisfied, that is for

$$(a')^2 \ll (ka)^2 \left[1 - \left(\frac{x_m}{ka}\right)^2\right]^3 \qquad (6.4.3b)$$

Since eqn. 6.4.3*b* is less restrictive than eqn. 6.4.3*a*, the horn mode analysis and eqn. 6.4.1 holds more generally than eqn. 6.3.16. It serves particularly well when waveguide transitions are to be analysed at frequencies for which eqn. 6.4.3*a* fails.

As an example, to compare the performance of waveguide transitions which operate at various distances above cutoff, we consider again the rectangular waveguide transition of Fig. 6.1*b*. We let $a_2/a_1 = 4$ and choose K_0 according to eqn. 6.3.18. The transition contour $A(z) = a(z)/a_1$ is then found from eqns. 6.3.24 and 6.3.21. The overall mode conversion of such transitions as a function of $\eta = f_0/f$ was calculated for different values of the highest design frequency f_0 and the respective wave number k_0. Note that according to eqn. 6.3.24*b* the geometrical length of the transition is inversely proportional to f_0.

Fig. 6.7 shows the over-all mode conversion as it was obtained by applying eqn. 6.4.1 with $\kappa_{31}^{(++)}$ and $\alpha_3^{(+)} - \alpha_1^{(+)}$ according to eqns. 3.1.42 with 3.1.41*c*. Because of eqn. 6.4.3*b* these expressions are only valid for $f > f_c$, where f_c is the cutoff frequency of the H_{30}-mode in the narrowest cross-section. Therefore the conversion curves of Fig. 6.7 apply only up to the corresponding value of η.

The results of Fig. 6.7 are quite typical in that they demonstrate how the conversion characteristics deteriorate in particular at low values of η. By utilising eqn. 6.3.24 the specified coupling distribution function is realised exactly only for very small values of η in the limit of $f \to \infty$. For the parameters v_1 and v_3 of the coupling distribution functions first-order approximations were used which are valid only for $m\pi/(2ka) \ll 1$. With these approximations instead of the exact values for v_1 and v_3 a coupling distribution function is obtained which deviates from the specified and wanted distribution if η is small.

Better performance of the waveguide transition, with less degradation of its conversion characteristics is obtained, at any frequency within the operating band, if the specified coupling distribution function is realised not for small values of η, but for $\eta \approx 1$. Since the corresponding design procedure is of general practical interest it will briefly be outlined here.

We start again from eqn. 6.4.1 and assume that

$$\Delta\beta = \alpha_n^{(+)} - \alpha_m^{(+)} \tag{6.4.4b}$$

as well as

$$\kappa = \kappa_{mn}^{(++)} \tag{6.4.4a}$$

are specified as functions of a'', a', a and k. While $\Delta\beta$ according to eqn. 3.1.38 is a real quantity, the value of κ is usually complex

$$\kappa = |\kappa| \exp\,[j \arg(\boldsymbol{\kappa})] \tag{6.4.5}$$

It facilitates the design procedure when we write eqn. 6.4.1 in the form

$$|\hat{A}_m| = \left| \int_0^L |\kappa| \exp\left[j \int_z^L \Delta\beta dz' + j \arg(\kappa) \right] dz \right| \tag{6.4.6}$$

Introducing

$$\hat{\xi} + \hat{\theta} = \int_0^z \Delta\beta dz' + \arg[\kappa(z)] - \arg[\kappa(z = 0)] \qquad (6.4.7a)$$

$$2\hat{\theta} = \int_0^L \Delta\beta dz' + \arg[\kappa(z = L)] - \arg[\kappa(z = 0)] \qquad (6.4.7b)$$

$$\hat{K} = \frac{|\kappa|}{\Delta\beta + \dfrac{d}{dz}[\arg(\kappa)]} \qquad (6.4.7c)$$

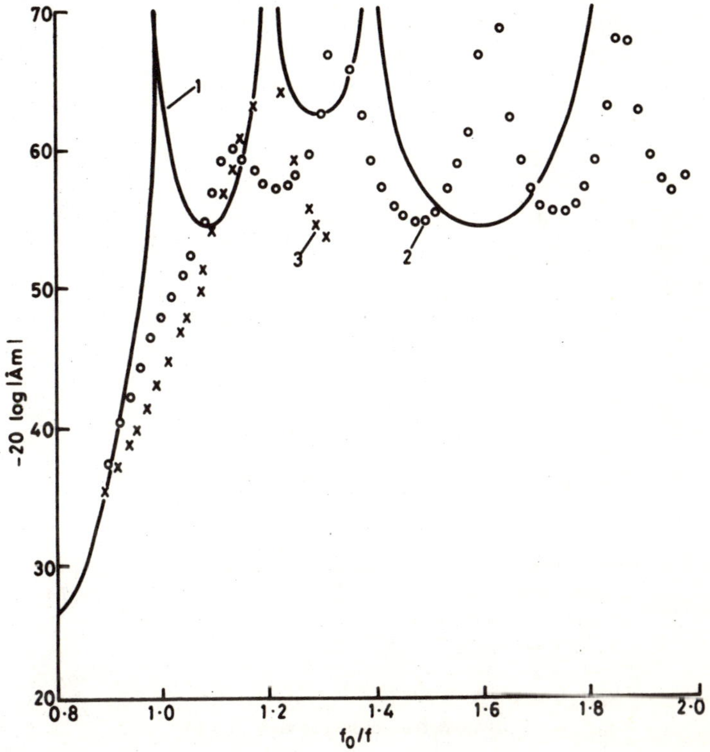

Fig. 6.7 *Mode conversion of a rectangular waveguide transition designed according to eqns. 6.3.24, 6.3.21, 6.3.18 with $a_2/a_1 = 4$*

Mode conversion curves

1 $f_0/f_c \to \infty$

2 $f_0/f_c = 2$

3 $f_0/f_c = 4/3$

f_c = cutoff frequency of the H_{30}-mode at a_1

yields

$$|\hat{A}_m| = \left| \int_{-\theta}^{\theta} \hat{K} \exp(-j\hat{\xi}) d\hat{\xi} \right| \tag{6.4.8}$$

When we now compare eqn. 6.4.7*a* with eqn. 6.2.6*c*, we find that the variable $\hat{\xi}$ corresponds to ξ according to

$$\hat{\xi} \to \eta\xi$$

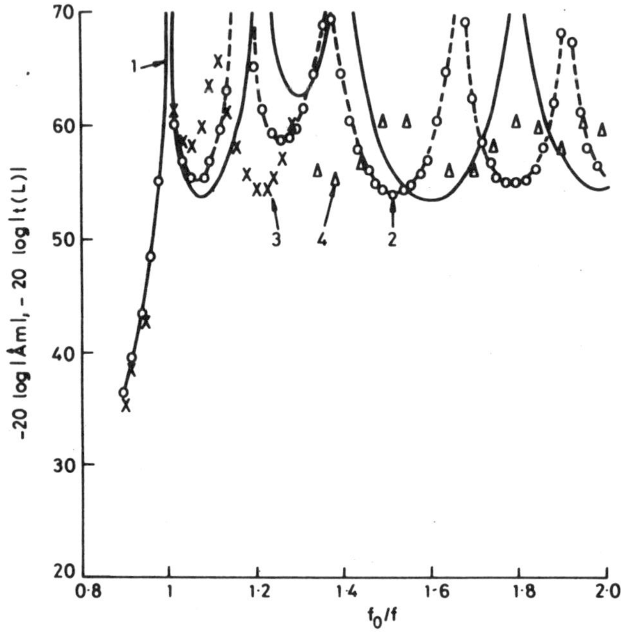

Fig. 6.8 *Mode conversion of a rectangular waveguide transition with $a_2/a_1 = 4$ designed according to eqns. 6.4.7, 6.4.10 with $K_0(\xi)$ of eqn. 6.3.18*
Mode conversion curves
1 $f_0/f_c \approx \infty$
2 $f_0/f_c = 2$
3 $f_0/f_c = 4/3$
4 $f_0/f_c = 4/3$ calculated according to eqns. 6.5.14

and in a first-order approximation $\hat{K}(\hat{\xi})$ of eqn. 6.4.7*c* corresponds to $C dK(\xi)/d\xi$ of eqn. 6.2.6*e*

$$\hat{K}(\hat{\xi}) \approx C \frac{dK(\xi)}{d\xi} \tag{6.4.9}$$

We now need to find for $K(\xi)$ a coupling distribution function at the upper frequency limit $f=f_0$, corresponding to $\eta=1$, that keeps spurious mode generation low over a wide frequency band. Also, to satisfy boundary conditions at both ends of the taper we introduce $\hat{\xi}=\xi$ for the upper frequency limit and let

$$\hat{K}(\hat{\xi}) = C_1 \frac{dK_0(\xi)}{d\xi} + C_2 K_0(\xi) \tag{6.4.10}$$

in accordance with eqn. 6.3.21. $K_0(\xi)$ represents one of the standard coupling distribution functions for low spurious mode generation. $\hat{K}(\hat{\xi})$ is with $k=k_0$ uniquely related to the transition contour $a(z)$ according to eqns. 6.4.7. The integration, which is required to determine $a(z)$ from $\hat{K}(\hat{\xi})$, has to be performed numerically. The integration constants C_1 and C_2 of eqn. 6.4.10 must be chosen to satisfy the boundary conditions

$$a = a_1; \quad a' = 0 \quad \text{at} \quad z = 0$$

$$a = a_2; \quad a' = 0 \quad \text{at} \quad z = L \tag{6.4.11}$$

By way of example this calculation was again carried out for the rectangular transition with $a_2/a_1=4$ and $K_0(\xi)$ according to eqn. 6.3.18. $\kappa=\kappa_{31}^{(++)}$ and $\alpha_1-\alpha_3=\alpha_1^{(+)}-\alpha_3^{(+)}$ were taken from eqns. 3.1.42. Fig. 6.8 shows the mode conversion. Compared with Fig. 6.7 the mode conversion characteristics are substantially improved near $\eta=1$ and yet do not deteriorate at larger values of $\eta=f_0/f$.

Although this excellent performance may not be altogether typical, it shows nevertheless that significant improvement is almost always achievable.

6.5 Coupling of modes below cutoff

We consider here the nonuniform waveguide of Fig. 6.9 from $z=0$ to $z=L$. The amplitudes of waves incident at $z=0$ are regarded as components of the vector $A^{(+)}(0)$. By the same token $A^{(-)}(0)$ is the vector of reflected wave amplitudes at $z=0$. $A^{(+)}(z)$ and $A^{(-)}(z)$, as well as $A^{(+)}(L)$ and $A^{(-)}(L)$ denote corresponding wave amplitude vectors at z along the waveguide transition and at $z=L$, respectively. The equations will restrict us to a finite number of modes and hence components of these vectors.

The wave amplitude vectors $A^{(\pm)}(z)$ at z relate to the corresponding vectors $A^{(\pm)}(z_1)$ at a different location z_1 along the taper by way of the transmission matrix $\mathbf{T}(z_1, z)$

$$\begin{bmatrix} A^{(+)}(z) \\ A^{(-)}(z) \end{bmatrix} = [\mathbf{T}(z_1, z)] \begin{bmatrix} A^{(+)}(z_1) \\ A^{(-)}(z_1) \end{bmatrix} \tag{6.5.1}$$

Any column of $\mathbf{T}(z_1, z)$ represents a solution of the coupled wave equations

$$\frac{d}{dz}\begin{bmatrix} A^{(+)}(z) \\ A^{(-)}(z) \end{bmatrix} = \begin{bmatrix} \mathbf{X}^{(++)}(z) & \mathbf{X}^{(+-)}(z) \\ \mathbf{X}^{(-+)}(z) & \mathbf{X}^{(--)}(z) \end{bmatrix} \cdot \begin{bmatrix} A^{(+)}(z) \\ A^{(-)}(z) \end{bmatrix} \tag{6.5.2}$$

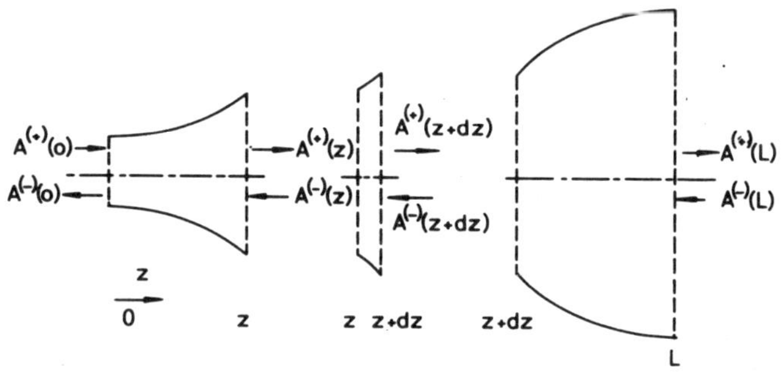

Fig. 6.9 *Arrays of wave amplitudes*

We can therefore state

$$\frac{d}{dz}[\mathbf{T}(z_1, z)] = \begin{bmatrix} \mathbf{X}^{(++)}(z) & \mathbf{X}^{(+-)}(z) \\ \mathbf{X}^{(-+)}(z) & \mathbf{X}^{(--)}(z) \end{bmatrix} \cdot [\mathbf{T}(z_1, z)] \tag{6.5.3}$$

as the differential equation system that defines $\mathbf{T}(z_1, z)$.

When we solve eqns. 6.5.3 by integration we must use the unitary matrix for $\mathbf{T}$ as the initial condition

$$\mathbf{T}(0, 0) = \mathbf{1} \tag{6.5.3a}$$

Note that in general eqns. 6.5.3 can be integrated only numerically, and only for a finite number of normal modes. For a given waveguide transition this integration will lead to the transmission matrix

$$\mathbf{T}(0, L) = \begin{bmatrix} \mathbf{T}^{(++)} & \mathbf{T}^{(+-)} \\ \mathbf{T}^{(-+)} & \mathbf{T}^{(--)} \end{bmatrix} \tag{6.5.4}$$

which represents the general solution of the coupled wave equations 6.5.2.

As long as the coupling matrix in eqns. 6.5.2 has off-diagonal elements which are all small in absolute value compared with the smallest difference between any of its relevant diagonal elements, that is as long as

$$|\chi_{mn}^{(p,p)}| \ll |\chi_{mm}^{(pp)} - \chi_{nn}^{(pp)}|$$

and

$$|\chi_{mn}^{(p,-p)}| \ll |\chi_{mm}^{(pp)} - \chi_{nn}^{(-p,-p)} \tag{6.5.5}$$

is satisfield for all normal modes under consideration, we can use the following approximate solution for the elements of $\mathbf{T}(0, L)$ according to eqns. 6.1.4

$$T_{mm}^{(pp)} = \exp\left[\int_0^L \chi_{mm}^{(pp)}(z)dz\right]$$

$m \neq n$:

$$T_{mn}^{(pp)}/T_{nn}^{(pp)} = \int_0^L \chi_{mn}^{(pp)}(z) \exp\left[\int_z^L (\chi_{mm}^{(pp)}(z') - \chi_{nn}^{(pp)}(z'))dz'\right] dz$$

$$T_{mn}^{(p,-p)}/T_{nn}^{(-p,-p)} = \int_0^L \chi_{mn}^{(p,-p)}(z) \exp\left[\int_z^L (\chi_{mm}^{(pp)}(z') - \chi_{nn}^{(-p,-p)}(z'))dz'\right] dz \tag{6.5.6}$$

For practical waveguide transitions these approximations are nearly always adequate if only all normal modes which need to be taken into account are above cutoff everywhere along the transition, including both ends.

Under these conditions the scattering matrix of the transition may also be evaluated to the same approximation without any difficulty. This scattering matrix is defined according to

$$\begin{bmatrix} A^{(-)}(0) \\ A^{(+)}(L) \end{bmatrix} = [\mathbf{S}(0, L)] \begin{bmatrix} A^{(+)}(0) \\ A^{(-)}(L) \end{bmatrix} \tag{6.5.7}$$

and it follows from eqns. 6.5.1 by exchanging in it $A^{(-)}(z)$ for $A^{(-)}(z)$, and letting $z = L$ as well as $z_1 = 0$. Thus, with the notations of eqns. 6.5.4, the scattering matrix may be written as

$$\mathbf{S}(0, L) = \begin{bmatrix} -(\mathbf{T}^{(--)})^{-1}\mathbf{T}^{(-+)} & (\mathbf{T}^{(--)})^{-1} = \mathbf{t}(L, 0) \\ \mathbf{T}^{(++)} - \mathbf{T}^{(+-)}(\mathbf{T}^{(--)})^{-1}\mathbf{T}^{(-+)} & \mathbf{T}^{(+-)}(\mathbf{T}^{(--)})^{-1} \\ = \mathbf{t}(0, L) & \end{bmatrix} \tag{6.5.8}$$

Under conditions for which eqns. 6.5.5 are valid the interaction between waves travelling in the same direction will be much stronger than between waves travelling in opposite directions. We then have

$$\mathbf{t}(0, L) = \mathbf{T}^{(++)}$$

and the approximations in eqns. 6.5.6 apply; $\mathbf{t}(L, 0)$ follows from reciprocity considerations.

In the case of waves which go through cutoff within the transition the conditions of eqns. 6.5.5 are not satisfied at all. On the contrary, there are even singularities in the system of coefficients χ. For a presentation in terms of modes with plane phase fronts they occur at the respective cutoff cross-section, while for a representation in terms of horn waves with curved phase fronts they occur at the junction of the transition and the cylindrical waveguide at the narrow end. Furthermore, and in particular, in the case of an analysis with plane-phase-front modes below cutoff the system of coupled wave eqns. 6.5.3 contains in all columns of $\mathbf{T}$ exponentially growing terms, which are due to the somewhat arbitrary initial conditions in eqns. 6.5.3. These terms will rather soon let the complete numerical integration deteriorate in its accuracy. Because there is strong interaction between forward and backward travelling waves in this case any analysis which neglects this interaction would lead to completely erroneous results.

All these difficulties may in principle be avoided by not solving the system of differential equations directly to obtain the transmission matrix, but by considering instead the corresponding system of equations for the scattering matrix. The scattering matrix has at least in its significant elements, because of their physical meaning, no objectionable irregularities.

Under the same condition as in eqns. 6.5.1 and for the same wave amplitude vectors at z and $z_1 \leqslant z$ the scattering matrix is defined by

$$\begin{bmatrix} A^{(-)}(z_1) \\ A^{(+)}(z) \end{bmatrix} = [\mathbf{S}(z_1, z)] \begin{bmatrix} A^{(+)}(z_1) \\ A^{(-)}(z) \end{bmatrix} \qquad (6.5.9)$$

If, in accordance with this definition, we let

$$\mathbf{S}(z_1, z) = \begin{bmatrix} \mathbf{R}^{(-)} & \mathbf{Q}^{(-)} \\ \mathbf{Q}^{(+)} & \mathbf{R}^{(+)} \end{bmatrix} \qquad (6.5.9a)$$

the differential equation for the scattering matrix follows by differentiating eqn. 6.5.9 with respect to z and by introducing eqn. 6.5.2

$$\frac{d}{dz}\mathbf{S}(z_1, z) = \begin{bmatrix} -\mathbf{Q}^{(-)}\mathbf{X}^{(-+)}\mathbf{Q}^{(+)} & -\mathbf{Q}^{(-)}[\mathbf{X}^{(--)} + \mathbf{X}^{(-+)}\mathbf{R}^{(+)}] \\ [\mathbf{X}^{(++)} - \mathbf{R}^{(+)}\mathbf{X}^{(-+)}]\,\mathbf{Q}^{(+)} & [\mathbf{X}^{(+-)} + \mathbf{X}^{(++)}\mathbf{R}^{(+)} \\ & -\mathbf{R}^{(+)}\mathbf{X}^{(--)} - \mathbf{R}^{(+)}\mathbf{X}^{(-+)}\mathbf{R}^{(+)}] \end{bmatrix} \qquad (6.5.10)$$

The interaction between waves travelling in opposite directions gives rise to the elements of $\mathbf{R}^{(\pm)}$ in $\mathbf{S}$. Under conditions 6.5.5 this interaction may be neglected. In this case the remaining matrices $\mathbf{Q}^{(\pm)}$ for interaction between waves travelling in the same direction obey much simpler differential equations. The interaction between forward travelling waves is described by

$$\frac{d}{dz}\mathbf{Q}^{(+)} = \mathbf{X}^{(++)}\mathbf{Q}^{(+)}$$

This equation system appears in all respects equivalent to what follows under conditions 6.5.5 from eqns. 6.5.3.

We will now employ eqn. 6.5.10 to analyse a rectangular waveguide taper for the fundamental H_{10}-mode according to Fig. 6.1*b*, and in particular calculate the excitation of the H_{30}-mode under the condition that this mode is below cutoff at the narrow end of the taper. We will represent the field in terms of local horn modes but take into account only the H_{10}-mode with the amplitudes $A_1^{(+)}$ and $A_1^{(-)}$ of its forward and backward travelling waves and the H_{30}-mode with the forward and backward travelling wave amplitudes $A_3^{(+)}$ and $A_3^{(-)}$. These amplitudes are assumed to be normalised according to eqns. 3.1.33, so that each mode transmits the net real power

$$\mathrm{Re}\,(P) = A^{(+)}(A^{(+)})^* - A^{(-)}(A^{(-)})^* \qquad (6.5.11)$$

in forward direction.

The wave amplitudes form forward and backward wave amplitude vectors which are given by

$$A^{(+)} = \begin{bmatrix} A_1^{(+)} \\ A_3^{(+)} \end{bmatrix}; \quad A^{(-)} = \begin{bmatrix} A_1^{(-)} \\ A_3^{(-)} \end{bmatrix}$$

To integrate eqn. 6.5.10 we start at $z_1 = 0$, the narrow end of the taper. With z_1 fixed, the elements of the scattering matrix depend only on z

$$\begin{bmatrix} A^{(-)}(0) \\ A^{(+)}(z) \end{bmatrix} = [\mathbf{S}(z)] \cdot \begin{bmatrix} A^{(+)}(0) \\ A^{(-)}(z) \end{bmatrix}.$$

$$\mathbf{S}(z) = \begin{bmatrix} \mathbf{R}^{(-)}(z) & \mathbf{Q}^{(-)}(z) \\ \mathbf{Q}^{(+)}(z) & \mathbf{R}^{(+)}(z) \end{bmatrix} \qquad (6.5.12)$$

For weak interaction between modes and good performance of the transition the waveguide dimensions must be tapered quite smoothly.

Moreover for such waveguide tapers we are mainly interested to limit the excitation of the H_{30}-mode, when the H_{10}-mode is incident from either end of the taper. We will therefore restrict our attention to the elements of the first column of $\mathbf{Q}^{(+)}(z)$ for which we write

$$Q_{11}^{(+)}(z) = T(z)$$
$$Q_{31}^{(+)}(z) = t(z), \tag{6.5.13a}$$

and to $\mathbf{R}^{(+)}(z)$ in which we neglect $R_{11}^{(+)}(z)$

$$\mathbf{R}^{(+)}(z) = \begin{bmatrix} 0 & d(z) \\ d(z) & R(z) \end{bmatrix} \tag{6.5.13b}$$

Because of the normalization with respect to power as expressed by eqn. 6.5.11, we have

$$|T| \approx 1 \quad |R| \approx 1 \tag{6.5.13c}$$

Well designed tapers excite only little parasitic mode power. For them we have

$$|t| \ll 1 \quad |d| \ll 1 \tag{6.5.13d}$$

everywhere along the taper. For such tapers we may neglect all terms which are proportional to d and thus extract the following approximations from eqn. 6.5.10:

$$\frac{dR(z)}{dz} = \chi_{33}^{(+-)} - \chi_{33}^{(-+)}[R(z)]^2 + [\chi_{33}^{(++)} - \chi_{33}^{(--)}]\,R(z) \tag{6.5.14a}$$

$$\frac{dt(z)}{dz} = [\chi_{31}^{(++)} - R(z)\chi_{31}^{(-+)}]\,T(z) + [\chi_{33}^{(++)} - R(z)\chi_{33}^{(-+)}]\,t(z) \tag{6.5.14b}$$

$$\frac{dT(z)}{dz} = \chi_{11}^{(++)}T(z) \tag{6.5.14c}$$

Because higher-order coupling terms have been neglected in these approximations they correspond in accuracy to eqns. 6.5.6. Eqns. 6.5.6 follow from eqns. 6.5.14 if we let $R = 0$.

For the integration of eqns. 6.5.14 the initial values for T and t at $z = 0$ are easily visualised to be

$$t(0) = 0; \quad T(0) = 1 \tag{6.5.15a}$$

To decide on the initial value $R(0)$ we refer to the spurious mode propagation through the abrupt transition from the cylindrical waveguide to the horn in Fig. 6.10, in which we let the flare angle 2ϕ go to zero. On the surface S at the junction and for $\phi \to 0$ the fields of the cylindrical waves with amplitudes $a_3^{(+)}$ and $a_3^{(-)}$ from Section 2.6. are given by

$$E_y = \hat{E}_y(a_3^{(+)} + a_3^{(-)})$$

$$H_\eta = \frac{\gamma_3}{j\omega\mu}\hat{E}_y(-a_3^{(+)} + a_3^{(-)})$$

$$\hat{E}_y = -\cos\left(\frac{3\pi}{2}\eta\right)\frac{1}{\sqrt{ab}}$$

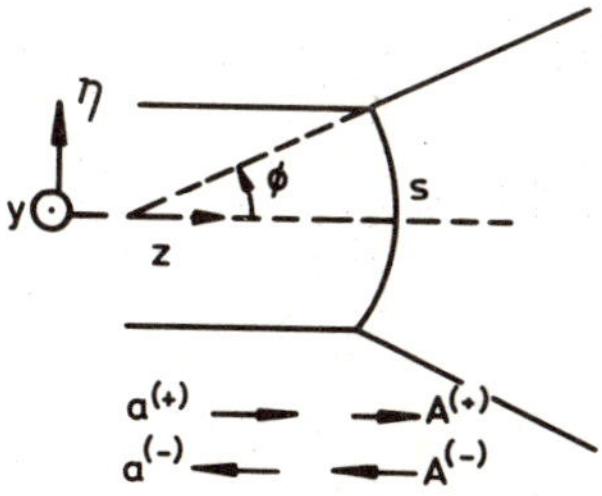

Fig. 6.10 *Abrupt transition between waveguide and horn*

The fields of the horn waves on this surface follow from eqns. 3.1.21, 3.1.18 and 3.1.32 and under present conditions appear as

$$E_y = \hat{e}_y(A_3^{(+)} + A_3^{(-)})$$

$$H_\eta = \hat{e}_y(-Y_3^{(+)}A_3^{(+)} + Y_3^{(-)}A_3^{(-)})$$

$$\hat{e}_y = -\cos\left(\frac{3\pi}{2}\eta\right)\frac{1}{\sqrt{ab}}\sqrt{\frac{2}{Y_3^T}}$$

Matching the one field representation to the other leads to

$$\lim_{\phi\to 0} Y_3^{(+)} = -\lim_{\phi\to 0} Y_3^{(-)} = \frac{\gamma_3}{j\omega\mu}$$

$$\begin{bmatrix} a_3^{(-)} \\ A_3^{(+)} \end{bmatrix} = \begin{bmatrix} 0 & 0 \\ \dfrac{2}{\sqrt{Y_3^T}} & -1 \end{bmatrix}\begin{bmatrix} a_3^{(+)} \\ A_3^{(-)} \end{bmatrix}$$

and yields

$$R(0) = -1 \tag{6.5.15b}$$

To integrate eqns. 6.5.14 we need to know the coefficients χ. By comparing eqns. 6.5.2 with eqns. 3.1.35 and by taking eqns. 3.1.28 and 3.1.3 into consideration we obtain

$$\chi_{nn}^{(++)} = -\chi_{nn}^{(--)} = -j\left[k\,\mathrm{Re}\,(y_n^{(+)}) - \phi'\left\{-\left(\frac{1}{3}+\frac{1}{n^2\pi^2}\right)ka\left(\frac{y_n^{(T)}}{2}-\phi\frac{\partial\, Im(y_n^{(+)})}{y_n^{(T)}\partial ka}\right) - \frac{\partial\, Im(y_n^{(+)})}{y_n^{(T)}\partial\phi} - \frac{1}{y_n^{(T)}2ka}\right\}\right]$$

$$\chi_{33}^{(-+)} = (\chi_{33}^{(+-)})^* = \frac{\phi'}{y_3^T}\left\{-\left(\frac{1}{3}+\frac{1}{9\pi^2}\right)ka\phi\frac{\partial y_3^{(+)}}{\partial ka}+\frac{\partial y_3^{(+)}}{\partial\phi}+\frac{j}{2ka}\right\}$$

$$\chi_{31}^{(++)}_{(-+)} = \frac{\phi'}{\sqrt{y_1^{(T)}y_3^{(T)}}}\frac{3}{8}\left\{\frac{2}{\pi^2}jka\left[\frac{y_1^{(+)}y_3^{(T)}}{2}(1\pm 1)\pm j\phi\frac{\partial y_1^{(+)}}{\partial ka}\right] + \frac{3}{2\pi^2}\phi(y_3^{(\mp)}\pm y_1^{(+)})\mp j\frac{2}{ka}\right)\right\} \tag{6.5.16}$$

$$\phi \approx a' = da/dz; \quad \phi' = d\phi/dz$$

The first of these expressions, with $y_n^{(+)}$ from eqns. 3.1.41c, allows us to determine $\chi_{11}^{(++)}$ and thus evaluate

$$T(z) = \exp\left[\int_0^z \chi_{11}^{(++)}(z')dz'\right]$$

for any given taper contour.

Eqns. 6.5.14a and 6.5.14b, however, contain coefficients χ which have a singular character at the input end of the taper. These singularities arise at the narrow end of the taper. When we evaluate eqn. 3.1.20a for the admittance parameter Y_n^T at this end of the taper, and when we use eqns. 9.3.7 and 9.3.8 of Reference 5 in this evaluation we obtain, in accordance with Fig. 3.5 for $ka < n\pi/2$.

$$\lim_{\phi\to 0} y_n^{(T)} = \sqrt{\left(\frac{n\pi}{2ka}\right)^2 - 1}\,\exp\left[-\frac{n\pi}{\phi}\left(\mathrm{ar\,cosh}\left(\frac{n\pi}{2ka}\right) - \sqrt{1-\left(\frac{2ka}{n\pi}\right)^2}\right)\right](1+O(\phi)) = 0 \tag{6.5.17a}$$

Despite these singularities we can nevertheless arrive at a simple solution for eqns. 6.5.14*a* and 6.5.14*b*, at least for small flare angles ϕ. To this end we let

$$R(z) = -\exp[j\varphi(z)] \quad \text{with} \quad \varphi(0) = 0 \tag{6.5.18}$$

and obtain instead of eqns. 6.5.14*a*

$$\frac{d}{dz}\varphi(z) = 2[\mathrm{Im}\,[\chi_{33}^{(-+)}\exp(j\varphi(z))] + \mathrm{Im}\,(\chi_{33}^{(++)})] \tag{6.5.18a}$$

As long as

$$|\varphi| \ll \pi/2 \tag{6.5.17b}$$

the exponential on the left-hand side of eqns. 6.5.18*a* may be expanded into a power series, yielding

$$\begin{aligned}\frac{d}{dz}\varphi(z) &= 2\{\mathrm{Im}\,(\chi_{33}^{(-+)}) + \mathrm{Im}\,(\chi_{33}^{(++)}) + \varphi\,\mathrm{Re}\,(\chi_{33}^{(-+)})\}\\ &= -ky_3^T - (1 + 1/(3\pi^2))kay_3^T\phi'/3\\ &\quad + \varphi\phi'\left[\frac{\partial(\ln y_3^T)}{\partial\phi} - \frac{2}{3}(1 + 1/(3\pi^2))\,ka\phi\frac{\partial(\ln y_3^T)}{\partial ka}\right]\end{aligned} \tag{6.5.19}$$

If now we consider eqns. 6.5.17*a* in the limit $\phi \to 0$ we find that

$$\frac{\partial[\ln(y_3^T)]}{\partial ka} \sim \frac{1}{\phi}$$

as well as

$$\frac{\partial[\ln(y_3^T)]}{\partial\phi} \sim \frac{1}{\phi^2}$$

Eqn. 6.5.19 may hence be simplified by dropping the last term within the brackets on its right-hand side. For the remaining term within these brackets we consider the derivative with respect to z

$$\begin{aligned}\frac{d[\ln y_3^T]}{dz} &= \phi'\frac{\partial[\ln y_3^T]}{\partial\phi} + k\phi\frac{\partial[\ln y_3^T]}{\partial ka}\\ &\approx \phi'\frac{\partial[\ln y_3^T]}{\partial\phi}\end{aligned} \tag{6.5.20}$$

which for

$$|\phi' a| \gg \phi^2 \tag{6.5.17c}$$

may be approximated just by

$$\frac{d(\ln y_3^T)}{dz} \approx \phi' \frac{\partial(\ln y_3^T)}{\partial\phi}$$

Transforming now to the new variable

$$\hat{\varphi}(z) = \varphi(z)/y_3^T(z),$$

the solution of eqns. 6.5.19 appears as

$$\hat{\varphi}(z) = \varphi(z)/y_3^T(z)$$

$$= -k\int_0^z \left[1 + \phi'(z')\left(\frac{1}{3} + \frac{1}{9\pi^2}\right)ka\right] dz'(1 + O(\phi^2)) \tag{6.5.21a}$$

When corresponding approximations are made in eqns. 6.5.14*b* they lead to

$$\frac{dt(z)}{dz} = \phi' T(z)\frac{3}{8}\sqrt{\frac{y_3^T}{y_1^T}}\left(\frac{2}{\pi^2}\, jkay_1^{(+)} + \frac{\hat{\varphi}(z)}{2ka}\right) + t(z)\frac{dy_3^T}{2y_3^T dz} \tag{6.5.22}$$

which may be integrated to yield as the solution

$$t(z) = \sqrt{y_3^T(z)}\,\frac{3}{8}\int_0^z \left(\frac{2}{\pi^2} jkay_1^{(+)} + \frac{\hat{\varphi}(z')}{2ka}\right)\frac{\phi'(z')T(z')}{\sqrt{y_1^T}}\, dz'(1 + O(\phi^2)) \tag{6.5.21b}$$

For comparison we list the approximate solution for an abrupt transition from the rectangular waveguide to a horn as it obtains when field components are matched at the junction. To first order in ϕ, and with initial values according to eqns. 6.5.15*b*, this solution is given by

$$t(\phi) = \sqrt{y_3^T(\phi)}\; jka\,\frac{3}{4\pi^2}\,\frac{y_1^{(+)}}{\sqrt{y_1^T}}\;\phi(1 + O(\phi))$$

Eqns. 6.5.21*a* and *b* serve well to determine φ and t as functions of z up to those values of z for which the conditions 6.5.17*a*, *b*, *c* are still satisfied. For larger values of z we need to integrate eqns. 6.5.14*a* and *b* directly. In this process, and as soon as we progress to values of z for which

$$\left|\frac{d\varphi}{dz}\right| \gg |\chi_{33}^{(++)} - \chi_{11}^{(++)}|$$

is satisfield we may, however, start to disregard those terms in eqns. 6.5.14*b* which are proportional to R. For more accurate results we should not just drop these terms abruptly from a certain z-value on, but rather replace $R(z)$ by $R(z)\exp(-c\varphi^2)$, with the coefficient c small enough to still account adequately for these terms.

The procedure which we just described was adopted to obtain results which are plotted as curve 4 in Fig. 6.8. Values of $y_\xi^{(+)}$ which are needed for this numerical solution were obtained from the differential eqns. 3.1.40 according to

$$(y_n^{(+)})^2 = 1 - \left(\frac{n\pi}{2ka}\frac{\sin\phi}{\phi}\right)^2 - j\frac{\sin\phi}{ka}\frac{\partial(kay_n^{(+)})}{\partial(ka)} \qquad (6.5.23a)$$

They were computed by keeping ϕ at a constant value and integrating over ka. The initial condition which in this case was used for ka at

$$ka = \frac{n\pi}{2}\frac{\sin\phi}{\phi} \qquad (6.5.23b)$$

was the solution

$$y_n^{(+)} = 0{\cdot}79014(-j\phi/n)^{1/3}(1 + 0{\cdot}0045(\phi/n)^2), \qquad (6.5.23c)$$

which follows from eqn. 3.1.20*a* for $\nu = k\rho$ and eqns. 9.3.31–9.3.34 of Reference 5.

Fig. 6.8 shows results from this evaluation in the form of curve 4. This curve demonstrates, by way of example, that the results of the previous Section may obviously be generalised to also include interaction with spurious modes which are not any more far above cutoff, and even with those modes which are at or below cutoff.

Waveguide tapers with low spurious mode conversion over wide frequency bands must be designed such that they have, at the upper limit f_0 of the band of operating frequencies, a coupling distribution function $K(\xi)$ with wide-band characteristics according to Section 5.2. Since at frequencies $f < f_0$ the condition 6.2.3*a* is in general not satisfied, the coupling distribution function $K(\xi)$ will deviate from the nominal distribution. This deviation will be more pronounced for operating frequencies that are substantially smaller than f_0. Experience shows, however, that such a deviation only shifts the zeros of the mode conversion characteristic, or even eliminates some of these zeros, but it hardly changes the maximum values of mode conversion as a function of frequency.

The typical example of curve 4 in Fig. 6.8 demonstrates, moreover, that the mode conversion characteristic approximately retains these same maximum values, when the operating frequency is so low, that the spurious mode goes below cutoff in the more narrow region of the waveguide taper. Under these conditions, even though the coupling distribution function cannot directly be calculated any more for such low operating frequencies, the mode conversion retains nevertheless its low values. The design procedure of Table 6.1 together with its refinements according to eqn. 6.3.21 or eqns. 6.4.7 and 6.4.10, respectively, may therefore be applied under more general conditions than the somewhat restrictive assumptions for which they were derived.

The formulas of the present Section for the generation of spurious modes which are below cutoff in the narrow region of the waveguide taper may hence also be applied to design suitable taper contours in which critically coupled spurious modes are at or even below cutoff. This application makes it necessary to simplify the mode conversion formulas somewhat by way of approximations, to arrive at a definite design procedure. It leads, however, to transition contours which curve rather strongly near the narrow end of the taper. We will therefore not pursue this design procedure any further here and instead derive a transition design in the next Section, that avoids such sharp curvatures of the transition contour, at the expense of a somewhat longer transition.

6.6 Horn transitions

All waveguide transitions which we considered in previous sections were between cylindrical waveguides. The different waveguides to which these transitions joined at both ends were uniform with a cross-section independent of the longitudinal co-ordinate. As a generalisation of these previous considerations we will now discuss gradual transitions between uniform horns. As before, we call a waveguide horn uniform when it changes its cross-sectional dimensions along the axis of the horn at a constant rate. Gradual transitions between such uniform horns constitute nonuniform horns which at both ends join to uniform horns without any discontinuity of horn slope, that is with a continuous function of the longitudinal co-ordinate for their rate of change of cross-sectional dimensions.

Fig. 6.11 shows a longitudinal section through such a gradual horn transition between $z = 0$ and $z = L$ with its contour given by $a(z)$. The initial and end values for $a(z)$ and its rate of change da/dz are determined by the uniform horns at both ends. They are at

$$z = 0: \quad a = a_1, \quad \frac{da}{dz} = a_1' \qquad (6.6.1a)$$

and at

$$z = L: \quad a = a_2, \quad \frac{da}{dz} = a_2' \tag{6.6.1b}$$

Such gradual transitions between uniform horns occur in feed systems for antennas, where one uniform horn of flare angle 2 arctan $(a_1') \neq 0$ must be connected to another uniform horn of a different flare angle 2 arctan (a_2').

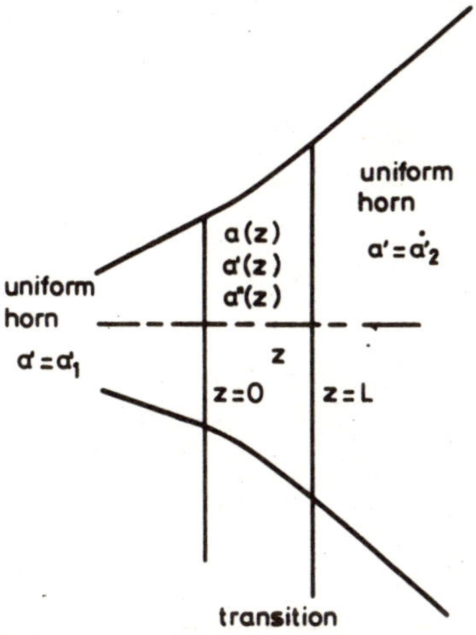

Fig. 6.11 *Nonuniform horn as transition between two uniform horns*

Sometimes it is also appropriate to compose a transition between two cylindrical waveguides of two horn transitions one from the narrow waveguide to a uniform horn and the other from this uniform horn to a wider waveguide. Such a combination of two transitions and a uniform horn in between serves particularly well, when two waveguides are to be joined, which differ widely in size. The first transition transforms a normal mode of the cylindrical input waveguide to the corresponding horn mode; the uniform horn transmits the horn mode to its wide end, and the transition there transforms the horn mode back to the normal mode of the wide cylindrical waveguide.

Fig. 6.12 shows a special case for such a combination where the two waveguide-to-horn transitions join directly, without any uniform horn in between. Although this transition does not differ in principle from those which we have treated before, we may still gain certain advantages when we consider it as composed of two waveguide-to-horn transitions.

Quite often, in practical applications, the cross-sectional dimension

a_0 of the more narrow waveguide is small enough for only the wanted mode to be above cutoff, but all other modes, which interact with it due to the nonuniformity of the transition, to still be below cutoff with evanescent fields. If we design such an over-all transition between cylindrical waveguides to perform well only with respect to unwanted mode suppression in its wide regions, and choose a contour with a favourable coupling distribution according to Section 5.2, then the most critical frequency for this design will be at the upper limit of the operating band. But in the narrow region of the transition, where even at the upper limit of the band of operating frequencies none of the unwanted modes is above cutoff, and hence not interacting significantly with the wanted mode, the transition will change its slope da/dz rather suddenly, with a sharp curvature of the contour function $a(z)$.

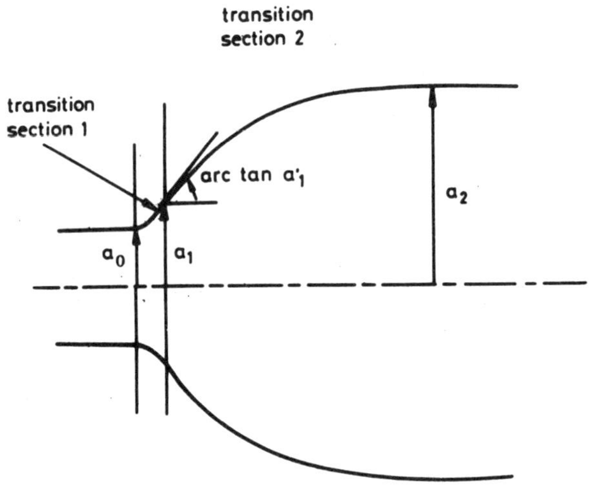

Fig. 6.12 *Transition between cylindrical waveguides composed of two waveguide-to-horn transitions*

At this nearly abrupt transition from a cylindrical waveguide to a horn of substantial flare angle a certain amount of the power which is incident in the wanted mode will be reflected. This reflection is strongest at the lower limit of the band of operating frequencies and may, in fact, exceed a specified value. Such reflections in the waveguide mode of operation are most conveniently avoided even for wide frequency bands by smoothing the contour at the transition from the narrow waveguide to the horn region. In such a case it appears appropriate to compose the complete transition of three sections, two waveguide-to-horn transitions and a uniform horn in between.

In designing the first of these transition sections from the narrow waveguide to the uniform horn we may neglect any interaction of the wanted mode with unwanted modes and consider only reflection of the wanted mode. In the wider sections of the transition, on the other hand, any such reflection into the wanted mode may be neglected, and only the interaction with forward travelling unwanted modes be considered. In most practical cases it is really not too difficult to examine by inspection whether these approximations are allowed. Both waveguide-to-horn transitions may then be designed for best performance over the specified band of operating frequencies, one independently from the other.

A quite general method to design horn transitions has already been presented in Section 6.4. Instead of eqns. 6.4.11 as boundary conditions we only need introduce the more general conditions of eqns. 6.6.1. We must, however, be aware that the inequality $|C_1| \gg |C_2|$ for the coefficients of eqn. 6.4.10 is under normal conditions only satisfied for transitions between cylindrical waveguides, but not for horn transitions. This will change the procedure by which we solve eqn. 6.4.10.

Before we discuss this solution, let us first find the local normal modes on which the representation of Chapter 6.4 is based and which must be known, to set up coupled wave equations for horn transitions. For round waveguide transitions in the form of nonuniform conical horns, as well as for rectangular waveguide transitions in form of sector horns, we found the local normal modes in form of the horn waves in Chapter 3.

For horn transitions with other cross-sections such horn waves are not readily available. Under certain assumptions however we may again resort to the procedure of Section 6.3 and obtain the coupled wave equations for local normal modes of the horn from the coupled wave equations for the normal modes of the local cylindrical waveguide by quasidiagonalization. To this end we need to modify the algorithm of Section 6.3 somewhat, and therefore summarise this procedure here.

The coupled wave equations for normal modes of the locally cylindrical waveguide are written in the form

$$\frac{d}{dz}A = [-\boldsymbol{\gamma} + a'\mathbf{X}]A \qquad (6.6.2)$$

where

A = the vector of normal mode amplitudes

z = the axial co-ordinate

$\boldsymbol{\gamma}$ = the diagonal matrix of propagation constants

$a'\mathbf{X}$ = the coupling matrix with $a' = da/dz$

The elements of $\boldsymbol{\gamma}$ and $\boldsymbol{\chi}$ depend for any given frequency only on the local dimension $a(z)$ and hence on z. Eqn. 6.6.2 may include not only forward but also backward travelling modes. For the sake of simplicity we let this system of equations contain only a finite number of normal mode amplitudes.

We now denote by **B** the amplitude vector of local normal modes of the horn and relate it to **A** by

$$A = [\mathbf{1} + \mathbf{w}]\, B \tag{6.6.3}$$

1 denotes again the unit matrix while **w** represents a quadratic matrix with its elements following from

$$\frac{d}{dz}\mathbf{w} = a''\frac{\partial}{\partial a'}\mathbf{w} \tag{6.6.4}$$

where

$$a'' = d^2a/dz^2.$$

If we substitute for A from eqn. 6.6.3 into eqn. 6.6.2 and utilise eqn. 6.6.4 we obtain

$$\frac{d}{dz}B = \left\{-[\boldsymbol{\gamma}+\boldsymbol{\epsilon}] - a''[\mathbf{1}+\mathbf{w}]^{-1}\cdot\frac{\partial}{\partial a'}\mathbf{w}\right\}B \tag{6.6.5}$$

where the transformation matrix $\mathbf{1}+\mathbf{w}$ and the diagonal matrix $\boldsymbol{\gamma}+\boldsymbol{\epsilon}$ with its diagonal and only elements $\gamma_i + \epsilon_i$, must satisfy the following condition:

$$[\mathbf{1}+\mathbf{w}]\,[\boldsymbol{\gamma}+\boldsymbol{\epsilon}] = [\boldsymbol{\gamma}-a'\boldsymbol{\chi}]\,[\mathbf{1}+\mathbf{w}] + a'\frac{\partial}{\partial a}\mathbf{w} \tag{6.6.6}$$

This matrix equation has as a typical element equation for an off-diagonal position mn with $m \neq n$

$$w_{mn}[\gamma_n - \gamma_m + \epsilon_n] = -a'\chi_{mn} - a'\cdot\sum_i \chi_{mi}w_{in}$$
$$+ a'\frac{\partial}{\partial a}w_{mn} \tag{6.6.7a}$$

In a diagonal position nn the element equation appears as

$$\epsilon_n(1 + w_{nn}) = -a'\chi_{nn} - a'\sum_i \chi_{ni}w_{in} + a'\frac{\partial}{\partial a}w_{nn} \tag{6.6.7b}$$

In addition to these equations there are relations for the elements of **w**, which follow from conservation of power. A normal mode i of the local cylindrical waveguide, which at a particular location of a lossless transition is above cutoff and propagates there, carries with its forward and backward travelling components the net power

$$P_i = A_i A_i^* - A_{(-i)} A_{(-i)}^*$$

Let us assume now that all components of the vector A are amplitudes of propagating modes. The power, which all these modes together transmit, is then given by

$$P = A^T \boldsymbol{\delta} A^* \tag{6.6.8a}$$

Here A^T denotes the transpose of A and A^* its complex conjugate. $\boldsymbol{\delta}$ is a diagonal matrix with its diagonal and only elements defined by

$$\delta_{nn} = \delta_n = 1 \quad \text{for} \quad n > 0$$

$$\delta_{nn} = \delta_n = -1 \quad \text{for} \quad n < 0 \tag{6.6.9}$$

In view of eqn. 6.6.8*a* let us also normalise the amplitudes of the local normal modes of the horn such that the net power, which together they transmit, is given by

$$P = B^T \boldsymbol{\delta} B^* \tag{6.6.8b}$$

By substituting from eqn. 6.6.3 into 6.6.8*a* we find that this requires the transformation matrix $\mathbf{1} + \mathbf{w}$ to obey the following equation:

$$[\mathbf{1} + \mathbf{w}]^T \boldsymbol{\delta} [\mathbf{1} + \mathbf{w}^*] = \boldsymbol{\delta} \tag{6.6.6b}$$

We evaluate its matrix product and obtain as further conditions for the elements w_{mn} of $\mathbf{w}$

$$w_{nm}\delta_n + \delta_m w_{mn}^* + \sum_i w_{im}\delta_i w_{in}^* = 0 \tag{6.6.7c}$$

To determine the elements of $\mathbf{w}$ from eqn. 6.6.7*c*, as well as from the other two eqns. 6.6.7, we try a Taylor-series expansion in terms of a' for them, as well as for the elements of the diagonal matrix $\boldsymbol{\epsilon}$. Since all the elements of $\mathbf{w}$ and of $\boldsymbol{\epsilon}$ vanish for $a' = 0$, these Taylor-series expansions have the following form:

$$w_{mn} = a' w_{mn}^{(1)} + (a')^2 w_{mn}^{(2)} + \dots \tag{6.6.10a}$$

$$\epsilon_n = a' \epsilon_n^{(1)} + (a')^2 \epsilon_n^{(2)} + \dots \tag{6.6.10b}$$

Its expansion coefficients $w_{mn}^{(k)}$ and $\epsilon_n^{(k)}$ depend for any given frequency only on the cross-sectional dimension $a(z)$ of the transition.

We introduce the Taylor-series representation of eqns. 6.6.10 into eqns. 6.6.7 and compare coefficients of equal powers of a'. The diagonal elements of **w** have imaginary components, which we are still at liberty to choose. If we let them all be equal to zero we obtain in case of $k = 1$, and for $m \neq n$

$$w_{mn}^{(1)} = \chi_{mn}/(\gamma_m - \gamma_n) \tag{6.6.11a}$$

while the diagonal elements in the case of $k = 1$ are

$$w_{nn}^{(1)} = 0 \tag{6.6.11b}$$

$$\epsilon_n^{(1)} = -\chi_{nn} \tag{6.6.11c}$$

In the case of $k > 1$ the off-diagonal elements of **w** follow as

$$w_{mn}^{(k)} = \frac{1}{\gamma_m - \gamma_n}\left[\sum_{j=1}^{k-1} \epsilon_n^{(j)} w_{mn}^{(k-j)} + \sum_i \chi_{mi} w_{in}^{(k-1)} - \frac{\partial}{\partial a} w_{mn}^{(k-1)}\right] \tag{6.6.11d}$$

while the diagonal elements for $k > 1$ are given by

$$w_{nn}^{(k)} = -\frac{1}{2}\sum_i \frac{\delta_i}{\delta_n} \sum_{j=1}^{k-1} w_{in}^{(j)}(w_{in}^{(k-j)})^* \tag{6.6.11e}$$

$$\epsilon_n^{(k)} = -\sum_{j=1}^{k-1} \epsilon_n^{(j)} w_{nn}^{(k-j)} - \sum_i \chi_{ni} w_{in}^{(k-1)} + \frac{\partial}{\partial a} w_{nn}^{(k-1)} \tag{6.6.11f}$$

These results allow us to estimate, up to what limits of a and a' the Taylor-series expansions 6.6.10 converge and how many terms of them we need to consider. We must, however, be aware that eqn. 6.6.2 takes into account only a finite number of modes and that it is difficult to make any estimate as to how higher order nonpropagating modes influence the transmission and interaction of the local normal modes in the nonuniform horn.

We will not pursue these questions any further here, and consider only a first-order approximation of eqn. 6.6.5. For this approximation we need to take into account only the first-order terms of eqns. 6.6.10. With them we obtain

$$w_{mn} = a' w_{mn}^{(1)} = a' \chi_{mn}/(\gamma_m - \gamma_n) \tag{6.6.12a}$$

while, because of eqn. 2.6.11b, we have $\chi_{nn} = 0$ for any gradual waveguide transition, and hence from eqn. 6.6.11c

$$\epsilon_n = 0 \tag{6.6.12b}$$

This first order approximation of eqn. 6.6.5, which considers coupling terms only that are all proportional to a', leads therefore to the following coupled wave equations for the amplitudes B of local normal modes of the horn

$$\frac{d}{dz}B = [-\gamma - a''\mathrm{w}^{(1)}]\,B \tag{6.6.13}$$

Let us designate the wanted mode by the index n and a typical unwanted mode by m. The output amplitude of such an unwanted mode at the end $(z = L)$ of the transition follows under present conditions from eqn. 6.1.4*c*. With the propagation constants and coupling coefficients of eqn. 6.6.13 its magnitude is given by

$$|\hat{B}_m| = \left|\int_0^L a''\mathrm{w}_{mn}^{(1)} \exp\left[-\int_z^L (\gamma_m - \gamma_n)\,dz'\right]dz\right|$$

$$= \left|\int_0^L a'' \frac{\chi_{mn}}{\gamma_m - \gamma_n} \exp\left[-\int_z^L (\gamma_m - \gamma_n)dz'\right]dz\right| \tag{6.6.14}$$

This solution for unwanted mode excitation may for small values of a' be compared with the results which in Section 6.2 were applied to design transitions between highly overmoded cylindrical waveguides. It has the same degree of accuracy as those results. For a first estimate of unwanted mode excitation this accuracy appears to be adequate.

If, for example, we consider as the main spurious mode the reflected and backward travelling component of the incident wanted mode, and evaluate eqn. 6.6.14 for an abrupt transition between a cylindrical waveguide and a uniform horn, then we obtain the same expression for the reflection coefficient as from eqns. 4.2.6 or 4.2.8, respectively.

To design horn transitions for low spurious mode excitation with the help of eqn. 6.6.14 we may apply the same method as it was developed in Section 6.4 for the quite general form of coupling integral in eqn. 6.4.6. Because of the more general conditions (eqn. 6.6.1) at both ends of the transition it is, however, difficult to decide with what initial values for the coefficients C_1 and C_2 to start the solution of eqn. 6.4.10. The choice of coupling function $K_0(\xi)$ in eqn. 6.4.10 depends likewise on these boundary conditions. Horn transitions for a wide band of operating frequencies may require coupling functions $K_0(\xi)$ that differ significantly from those in Section 5.2 for transitions between cylindrical waveguides.

References

1 UNGER, H.-G.: 'Circular waveguide taper of improved design' *Bell Syst. Tech. J.*, 1958, **37**, pp. 899–912
2 TANG, C.C.H.: 'Optimization of waveguide tapers capable of multimode propagation', *IEEE Trans.*, 1961, **MMT-11**, pp. 442–452
3 HECKEN, R.P., and ANUFF, A.: 'On the optimum design of tapered waveguide transitions', *ibid.*, 1973, **MTT-21**, pp. 374–380
4 SPORLEDER, F.: 'Waveguide transition design of improved accuracy', *AEÜ*, 1976, **30**, pp. 289–296
5 ABRAMOWITZ, M., and STEGUN, I.A.: 'Handbook of mathematical functions'. National Bureau of Standards, Applied Mathematics Series 55, March 1965

Chapter 7

Waveguide tapers with spurious mode degeneracies

We consider gradual waveguide transitions in this Section in which the wanted mode interacts with other unwanted modes, but which are to transmit the wanted mode with little mode conversion loss. In contrast to previous chapters, however, we will include in our considerations the case where, at some location along the transition, an unwanted coupled mode has the same phase constant as the wanted mode and is hence degenerate with the wanted mode at this particular location. The degeneracy is assumed to be confined to this location so that on one side of it the difference in phase constants between the wanted and this unwanted mode is of opposite sign than on the other. At the location of degeneracy the phase constants of both modes merely cross over.

In general, in such situations, the wanted mode also interacts with other modes, and we may even be confronted with the case that several of these unwanted modes become locally degenerate with the wanted mode at different points along the transition. Wave propagation and interaction is then much harder to understand and to analyse than for the transitions in previous chapters where no such local degeneracies occurred. It is also much more difficult, if not impossible, in these situations to develop a general procedure with which to optimise the contour of such transitions with respect to mode conversion loss.

To recognise clearly the problems that arise in the analysis and design of transitions with several locally degenerate spurious modes, and to attempt their solution, we concentrate here on a specific example, which is not only typical, but also of some practical significance. This example is shown in Fig. 7.1 and represents a broadband transition from the fundamental H_{10}-mode of the rectangular waveguide to the circular electric H_{01}-mode of the round waveguide, in the form of a sector transducer.

Such a transducer is expediently composed of two sections. The first section, according to Fig. 7.1*a*, transforms the H_{10}-mode of the rectangular waveguide to the H_{01}-mode of a sector waveguide. For all sector waveguides with sector angles α_1 smaller than 130° this H_{01}-mode has the lowest cutoff frequency and is hence the fundamental mode.

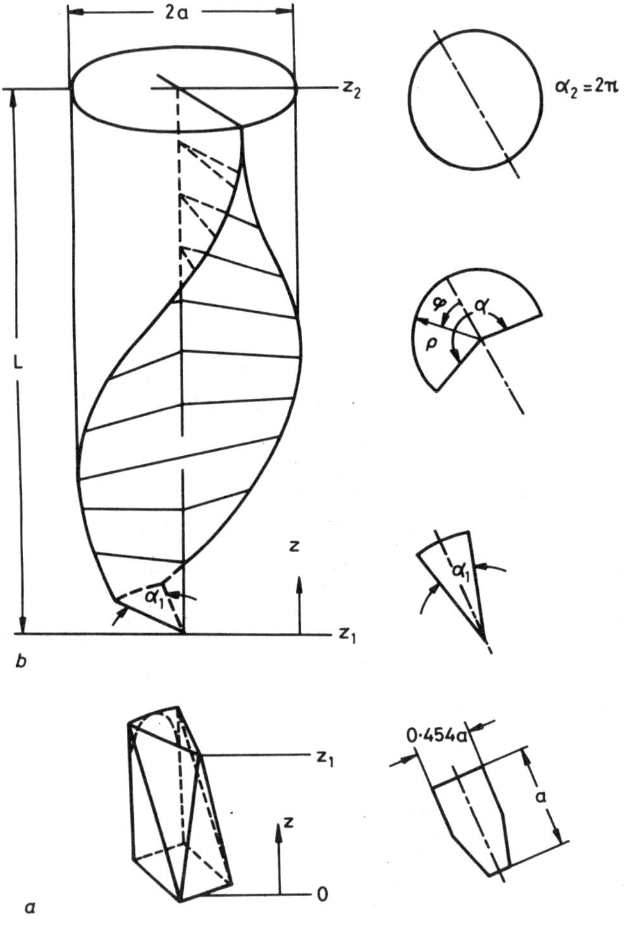

Fig. 7.1 *a Transducer from rectangular waveguide to sector waveguide*
b Sector transducer

The second section according to Fig. 7.1*b* represents the sector transducer. It transforms the H_{01}-wave of the sector waveguide to the H_{01}-wave of the round waveguide.

The problem is to find transition contours for both these transducer sections that transform the wanted mode from one waveguide to the other with a minimum of mode conversion loss, within a specified band of operating frequencies $f_1 < f < f_2$, for a given over-all length of the transducer. The power conversion to one or other of the spurious modes is in this case of not too much concern, as long as the net power loss of the wanted mode stays small enough. Mode filters can always be inserted at both ends of the transducer to absorb any of the unwanted modes without any significant loss to the wanted mode. We will also exclude heat loss due to finite conductivity of the transducer walls from our considerations.

7.1 Sector-transducer

The transition from the rectangular waveguide to the sector waveguide has, on the average, local cross-sections of much smaller area than the transition from the sector waveguide to the round waveguide. This latter section has therefore more modes above cutoff interacting with the wanted mode than the first section. The interaction in the second section leads to more mode conversion than in the first section. Unwanted modes that are locally degenerate with the wanted mode are also more liable to occur in this second section. We will therefore concentrate on this section first, and in particular on the interaction with locally-degenerate modes in the sector transducer.[1] To keep the analysis as simple as possible, and yet include all the significant characteristics, we make a number of assumptions as follows

(*a*) First, we assume the radius a of the local sector waveguide to remain constant along the sector transducer in Fig. 7.1*b*.
(*b*) Secondly, we let the smallest sector angle α_1 at the sector waveguide input of the transducer be smaller than 30°. For sector waveguide inputs of such small sector angles the spurious mode excitation does for all practical purposes not depend on the angle α_1.
(*c*) Thirdly, we let the total power which converts maximally to all unwanted modes be much lower than the power which is incident in the wanted mode. Such an assumption is well justified as long as the sector angle α changes sufficiently gradual with axial distance z along the transducer.
(*d*) Finally, we assume that the operating frequency f is large compared to the cutoff frequency f_{co} of the wanted H_{01}-mode.

The radius a of the sector transducer must of course in any case be chosen large enough for the H_{01}-mode to be above cutoff at the lower limit f_1 of the band of operating frequencies. For a wide band of operating frequencies the upper limit f_2 will then be large compared to f_{co}, and our final assumption will automatically be satisfied at the more critical frequencies near the upper band limit. We will even assume now that those unwanted modes, that become degenerate with the wanted mode at some point along the transducer are also far above cutoff, even in the narrower cross-sections towards the sector waveguide end of the transducer. Transducer regions in which this further assumption should not be satisfied have a sufficiently large phase difference between wanted and unwanted modes and enough destructive interference to not contribute very much to mode conversion.

When the wanted mode is far above cutoff it interacts mainly with modes travelling in the same direction, and hardly at all with modes in opposite direction. The coupled wave equations need therefore to account only for forward travelling waves with amplitudes $A_i(z)$. The coupling coefficients for interaction between them are proportional to the rate of change of the sector angle α. We denote the wanted mode by the mode index 0, and let m and n be mode indicies of critical unwanted modes. The coupled wave equations for their amplitudes are

$$\frac{d}{dz}\begin{bmatrix} A_0(z) \\ A_m(z) \\ A_n(z) \end{bmatrix} = \begin{bmatrix} -j\beta_0 & +\chi_{om}\dfrac{d\alpha}{dz} & +\chi_{on}\dfrac{d\alpha}{dz} \\ -\chi_{om}^*\dfrac{d\alpha}{dz} & -j\beta_m & +\chi_{mn}\dfrac{d\alpha}{dz} \\ -\chi_{on}^*\dfrac{d\alpha}{dz} & -\chi_{mn}^*\dfrac{d\alpha}{dz} & -j\beta_n \end{bmatrix}\begin{bmatrix} A_0(z) \\ A_m(z) \\ A_n(z) \end{bmatrix} \tag{7.1.1}$$

We let the amplitudes A_i be again so normalised that the power P_i, which any of these modes transmits, is given by

$$P_i = A_i A_i^* \tag{7.1.2}$$

The transducer constitutes a nonuniform waveguide which is either empty or homogeneously filled with isotropic dielectric material. Its walls may be assumed to be perfectly conducting. Formulas for propagation constants and coupling coefficients in such nonuniform waveguides were derived in Section 2.6. From them we obtain the phase constants β_i and the coupling factors χ_{ij} in eqn. 7.1.1.

The phase constants of the uniform waveguide follow from eqn. 2.6.3. Any nonuniformity, such as the changing sector angle α in the sector transducer changes this phase constant only to the extent, that k_{ci} changes. But, as we may see from eqn. 2.6.11*b*, no additional terms arise in the diagonal elements of the coupling matrix. Accordingly, we have for the phase constants in eqn. 7.1.1

$$\beta_i = \sqrt{k^2 - k_{ci}^2} \tag{7.1.3a}$$

Far above cutoff, for $k^2 \gg k_{ci}^2$, we approximate this expression by

$$\beta_i = k - k_{ci}^2/(2k). \tag{7.1.3b}$$

To evaluate the coupling factors of the sector transducer we need to know the transverse field distributions of its local normal modes. Table 2.1 lists the eigenfunctions T_i of the round waveguide. By replacing, in these eigenfunctions, the factor π in the denominator of the square root by $\alpha/2$ and the integer order $n = n_i$ of circumferential dependence and of the Bessel functions J_n by the noninteger number

$$\nu_i = 2n_i\pi/\alpha \tag{7.1.4}$$

we obtain the eigenfunctions T_i of the sector waveguide. The normal mode fields which we derive from them satisfy the boundary conditions of the sector cross-section and are normalised according to eqn. 2.6.4*d*.

The eigenfunction of the wanted H_{01}-mode follows with these substitutions as

$$T_0 = \frac{1}{x_0}\sqrt{\frac{2}{\alpha}}\,\frac{J_0(x_0\rho/a)}{J_0(x_0)} \tag{7.1.5a}$$

x_0 denotes the lowest-order zero of the derivative of the Bessel function $J_0(x)$ with respect to the argument x

$$J_0'(x_0) = 0 \tag{7.1.5b}$$

The transverse electric field of the H_{01}-mode derives from eqns. 2.6.4 and 2.6.5*a* and is given by

$$E_{to} = -n_\varphi\sqrt{Z_0}\,\frac{1}{a}\sqrt{\frac{2}{\alpha}}\,\frac{J_0'(x_0\rho/a)}{J_0(x_0)} \tag{7.1.5c}$$

The prime on the Bessel function denotes its derivative with respect to the argument. The H_{01}-mode has the cutoff wave number

$$k_{co} = x_0/a \tag{7.1.3c}$$

k_{co} does not depend on the sector angle. It remains constant along the transducer just as the phase constant β_0 according to eqn. 7.1.3*a* does.

To evaluate the coupling coefficients for the interaction of the H_{01}-mode with other modes of the sector guide owing to a change in its sector angle, we note first that in Fig. 7.1*b* the sector transducer has the longitudinal section $\varphi = 0$ as a plane of symmetry. The electric field of the H_{01}-mode is φ-directed and independent of φ. Such a mode, with an even distribution of its electric field E_φ with respect to the plane of symmetry $\varphi = 0$ of the transducer, couples only to those modes which have likewise an even distribution of E_φ with respect to $\varphi = 0$. They do not excite any modes with odd distributions of E_φ about $\varphi = 0$.

Of all the H-waves of the sector waveguide only those have an even distribution of E_φ about $\varphi = 0$ for which the normal mode function depends on φ according to $\cos(\nu_i\varphi)$. Their normal mode functions are hence given by

$$T_i = \left\{\frac{2\epsilon_{ni}\,\alpha}{x_i^2 - \nu_i^2}\right\}^{1/2} \frac{J_{\nu_i}(x_i\rho/a)}{J_{\nu_i}(x_i)} \cos(\nu_i\varphi) \tag{7.1.6a}$$

The electric field which, according to eqns. 2.6.4 and 2.6.5a, derives from these eigenfunctions has the components

$$E_{\varphi i} \sim J'_{\nu_i}(x_i\rho/a)\cos(\nu_i\varphi) \tag{7.1.6b}$$

$$E_{\rho i} \sim J_{\nu_i}(x_i\rho/a)\sin(\nu_i\varphi) \tag{7.1.6c}$$

The boundary condition $E_{\varphi i}(a) = 0$ requires

$$J'_{\nu i}(x_i) = 0 \tag{7.1.6d}$$

According to eqn. 7.1.4 the order ν_i of the Bessel function $J_{\nu i}$ depends on α. The zero points x_i of its derivative are therefore also functions of α and vary along the transducer. The cutoff wave numbers

$$k_{ci} = x_i/a \tag{7.1.3d}$$

vary likewise along the transducer.

Of the E-modes in the sector horn, only those have an even distribution of E_φ with respect to the plane of symmetry $\varphi = 0$, which derive from the eigenfunctions

$$T_i = \sqrt{\frac{2\epsilon_{ni}}{\alpha}} \cdot \frac{J_{\nu_i}(x_i\rho/a)}{x_i J_{(\nu_i-1)}(x_i)} \sin(\nu_i\varphi) \tag{7.1.7a}$$

Their transverse electric fields have the components

$$E_{\varphi i} \sim J_{\nu_i}(x_i\rho/a)\cos(\nu_i\varphi) \tag{7.1.7b}$$

$$E_{\rho i} \sim J'_{\nu_i}(x_i\rho/a)\sin(\nu_i\varphi) \tag{7.1.7c}$$

The boundary condition $E_{\varphi i}(a) = 0$ requires

$$J_{\nu i}(x_i) = 0 \tag{7.1.7d}$$

and, with ν_i depending on α, the zero points x_i as well as the cutoff wave number k_{ci} of these E-modes depend also on α.

To evaluate the coupling coefficients between the H_{01}-modes and these H- and E-modes we determine first the coupling factors C of Table 2.2. Their integrals along the boundary of the waveguide cross-section have the form

$$\oint \tan(\phi) Y(\rho, \varphi) dl_s$$

where $Y(\rho, \varphi)$ represents products of different eigenfunctions T_i and their derivatives.

For the sector transducer the integrand differs from zero only along the radii $\varphi = \pm\alpha/2$ where $\tan\phi = \pm(\rho/2)d\alpha/dz$. The integral is hence given by

$$\oint \tan(\phi) Y(\rho, \varphi) dl_s = \frac{d\alpha/2}{dz}\left[-\int_a^0 Y(\rho, \alpha/2)\rho d\rho + \int_0^a Y(\rho, -\alpha/2)\rho d\rho\right] \tag{7.1.8a}$$

and the gradients of T_i, perpendicular as well as parallel to these radii, are obtained by applying the operators

$$\frac{\partial}{\partial n_s} = \pm\frac{\partial}{\rho\partial\varphi}, \quad \frac{\partial}{\partial l_s} = \mp\frac{\partial}{\partial\rho}, \tag{7.1.8b}$$

respectively. The upper signs apply for integration along $\varphi = \alpha/2$ and the lower signs for $\varphi = -\alpha/2$.

Employing all these formulas we obtain for the coupling factors between the H_{01}- and other H-modes

$$C_{oi} = 0$$

$$\frac{1}{2}C_{io} = -\frac{1}{\alpha}\frac{d\alpha}{dz}\frac{x_0}{x_i^2 - x_0^2}\sqrt{\frac{2}{x_i^2 - \nu_i^2}}\frac{\nu_i^2}{J_0(x_0)J_{\nu_i}(x_i)}$$

$$\int_0^a J_0(x_0\rho/a)J_{\nu_i}(x_i\rho/a)\frac{d\rho}{\rho} \tag{7.1.9}$$

and for the coupling factors between the H_{01}- and E-modes

$$C_{oi} = 0$$

$$\frac{1}{2}C_{io} = -\frac{1}{\alpha}\frac{d\alpha}{dz}\sqrt{2}\,\frac{\nu_i/a}{x_i J_0(x_0)J_{(\nu_i-1)}(x_i)}$$

$$\int_0^a J_0'(x_0\rho/a)J_{\nu_i}(x_i\rho/a)d\rho \tag{7.1.10}$$

From eqn. 2.6.11 the coupling coefficients are determined by these coupling factors together with the ratio Z_i/Z_0 of wave impedances of coupled modes. Far above cutoff these wave impedances differ only little, one from another, and we may let $Z_i/Z_0 \simeq 1$. The coefficients χ_{oi} of the coupling coefficients in eqn. 7.1.1 relate under these conditions to the coupling factors C_{io} as

$$\frac{Z_i}{Z_0} = 1$$

$$\chi_{oi} = -\frac{1}{2}C_{io}\Big/\frac{d\alpha}{dz} \tag{7.1.11}$$

For the coupling factor in eqn. 7.1.9 we need to evaluate the expression

$$E = \frac{1}{x_i^2 - x_0^2}\int_0^1 J_0(x_0y)J_{\nu_i}(x_iy)dy/y \tag{7.1.12a}$$

For $x_i \to x_0$, when the two coupled modes are degenerate, this expression becomes an indeterminate form. We obtain a more suitable expression for this indeterminate form when we consider the following definite integral from eqn. 5.53 of Reference 2:

$$\int_0^1 [(x_i^2 - x_0^2)y - (\nu_i^2/y)]\,J_0(x_0y)J_{\nu_i}(x_iy)dy = x_0J_{-1}(x_0)J_{\nu_i}(x_i)$$

$$-x_iJ_0(x_0)J_{(\nu_i-1)}(x_i) + \nu_iJ_0(x_0)J_{\nu_i}(x_i) \tag{7.1.12b}$$

Under present conditions we have from eqn. 7.1.5*b*

$$J_0'(x_0) = -J_{-1}(x_0) = 0$$

and from eqn. 7.1.6*d*

$$x_iJ_{(\nu_i-1)}(x_i) - \nu_iJ_{\nu_i}(x_i) = J_{\nu_i}'(x_i) = 0$$

The definite integral of eqn. 7.1.12*b* is therefore equal to zero under these conditions, and eqn. 7.1.12*a* may be replaced by

$$E = \frac{1}{\nu_i^2}\int_0^1 J_0'(x_0y)J_{\nu_i}(x_iy)y\,dy \tag{7.1.12c}$$

When we introduce this expression instead of eqn. 7.1.12*a* into eqn. 7.1.9, we obtain for coupling between the H_{01}-mode and an unwanted H-mode the following coefficient:

$$\chi_{oi} = \frac{x_0}{\alpha}\sqrt{\frac{2}{x_i^2 - \nu_i^2}}\,\frac{1}{J_0(x_0)J_{\nu_i}(x_i)}\int_0^1 J_0(x_0y)J_{\nu_i}(x_iy)y\,dy \tag{7.1.13a}$$

For coupling between the H_{01}-mode and an unwanted E-mode, on the other hand, we obtain directly from eqns. 7.1.10 and 7.1.11 the coefficient

$$\chi_{oi} = \frac{\sqrt{2}\,\nu_i/\alpha}{x_i J_0(x_0) J_{(\nu_i-1)}(x_i)} \int_0^1 J_0'(x_0 y) J_{\nu_i}(x_i y) dy \qquad (7.1.13b)$$

All these coupling coefficients do not depend on frequency. This is a consequence of our previous assumption, that the coupled modes are far above cutoff. Likewise under this assumption the difference in phase constants may, from eqn. 7.1.3*b*, be approximated by

$$\beta_0 - \beta_i = \frac{1}{2ka^2}(x_i^2 - x_0^2) \qquad (7.1.14a)$$

and with

$$X_i = x_i^2 - x_0^2 \qquad (7.1.14b)$$

as well as with

$$\Lambda = L/(2ka^2) \qquad (7.1.14c)$$

as a wavelength parameter, be written as

$$\beta_0 - \beta_i = \Lambda X_i / L \qquad (7.1.14d)$$

Let us now find out under what conditions a coupled unwanted mode will draw a significant amount of power from the wanted mode. To interact to any extent with the wanted mode it must be coupled to it with a substantial value of $\chi_{oi} d\alpha/dz$. In addition, its phase constant β_i must not differ too much from β_0. Table 7.1 lists those unwanted modes which, in accordance with these conditions, interact most strongly with the H_{01}-mode. The mode indices in Table 7.1 correspond to those which the respective mode has after gradual transformation into the round waveguide

Table 7.1 *Modes which interact most strongly with the H_{01}-mode*

Mode	H_{01}	H_{11}	H_{21}	E_{11}	H_{31}
Index *n*	0	1	2	1	3
Index *i*	0	1	2	3	4

Fig. 7.2 shows from a numerical evaluation of eqns. 7.1.5*b*, 7.1.6*d* and 7.1.7*d* for these most critically coupled modes the normalised difference X_i in phase constants according to eqn. 7.1.14*b*. The X_i

depend only on α. For different values of α the H_{11}-, H_{21}- and E_{11}-modes become degenerate with the H_{01}-mode, and their differences in phase constants become zero and change signs at the respective points of degeneracy.

The integrals in eqns. 7.1.13 have also been numerically evaluated for these most critically coupled modes, and their coupling coefficients χ_{oi} have been plotted as a function of α in Fig. 7.3.

For the most critically coupled modes let us first examine what effect the degeneracy has on the mode conversion. To this end we

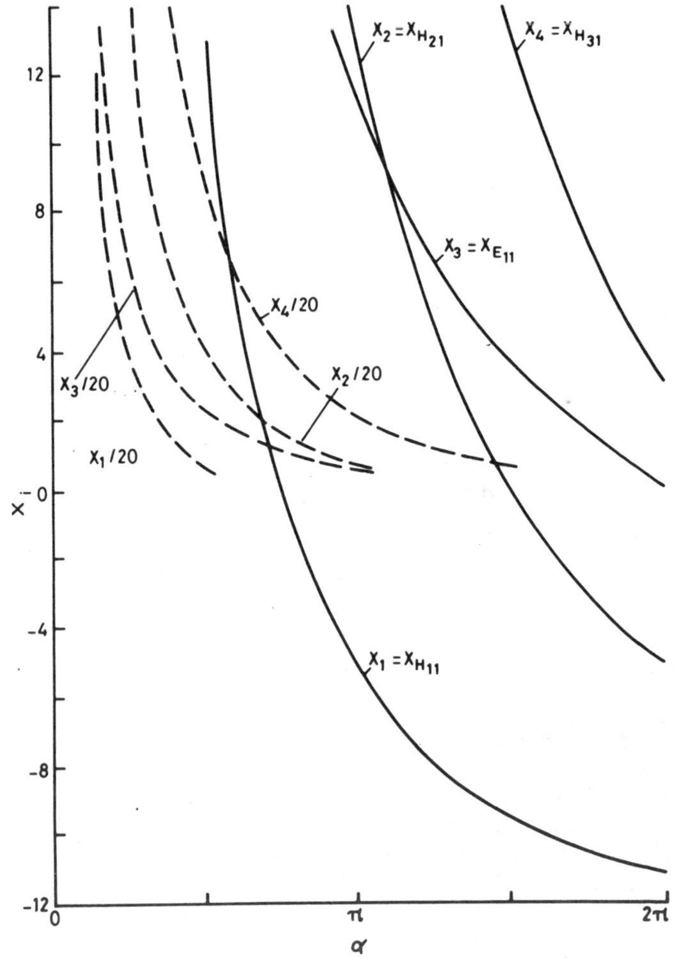

Fig. 7.2 *Normalised difference in phase constants X_i against angle α*

consider one of these modes at a time and the transducer region $z_1 < z < z_2$ that includes the point of degeneracy, where $X_i = 0$. We then assume the wanted mode to be incident at z_1 with unit power and the converted power to remain small compared with unity. Under these conditions the amplitude of the unwanted mode follows from eqn. 6.1.4*c* and, for $z = z_2$, is given by

$$\hat{A}_i = |A_i(z_2)| = \left| \int_{z_1}^{z_2} \chi_{oi} \frac{d\alpha}{dz} \exp\left[j \int_{z_0}^{z} (\beta_0 - \beta_i) dz' \right] dz \right| \qquad (7.1.15)$$

with the lower limit z_0 of integration in the phase factor remaining arbitrary.

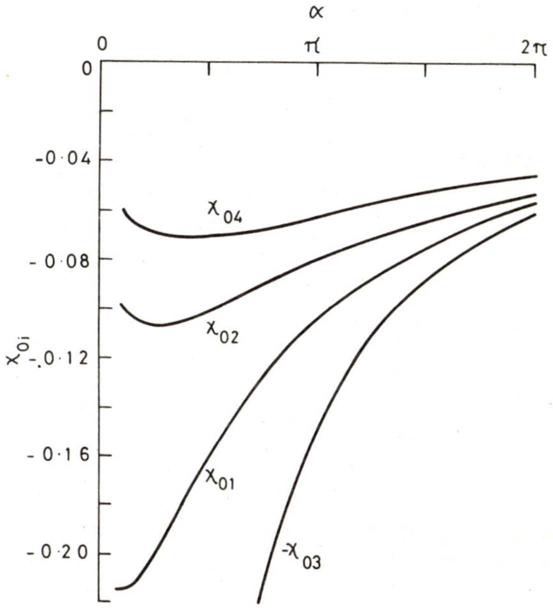

Fig. 7.3 *Normalised coupling coefficients against angle α*

If we consider interaction between the wanted mode and several such unwanted modes then the amplitude of each is given by an expression as in eqn. 7.1.15, and conservation of power leads to a transmission factor for the wanted mode with an absolute value given by

$$T = 1 - \sum_i \hat{A}_i^2 \qquad (7.1.16)$$

The sum includes all unwanted modes which the wanted mode excites to any significant extent.

To solve the integral in eqn. 7.1.15 and find coupling distributions with low over-all mode conversion we introduced

$$\xi = \int_{z_0}^{z} (\beta_0 - \beta_i) dz'$$

as the new variable of integration in Chapters 5 and 6. With it the coupling integral assumed the form

$$\int_{\xi_1}^{\xi_2} K(\xi) \exp(j\xi) d\xi$$

Such a change of variable is not feasible for unwanted modes which become degenerate with the wanted mode. The difference $\beta_0 - \beta_i$ changes its sign at the point of degeneracy and $\xi(z)$ would not have a one-to-one correspondence.

We therefore employ another change of variables which, for degeneracy and little destructive interference between the degenerate modes, leads to simple expressions for unwanted mode excitation. We assume that χ_{oi} and $d\alpha/dz$ do not change signs in the range $z_1 < z < z_2$, and let

$$y(\alpha) = \int_{z_0}^{z} |\chi_{oi}| \frac{d\alpha}{dz} dz' = \int_{\alpha_0}^{\alpha} |\chi_{oi}| d\alpha \qquad (7.1.17a)$$

y is a function of α with a one-to-one correspondence. By defining y with the absolute value of χ_{oi} as in eqn. 7.1.17a we arrive at a unified representation for all unwanted modes. The lower limit α_0 of integration, corresponding to z_0, must here be chosen within the range $\alpha_1 \leqslant \alpha \leqslant \alpha_2$ of the sector transducer. We denote the y-values at the limits of this range by

$$y_1 = y(\alpha_1) \quad \text{and} \quad y_2 = y(\alpha_2) \qquad (7.1.17b)$$

If now we abbreviate by $I(y)$ the integral

$$I(y) = \int_0^y dy' \exp\left[j \int_0^{y'} \Lambda K(y'') dy'' \right] \qquad (7.1.18a)$$

where

$$K(y) = \frac{X_i}{|\chi_{oi}|} \frac{dz/L}{d\alpha} \qquad (7.1.18b)$$

we can write the magnitude $\hat{A}_i$ of the amplitude of the unwanted mode i at $z = z_2$ from eqn. 7.1.15 as

$$\hat{A}_i = |I(y_2) - I(y_1)| \tag{7.1.19}$$

Except for a phase factor, the integral $I(y)$ in this expression represents the amplitude of the unwanted mode i as it is excited in the range between z_0, where $y = 0$, and $z(y)$.

The locus $I(y)$ in the plane of complex I-values can be traced by noting the following two simple characteristics:

The infinitesimal displacement dI of $I(y)$ when y changes by dy follows from the definition in eqn. 7.1.18*a* as

$$|I(y+dy) - I(y)| = dy \tag{7.1.20a}$$

The line which $I(y)$ traces in the complex I-plane has the curvature

$$\frac{d \arg [I(y)]}{dy} = \Lambda K(y) \tag{7.1.20b}$$

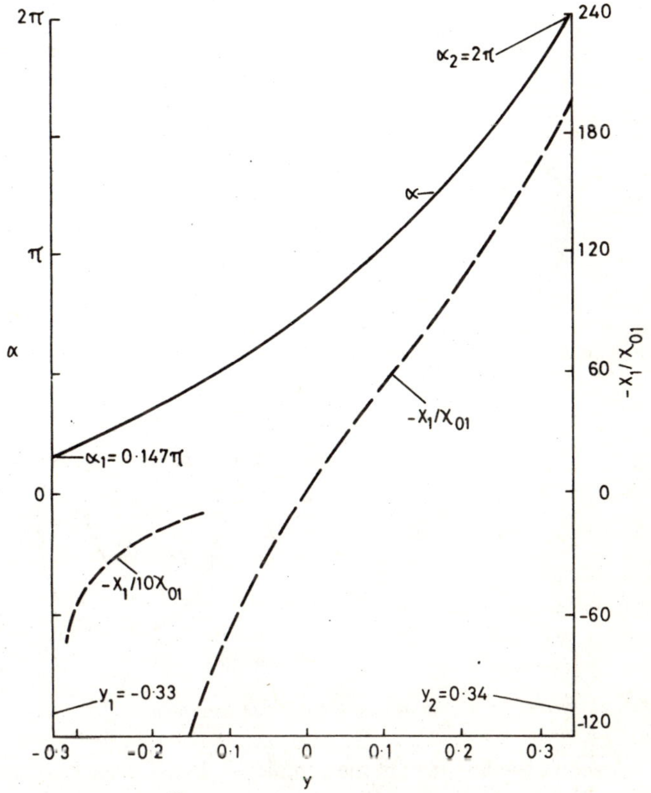

Fig. 7.4 *Angle α and value of X_1/X_{01} against y for the H_{11}-mode*

The length of this line $I(y)$ between any two points with y_1' and y_2' depends for any particular mode only on the values α_1' and α_2' at the respective locations along the transducer. $K(y)$ on the other hand depends only on the form of the transition, that is on the function $\alpha(z)$.

Fig. 7.4 shows as an example for the unwanted H_{11}-mode the parameter α as a function of y. The values y_1 and y_2 at both ends of the transducer may also be read from this diagram. In addition $X_1/|\chi_{01}|$ is plotted against y in Fig. 7.4 to determine $K(y)$ according to eqn. 7.1.18*b*. For this representation the origin of y was chosen to have $y = 0$ at the point of degeneracy, where $X_1/|\chi_{01}| = 0$. Because of this degeneracy we obtain zero curvature and a point of inflection for the line $I(y)$ in the complex I-plane at $y = 0$. As soon as we know $d(z/L)/d\alpha$ we can evaluate $K(y)$ and then, for any value of the wavelength parameter Λ, determine the curvature of the line $I(y)$.

In a linear transducer the sector angle α changes at a constant rate, so that

$$\frac{dz/L}{d\alpha} = \text{constant} = \frac{1}{\alpha_2 - \alpha_1} \tag{7.1.21}$$

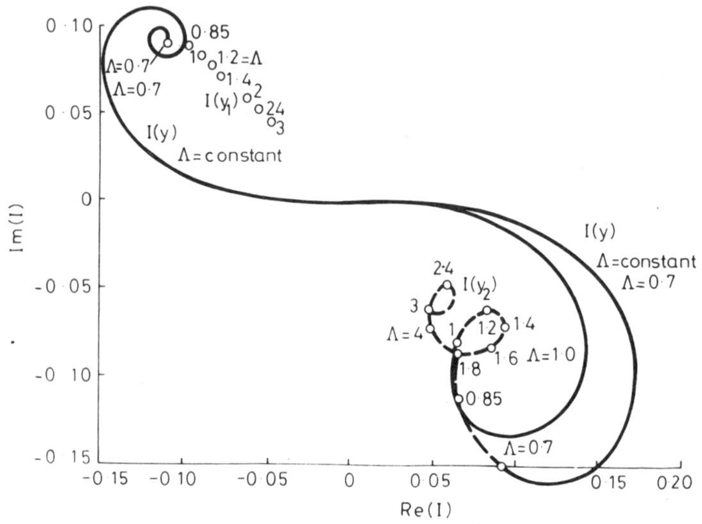

Fig. 7.5 *Loci I(y) for H_{11}-conversion in a linear transducer*

Fig. 7.5 shows the line $I(y)$ in the complex I-plane for the H_{11}-mode as the unwanted mode in such a linear transducer. The branch of $I(y)$ for positive y-values increases its curvature approximately linearly with line length. This branch follows therefore approximately a Cornu spiral, with both its real and imaginary components approximately given by Fresnel integrals. The branch of the line that $I(y)$ forms in the complex I-plane for negative values of y increases in curvature more than linearly with line length. This branch curls therefore more than the other branch. At any point along the line $I(y)$ the curvature increases also, proportional to the wavelength parameter Λ. As Λ becomes larger, both branches of the line $I(y)$ curl more quickly into their end points, and these end points move nearer to the origin of the I-plane, and closer together. In this process their spacing and with it the amplitude $\hat{A}_i$ of the unwanted mode amplitude (eqn. 7.1.19) diminish. Fig. 7.6 shows $\hat{A}_1^2$ for the H_{11}-mode as a function of Λ.

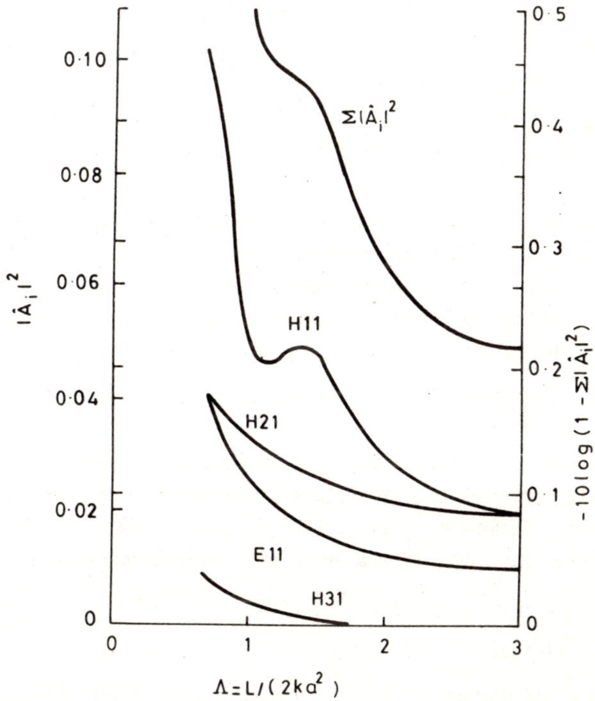

Fig. 7.6 *Mode conversion in the linear transducer*

With these general characteristics of $I(y)$ we are now able to choose a suitable transducer contour $\alpha(z)$, and with it a function $K(y)$ that keeps the over-all excitation of a particular unwanted mode low for the smallest and hence most critical value of the wavelength parameter

Λ. Such a contour and the corresponding function $K(y)$ should obey the following requirements:

(*a*) At $y = 0$ and for finite values of $d(z/L)/d\alpha$ there, $K(y)$ must also be zero.
(*b*) At larger absolute values of y the function $K(y)$ should be specified such that the branches of $I(y)$ in the complex I-plane curl with their end points as close as possible towards their origin at $I = 0$. Because of this the curvature of these branches should not change too much along them, and therefore the rate of change $d\alpha/dz$ of sector angle should be proportional to y. Curling the end points of $I(y)$ back close towards the origin of the I-plane means physically that destructive interference reconverts most of the converted power back to the wanted mode.
(*c*) To obtain effective reconversion the wavelength parameter Λ must at least be large enough, to make both branches of $I(y)$ change the direction of the tangents at their end points by at least $3\pi/2$ compared to their tangent at the origin.

$$\Lambda_{min} \left| \int_0^{y_{1,2}} K(y)dy \right| \geqslant 3\pi/2 \tag{7.1.22}$$

(*d*) Near their end points both branches of $I(y)$ must have a curvature distribution such that their mutual spacing does not vary greatly over the specified range of wavelength parameter values. We meet this requirement by letting both branches near their respective ends curl tightly around the end points.

When compared to the linear transducer (eqn. 7.1.21) a transducer with a distribution $K(y)$ according to Fig. 7.7 affords considerable reduction in mode conversion. This distribution more or less obeys the above requirements and guarantees a minimum of H_{11}-excitation for wavelenght parameter values $\Lambda > \Lambda_{min} = 1{\cdot}1$. Fig. 7.8 shows the H_{11}-power conversion $\hat{A}_1^2$ as a function of Λ. By integrating

$$\frac{K(y(\alpha))}{X_1/|\chi_{01}|} = \frac{dz/L}{d\alpha} \tag{7.1.23}$$

with the distribution of Fig. 7.7 for $K(y)$ we obtain the transducer contour in Fig. 7.9.

To improve the transducer as compared to the quasilinear transducer of eqn. 7.1.21, also for still lower values of Λ than the above specified Λ_{min} the requirement of eqn. 7.1.22 calls for the branch of $I(y)$ at negative y-values to curl less than in the above case.

At larger values of Λ_{min}, on the other hand, a judicious gradation

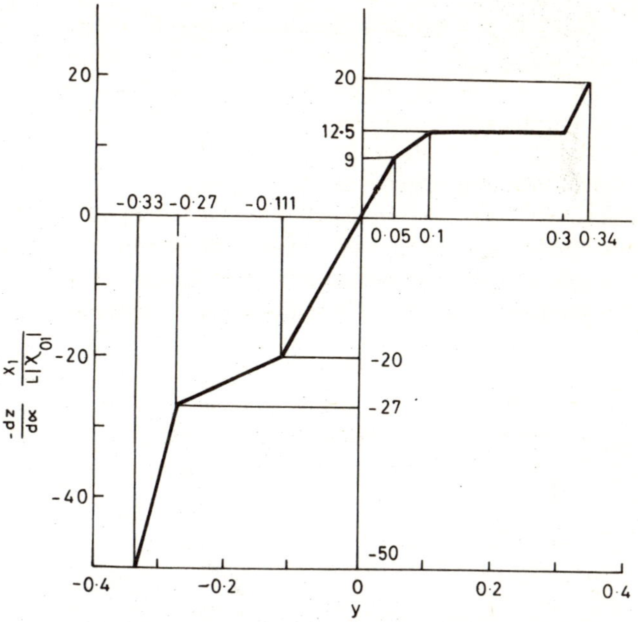

Fig. 7.7 *Curvature of $l(y)$ of an improved transducer for H_{11}-conversion*

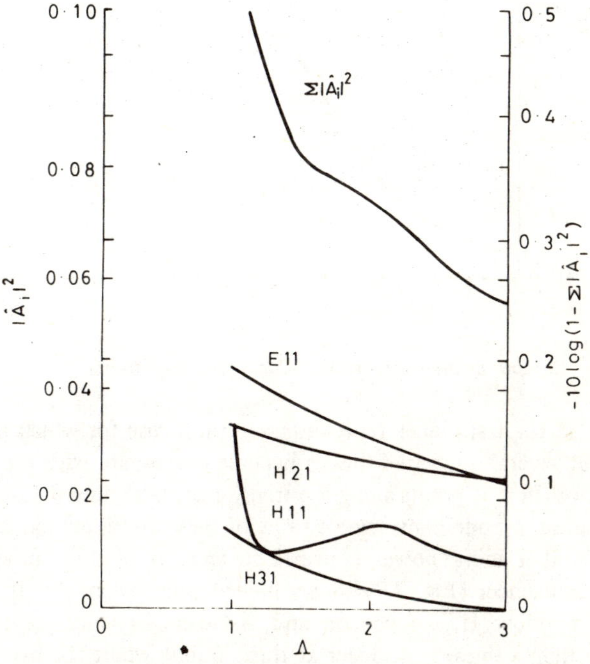

Fig. 7.8 *Mode conversion of improved transducer according to Fig. 7.7*

of $K(y)$ allows to improve the transducer to any desired extent. Near $y = 0$ both branches of $I(y)$ should start with a strong curvature, or at least rise to such strong curvature in a short distance. At the point of degeneracy for $y = 0$, and near it, the derivative $dK(y)/dy$ should hence be as large as possible in absolute value. The slope of the contour $d\alpha/dz$ should therefore be very small at this point.

The function $I(y)$ which from eqn. 7.1.19 determines the unwanted mode amplitude allows us thus to visualise unwanted mode excitation. It also reveals all significant requirements for a transducer contour that converts only a minimum of power to the respective unwanted mode.

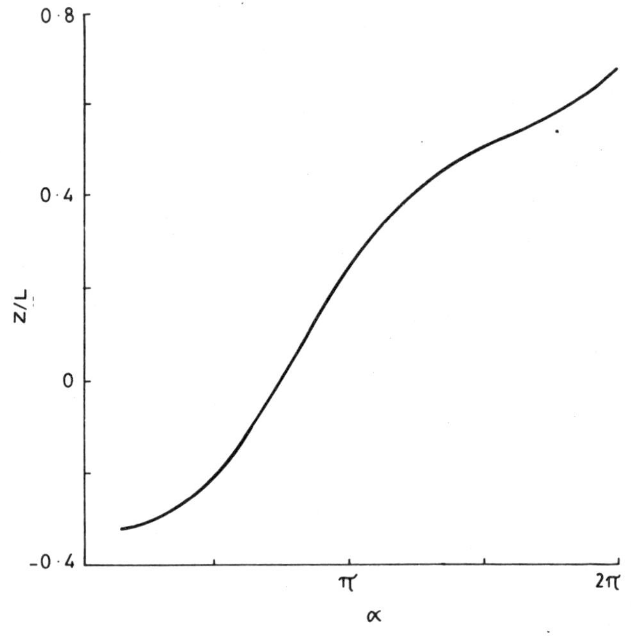

Fig. 7.9 *Contour of the improved transducer according to Fig. 7.7*

The sector transducer represents a typical case for which not only one, but several unwanted modes become degenerate with the wanted mode at different points along the transducer. Designing the transducer for minimum mode conversion to one of these unwanted modes makes it likely that more power converts to another of these modes. The sector transducer (Fig. 7.9) shows indeed such behaviour. It was designed for low H_{11}-conversion and, as a consequence, tapers more steeply than a linear transducer at those points where the E_{11}- and the H_{21}-mode pass through degeneracy with the wanted mode. As a consequence, more power converts to these modes than would convert to them in a linear transducer. The design of an improved transducer should therefore take all these modes into account simultaneously and reduce the slope $d\alpha/dz$ wherever an unwanted mode passes through degeneracy.

However, the straightforward linear transducer exhibits quite favourable conversion characteristics for the E_{11}-mode. Fig. 7.10 shows α and the coupling parameter X_i/χ_{oi} of $H_{01}-EH_{11}$ interaction as a function of y. For a linear transducer with $d(z/L)d\alpha =$ constant $K(y)$ (eqn. 7.1.18b) has very nearly the distribution which Fig. 7.7 shows for an improved design of the branch of $I(y)$ with $y > 0$. Considering, that of all unwanted modes, the E_{11}-mode has the longest line $I(y)$ in the complex I-plane and that in its point of degeneracy the absolute value of

$$\left|\frac{dK(y)}{dy}\Bigg/\left(\frac{d(z/L)}{d\alpha}\right)\right| = \left|\frac{dX_i}{d\alpha}\right| \tag{7.1.24}$$

tends to be smaller than for any of the other unwanted modes, the E_{11}-mode turns out to be the most critical unwanted mode and the hardest to suppress for wavelength parameter values in the range of $\Lambda \simeq 1$.

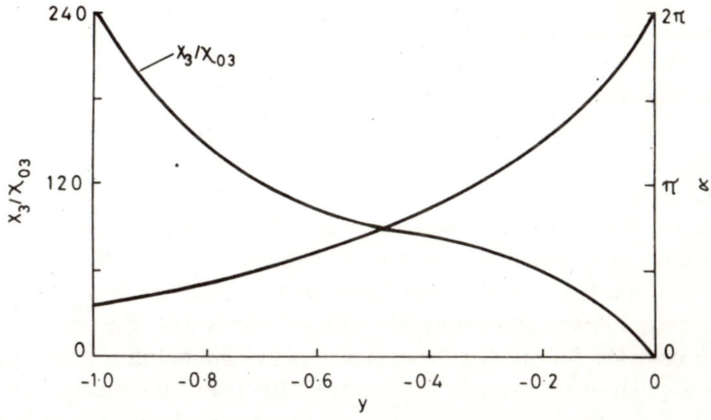

Fig. 7.10 *Angle α and value of X_3/χ_{03} against y for the E_{11}-mode*

It therefore appears quite appropriate to base the design of the transducer on the control of power conversion to the E_{11}-mode. The linear sector transducer seems to be a near-optimum design in particular when, at the upper limit f_2 of the band of operating frequencies, we require the H_{01}-mode to have less than 0·5 dB conversion loss. From

Fig. 7.6 for this specification, the wavelength parameter must be larger than $\Lambda_{min} = 1$, which together with f_2 fixes the minimum length of the transducer for a given radius a of the sector waveguide. This radius a in turn must be large enough for the H_{01}-mode to be above cutoff at the lower limit f_1 of the band of operating frequencies.

So far we have considered only the unwanted modes of Table 7.1 which the H_{01}-mode excites to any significant extent. All other unwanted modes draw only little power from the H_{01}-mode and may hence be neglected. Table 7.2 lists a number of these other unwanted modes, together with their respective values of the normalised phase difference X_i and coupling factor $|\chi_{oi}|$, both for $\alpha = \alpha_2 = 2\pi$ at the junction of the sector transducer to the round waveguide. We obtain an upper bound for the power conversion to these modes by tracing for them the locus of $I(y)$ in the complex I-plane. For the parameter y in this representation we let the lower limit of integration in eqn. 7.1.17*a* be located at $z_0 = z_2$ or $\alpha_0 = \alpha_2 = 2\pi$, so that $y = 0$ at $\alpha = 2\pi$.

Table 7.2 *Characteristics of some unwanted modes*

Mode	E_{21}	H_{41}	H_{12}	E_{31}	H_{22}	E_{41}
$\lvert\chi\rvert$	0·079	0·041	0·070	0·084	0·402	0·084
X	11·7	13·6	13·7	26·0	30·3	42·9
$\lvert\chi\rvert/X \cdot 10^3$	6·75	3·0	5·1	3·23	1·4	1·96
$\Lambda^2\lvert\hat{A}_i^2\rvert \cdot 10^3 <$	1·5	0·31	0·89	0·35	0·09	0·13

The curvature of the line $I(y)$ in the complex I-plane follows from eqn. 7.1.20*b* and may be evaluated by substituting for $K(y)$ from eqn. 7.1.18*b* with the respective values for X_i and $|\chi_{oi}|$. The unwanted modes of Table 7.2 are nowhere along the transducer degenerate with the H_{01}-mode. Their normalised phase differences X_i increase monotonically as α decreases from $\alpha = 2\pi$. Going along the transducer from $\alpha = 2\pi$ and $y = 0$ to smaller values of α the coupling parameter $K(y)$ increases likewise monotonically with increasing $|y|$. The line $I(y)$ in the complex I-plane curls ever more tightly, and eventually converges to a point that is closer to its origin $I(0)$ than the initial radius of curvature. We therefore obtain as an upper bound for the amplitude of such an unwanted mode

$$\hat{A}_i \leqslant \frac{1}{\Lambda|K_0|} = \frac{|\chi_{oi}|}{\Lambda|X_i|}\left|\frac{d\alpha}{d(z/L)}\right| \tag{7.1.25}$$

For X_i and χ_{oi} in this expression their values at $\alpha = 2\pi$ must be substituted. Table 7.2 lists in addition to these values their ratio $|\chi_{oi}|/X_i$

as well as $\Lambda^2 \hat{A}_i^2$ for the relative slope $L(d\alpha/dz) = \alpha_2 - \alpha_1 = 1{\cdot}85\pi$ of a linear transducer.

Fig. 7.11 compares calculated and measured results of the H_{01}-transmission loss in a linear sector transducer with a radius $a = 8{\cdot}5$ mm and a length $L = 280$ mm. The heat losses in the transducer wall were determined by assuming the H_{01}-mode to transmit all the incident power everywhere along the transducer. The measurements verify the theory to within the measuring accuracy of 0·1 dB.

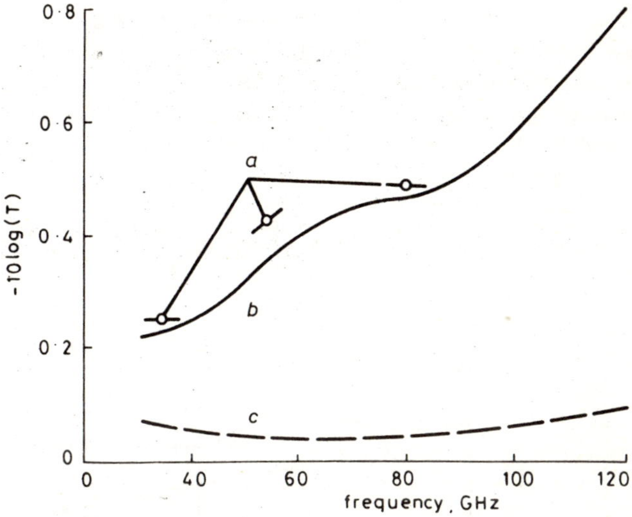

Fig. 7.11 *H_{01}-loss in a linear sector transducer*
$a = 8{\cdot}5$ mm $L = 280$ mm
a measured attenuation
b attenuation due to mode conversion (calculated)
c attenuation due to ohmic losses (calculated)

7.2 Rectangular-to-sector transducer

For the sake of completeness we will also discuss briefly the transition from the fundamental H_{10}-mode of the rectangular waveguide to the H_{01}-mode of the sector waveguide.

This transition represents a typical case for intermediate cross-sections that cannot readily be described in terms of separable co-ordinates. For an approximate analysis of fundamental mode propagation through this transducer, and its interaction with higher-order modes, we use for each mode the notation that it has upon transition into the rectangular waveguide. No higher-order mode becomes degenerate with the fundamental H_{10}-mode. Of all the unwanted modes the H_{20}-mode interacts most strongly with the H_{10}-mode.

For the field distribution of these modes and their propagation and coupling characteristics we assume here that all intermediate cross-sections according to Fig. 7.12 have a height b which is small compared to the width a, and that the angle α_1 is likewise small. The wanted H_{10}-mode will then excite electric fields of predominantly y-direction in Fig. 7.12. The normal mode fields of the H_{no}-modes are nearly independent of y, and to approximate their distributions by simple expressions, we divide the cross-section into two regions. Region 1 from $x = 0$ to $x = a(1-\Delta)$ has very nearly the form of a circular ring sector. The electric field of the H_{no}-modes in this region corresponds to the electric field of an H_{om}-mode in a coaxial waveguide, and may be expressed by the following combination of Bessel and Neumann functions of order one:

$$e_i(x) = D_1 J_1\left[\frac{x_i}{a}(x+\Delta a)\right] + D_2 N_1\left[\frac{x_i}{a}(x+\Delta a)\right] \quad (7.2.1a)$$

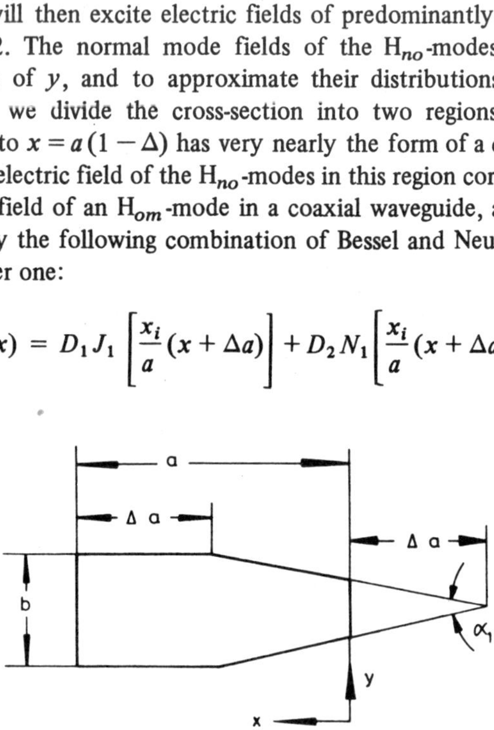

Fig. 7.12 *Intermediate cross-section of the rectangular-to-sector transducer*

Region 2 from $x = a(1-\Delta)$ to $x = a$ represents part of the cross-section of a rectangular waveguide with the electric field of its H_{no}-modes given by

$$e_i(x) = D_3 \sin\left[\frac{x_i}{a}(x-a)\right] \quad (7.2.1b)$$

We subject eqn. 7.2.1 to the boundary condition $e_i = 0$ at $x = 0$ and match both field expressions 7.2.1 as well as their derivatives $de_i(x)/dx$

at $x = a(1-\Delta)$. Matching these derivatives ensures continuity of the magnetic field component H_z. We thus obtain three homogenous linear equations for the coefficients D_1, D_2 and D_3. The condition for non-trivial solutions to these homogenous equations constitutes the characteristic equation for eigenvalues x_i of the boundary value problem.

From a more detailed investigation of this transducer we summarise here only the most important results, which, for a = constant and α_1 = constant along the transducer, have been computed numerically.

Just as for the sector transducer in eqn. 7.1.14 we define again a wavelength parameter

$$\Lambda_1 = z_1/(2ka^2)$$

for the rectangular-to-sector transducer of length z_1. The difference in phase constants between the wanted H_{10}-mode with $i = 0$ and the H_{20}-mode with $i = 1$ is related to this phase parameter by

$$\beta_0 - \beta_1 = \frac{1}{z_1} \Lambda_1 X$$

where

$$X = x_1^2 - x_0^2$$

represents again a normalised phase difference. Fig. 7.13 shows X as a function of Δ.

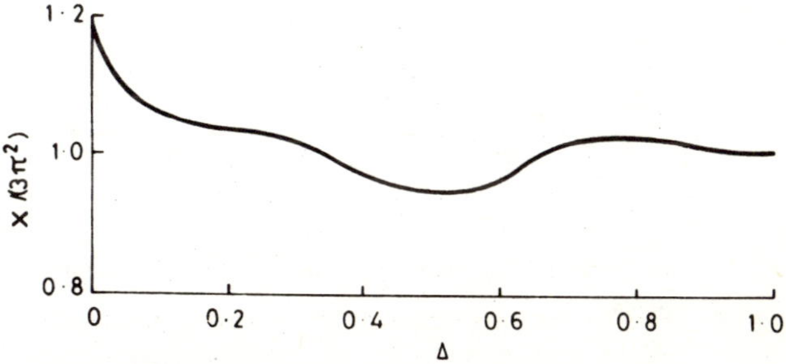

Fig. 7.13 *Normalised difference in phase constants between the H_{10}- and the H_{20}-mode in a rectangular-to-sector transducer*

Δ is the only parameter of the intermediate cross-section in Fig. 7.12 that changes along the transducer. Because of its change the local H_{20}-mode couples to the local H_{10}-mode. For the coupling factor $\kappa_{01}^{(++)}$ between forward travelling components of these modes in eqn. 2.4.1*a* we may therefore write

$$\kappa_{01} = \kappa_{01}^{(++)} = \chi \frac{d\Delta}{dz}$$

and use eqns. 2.6.11*a* and 2.6.12 to evaluate the factor χ in this expression. Sufficiently far above cutoff both coupled modes have nearly the same wave impedance, and κ_{01} as well as χ become independent of frequency. Fig. 7.14 shows χ as a function of Δ under this far-above-cutoff condition.

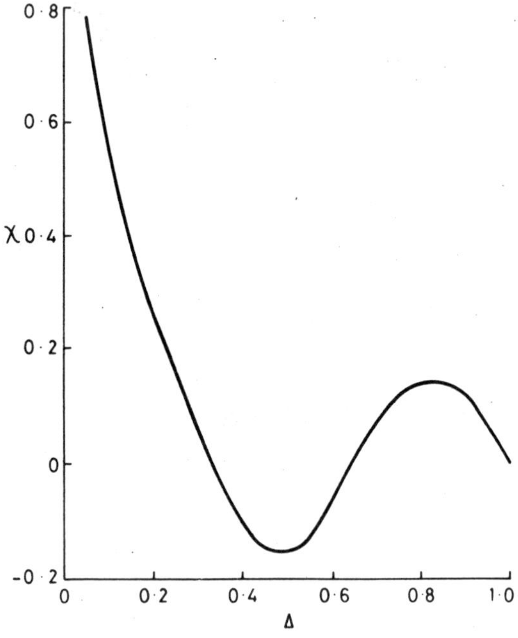

Fig. 7.14 *Coupling factor for $H_{10}-H_{20}$-interaction in a rectangular-to-sector transducer*

In the rectangular-to-sector transducer of Fig. 7.1*a* the parameter Δ changes at a constant rate. The amplitude $\hat{A}$ of the H_{20}-mode, which an incident H_{10}-mode of unit power excites in such a linear transducer, has been determined by applying eqn. 6.1.4*c* and is plotted in Fig. 7.15 as a function of the wavelength parameter Λ_1. According to this plot, mode conversion to the H_{20}-mode in the linear transducer remains at a fairly high level up to $\Lambda_1 = 0{\cdot}4$. Destructive interference and reconversion becomes effective only for $\Lambda_1 > 0{\cdot}4$. A more suitable choice of $d\Delta/dz$ as a function of z, however, leads to reconversion and less overall H_{10}-loss, already at much lower values of Λ_1.

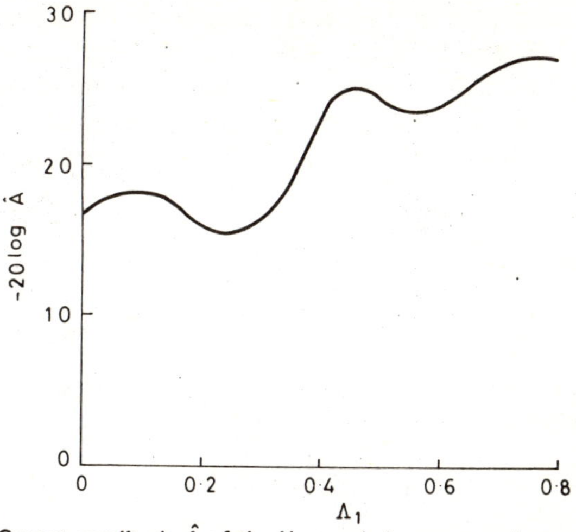

Fig. 7.15 *Output amplitude $\hat{A}$ of the H_{20}-mode in a rectangular-to-sector transducer for an H_{10}-mode incident with unit power. Transducer contour according to Fig. 7.1a*

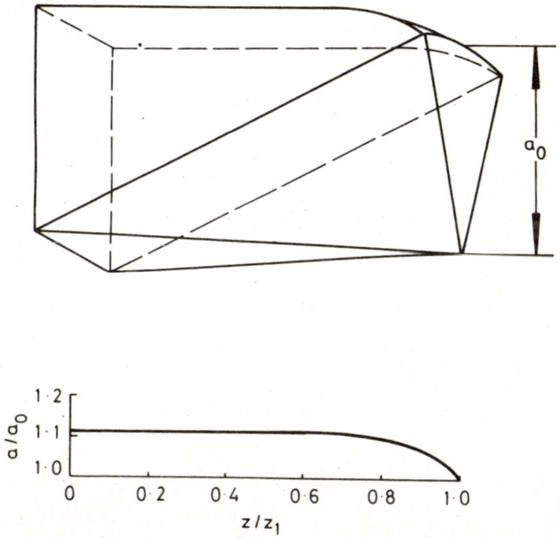

Fig. 7.16 *Improved rectangular-to-sector transducer for reduced unwanted mode excitation*

Still better performance for the rectangular-to-sector transducer is obtained when, in addition to the change of Δ, the radius a varies also along the transition. Variations of both Δ and a together, such as in Fig. 7.16, reduce the direct H_{10}–H_{20} coupling substantially. For a suitable combination of Δ and a as functions of z the power conversion to the H_{20}-mode remains smaller than −26 dB for all values of Λ_1. At small Λ_1-values such a transducer converts more of the incident H_{10}-power to the H_{30}-mode. But for $\Lambda_1 > 0{\cdot}15$ even the total H_{10}-loss due to conversion to both the H_{20}- and the H_{30}-modes is less than 0·05 dB.

References

1 SPORLEDER, F., and SAAD, L.: 'Broadband sector transducer', *AEÜ*, 1976, **30**, pp. 253–259

2 GRADSHTEYN, I.S., and RYSHIK, I.W.: 'Tables of integrals, series and products' (Academic Press, New York, 1965)

Chapter 8

Wave coupling with complete and nearly complete power transfer

The previous sections considered mode coupling in nonuniform or coupled waveguides as an undesired effect which interferes with normal mode propagation and degrades the transmission performance of these guiding structures. The taper contours of waveguide transitions and the coupling distribution along any waveguide section with mode interaction were accordingly designed to keep the over-all power conversion between a wanted mode and unwanted or spurious modes as low as possible.

In this Section we will consider coupling between modes where the aim is to achieve a certain amount of power conversion from one mode to another, possibly also complete power conversion. We will, in particular, treat two different design principles for waveguide couplers which allow the power of one mode to be completely transferred to another mode. The first of these design principles requires the coupled modes to propagate in phase synchronism with nearly the same phase constants for both modes. For complete power transfer it also requires the coupling strength and the length of the coupling section to be controlled within tight tolerances. By applying strong coupling between two modes complete power transfer takes place in a short coupling length.[1] If coupling occurs not only between these two modes in phase synchronism but also to other modes, power conversion to these other spurious modes may be kept low by destructive interference. For this purpose these modes must propagate with different phase constants so that along the coupling section they accumulate enough phase difference with respect to the two modes which are in phase synchronism. This particular design principle for coupling between modes in phase synchronism allows therefore to build couplers in overmoded structures for selective power transfer between just these two modes.

The other design principle which will be discussed in the second section of this Chapter allows the propagation and coupling parameters to vary over relatively wide ranges while still achieving complete power transfer. It only requires the two modes to be in phase synchronism at one point within the coupling region and their phase constants to cross over gradually near this point. Otherwise the phase constants may cross over at any location along the coupler and this location may shift with frequency or when other parameters are varied. Complete power transfer according to these design principles corresponds to the adiabatic fast passage for the transition in time of physical system between different states; it is therefore referred to as adiabatic coupling. Adiabatic coupling requires a large over-all length of the coupler but allows wide tolerances for the critical parameters. Because of these special features it is particularly well suited for complete power transfer between optical waveguides.

8.1 Mode-selective directional couplers

Directional couplers are 4-port arrangements consisting of two waveguides which may differ in cross-sectional form and size. For interaction these two waveguides run parallel to each other and the fields of waves in one waveguide are allowed to reach into the other waveguide.

In metallic waveguides a common wall separating both waveguides is provided with arrays of holes through which modes of one waveguide couple to modes of the other waveguide. Fig. 8.1 shows as a typical example a rectangular metallic waveguide which through an array of small holes is coupled to a metallic waveguide of circular cross-section. Optical waveguides which run parallel to each other interact with one another by way of their evanescent fields. In the latter case the coupling is continuous and, as analysed in Section 2.8, may be described by suitable coupling coefficients with which coupled wave equations are set up. The discrete coupling by small holes in the common wall of metallic waveguide couplers may likewise be described by effective coupling coefficients of the equivalent continuous coupling. Such a description fails only where the spacing of the coupling holes approaches half the guide wavelength and then only for the interaction between forward and backward travelling waves. Coupling holes which are spaced by half the guide wavelength couple a forward travelling wave quite strongly to its backward travelling counterpart. Any continuous coupling arrangement will not do this and can therefore not be equivalent to such widely-spaced holes.

Directional couplers for over-moded waveguides are usually required to transfer power only from one mode of the first waveguide to one mode of the second waveguide. A typical example is the rectangular-to-round waveguide coupler of Fig. 8.1 for the excitation of the H_{01}-mode in the round waveguide. This mode is not the fundamental mode of the round waveguide. The H_{11}-, E_{01}- and H_{21}-modes have lower cutoffs and the H_{31}-mode has a cutoff only slightly higher than that of the H_{01}-mode. If therefore the fundamental H_{10}-mode of the rectangular waveguide is to transfer its power to the H_{01}-mode of the round waveguide the coupler must be designed to convert only little of the incident H_{10}-power to the other propagating modes in the round waveguide.

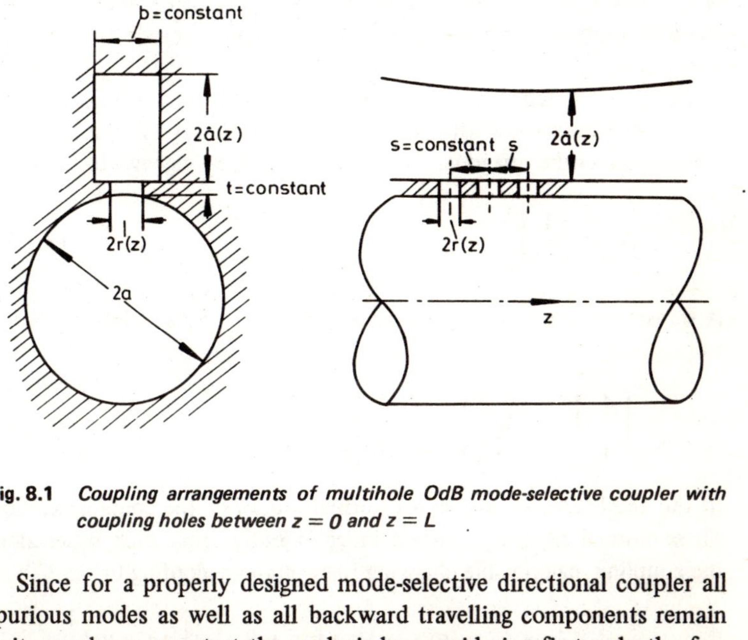

Fig. 8.1 *Coupling arrangements of multihole OdB mode-selective coupler with coupling holes between z = 0 and z = L*

Since for a properly designed mode-selective directional coupler all spurious modes as well as all backward travelling components remain quite weak we can start the analysis by considering first only the forward travelling components of the two wanted modes. With continuous but z-dependent coupling between these modes the coupled wave equations for their forward travelling wave amplitudes appear as

$$\frac{d}{dz}\begin{bmatrix} A_1 \\ A_2 \end{bmatrix} = \begin{bmatrix} -j\beta_1 + \kappa_{11} & \kappa_{12} \\ \kappa_{21} & -j\beta_2 + \kappa_{22} \end{bmatrix}\begin{bmatrix} A_1 \\ A_2 \end{bmatrix} \qquad (8.1.1a)$$

Both coupling coefficients κ_{12} and κ_{21} are equal to each other, and, for a lossless structure purely imaginary so that the effective coupling coefficient

$$\hat{\kappa} = -j\kappa_{12} = -j\kappa_{21} \tag{8.1.1b}$$

is a real quantity just as the effective phase constants

$$\hat{\beta}_1 = \beta_1 + j\kappa_{11} \tag{8.1.1c}$$

and

$$\hat{\beta}_2 = \beta_2 + j\kappa_{22} \tag{8.1.1d}$$

are real quantities for the propagating modes in the lossless waveguides.

For exact phase synchronism of both modes the effective phase constants must be equal to each other

$$\hat{\beta}_1 = \hat{\beta}_2 = \hat{\beta} \tag{8.1.2a}$$

Under this condition both waves interfere constructively with each other, and along the coupling section power transfers only from the incident mode in one waveguide to the wanted mode in the other waveguide.

For phase synchronism the coupling matrix in eqn. 8.1.1*a* becomes symmetrical, not only with respect to its main diagonal, but also with respect to its other diagonal. Such a matrix has the eigenvector matrix

$$\mathbf{V} = \frac{1}{\sqrt{2}}\begin{bmatrix} 1 & 1 \\ 1 & -1 \end{bmatrix} \tag{8.1.3a}$$

It transforms the wave amplitudes A_1 and A_2 of eqns. 8.1.1*a* according to

$$\begin{bmatrix} A_1 \\ A_2 \end{bmatrix} = \frac{1}{\sqrt{2}}\begin{bmatrix} 1 & 1 \\ 1 & -1 \end{bmatrix}\begin{bmatrix} a_1 \\ a_2 \end{bmatrix} \tag{8.1.3b}$$

to the amplitudes a_1 and a_2 of normal modes of the coupling section. These normal modes propagate independently from each other along the coupling region while changing their phases according to

$$a_1(z) = a_1(z = 0) \exp\left[-j\int_0^z \hat{\beta}\,dz'\right] \exp\left[j\int_0^z \hat{\kappa}\,dz'\right]$$

$$a_2(z) = a_2(z = 0) \exp\left[-j\int_0^z \hat{\beta}\,dz'\right] \exp\left[-j\int_0^z \hat{\kappa}\,dz'\right] \tag{8.1.3c}$$

The normal mode with amplitude a_1 belongs to the first eigenvector of $\mathbf{V}$. Its two wave amplitudes A_1 and A_2 are equal in absolute value and phase. Therefore this normal mode is referred to as the even mode of the coupling section. The other normal mode belongs to the second eigenvector of **V**. Its two wave amplitudes A_1 and A_2 are also equal in absolute value but opposite in phase. It is therefore referred to as the odd mode of the coupling section.

If we excite the coupler at $z = 0$ with unit power in the wanted mode of one of the two waveguides we have as initial conditions

$$A_1 = 1, \quad A_2 = 0 \quad \text{at} \quad z = 0 \tag{8.1.4a}$$

In this case we obtain

$$a_1 = \frac{1}{\sqrt{2}}, \quad a_2 = \frac{1}{\sqrt{2}} \quad \text{at} \quad z = 0$$

so that the even and odd mode of the coupling section are initially in phase and have equal amplitude. At the end of the coupling region at $z = L$ even and odd mode amplitudes follow from

$$a_1 \exp\left[j\int_0^L \hat{\beta} dz\right] = \frac{1}{\sqrt{2}} \exp\left[j\int_0^L \hat{\kappa} dz\right]$$

$$a_2 \exp\left[j\int_0^L \hat{\beta} dz\right] = \frac{1}{\sqrt{2}} \exp\left[-j\int_0^L \hat{\kappa} dz\right] \tag{8.1.4b}$$

and result in the following output amplitudes in both waveguides

$$|A_1| = \left|\cos\left(\int_0^L \hat{\kappa} dz\right)\right|; \quad |A_2| = \left|\sin\left(\int_0^L \hat{\kappa} dz\right)\right| \quad \text{at} \quad z = L \tag{8.1.4c}$$

For complete power transfer from one waveguide to the other the coupling coefficient integrated over the coupling section must satisfy

$$\int_0^L \hat{\kappa} dz = n\frac{\pi}{L}, \quad n = 1, 3, 5 \ldots \tag{8.1.4d}$$

In general $\hat{\kappa}$ depends on frequency. Complete power transfer occurs therefore only at the design frequency under the condition of eqn. 8.1.4*d*. In eqn. 8.1.4*c* the sine function for A_2 has, however, a fairly broad maximum so that nearly complete power transfer occurs within a correspondingly wide frequency band centered at the design frequency. By the same token any deviation of other design parameters from their nominal value will only degrade the power transfer as much as they shift the argument in the sine function for A_2 from its value for the maximum of $|A_2|$.

Much more critical for complete power transfer is the condition of expr. 8.1.2*a* for phase synchronism of both modes. Quite often it appears difficult to satisfy this condition. Typical consequences of a phase difference between both modes may be seen from the simple example of a uniform coupling section along which $\hat{\kappa}$ as well as $\hat{\beta}_1$ and $\hat{\beta}_2$ remain constant. For phase synchronism the coupling condition for complete power transfer reduces for uniform coupling to

$$\int_0^L \hat{\kappa}\,dz = \hat{\kappa}L = \pi/2$$

For $\hat{\beta}_1$ different from $\hat{\beta}_2$ we introduce their arithmetic mean value $\hat{\beta}$ and their difference $2\Delta\beta$:

$$\begin{aligned}\hat{\beta} &= (\hat{\beta}_1 + \hat{\beta}_2)/2 \\ \Delta\beta &= (\hat{\beta}_1 - \hat{\beta}_2)/2\end{aligned} \tag{8.1.5a}$$

and replace the wave amplitudes A_1 and A_2 in eqn. 8.1.1 by

$$\begin{aligned}A_1' &= A_1 \exp(-j\hat{\beta}z) \\ A_2' &= A_2 \exp(-j\hat{\beta}z)\end{aligned} \tag{8.1.5b}$$

The coupled wave equations for these new wave amplitudes appear as

$$\frac{d}{dz}\begin{bmatrix} A_1' \\ A_2' \end{bmatrix} = \begin{bmatrix} -j\Delta\beta & j\hat{\kappa} \\ j\hat{\kappa} & j\Delta\beta \end{bmatrix} \cdot \begin{bmatrix} A_1' \\ A_2' \end{bmatrix} \tag{8.1.5c}$$

To derive the general solution of these equations we diagonalise the coupling matrix by transforming with its eigenvector matrix from the wave amplitudes A_1' and A_2' to the amplitudes of normal modes of the coupling region, which propagate independently of each other within the coupling region. When we evaluate this general solution for the initial conditions according to eqn. 8.1.4*a* we obtain the z-dependence of the wave amplitudes A_1' and A_2' in the following form:

$$\begin{aligned}A_1'(z) &= \frac{\Delta\beta}{\sqrt{\hat{\kappa}^2 + \Delta\beta^2}} \exp\left(j\sqrt{\hat{\kappa}^2 + \Delta\beta^2}\,z\right) \\ &\quad + \left(1 - \frac{\Delta\beta}{\sqrt{\hat{\kappa}^2 + \Delta\beta^2}}\right) \cos\left(\sqrt{\hat{\kappa}^2 + \Delta\beta^2}\,z\right) \\ A_2'(z) &= j\frac{\hat{\kappa}}{\sqrt{\hat{\kappa}^2 + \Delta\beta^2}} \sin\left(\sqrt{\hat{\kappa}^2 + \Delta\beta^2}\,z\right)\end{aligned} \tag{8.1.6a}$$

The point of maximum power transfer is now given by

$$\sqrt{\hat{\kappa}^2 + \Delta\beta^2}\,z = \pi/2$$

but only the fraction

$$|A_2'|^2_{max} = |A_2|^2_{max} = \frac{1}{1+(\Delta\beta/\hat{\kappa})^2} \tag{8.1.6b}$$

of the power incident in mode 1 transfers maximally to mode 2. Nearly complete power transfer requires good enough phase synchronism between both modes for

$$2\Delta\beta = |\hat{\beta}_1 - \hat{\beta}_2| \ll |\hat{\kappa}| \tag{8.1.6c}$$

to be satisfied. According to this inequality the local phase difference between the two modes must be small compared to the local coupling coefficient everywhere along the coupler. However, this condition is somewhat too restrictive, in particular for couplers with nonuniform coupling distributions. For such couplers, and in phase synchronism, nearly complete power transfer requires

$$\int_0^L \hat{\kappa}\, dz \approx \pi/2$$

The tolerances for phase synchronism then need to be set only tight enough for

$$\left| \int_{z_1}^{z_2} (\hat{\beta}_1 - \hat{\beta}_2) dz \right| \ll \pi/2 \tag{8.1.7a}$$

to be satisfied for any values of z_1 and z_2 within $0 < z_1 < L$ and $0 < z_2 < L$. This condition for phase differences between the two modes is clearly less restrictive than eqn. 8.1.6*c*. To be more specific about tolerances on

$$\hat{\beta}_1 - \hat{\beta}_2 = \beta_1 - \beta_2 + j\kappa_{11} - j\kappa_{22} \tag{8.1.7b}$$

we need to know the distribution of $\hat{\kappa}(z)$ as well as how much the over-all power conversion is allowed to deviate from its nominal value. Quite generally we can state that complete power transfer, or any other specified amount of overall coupling, imposes strict tolerances on the waveguide dimensions as well as on the coupling gap or coupling holes between them. If, for reasons of mode selectivity, a certain distribution of $\hat{\kappa}(z)$ is specified these dimensions vary along the coupler and any tolerances on them are then even more difficult to hold.

We will now consider the excitation of spurious and unwanted modes in such directional couplers and see what length L and coupling distribution $\hat{\kappa}(z)$ the coupler requires to keep such spurious modes below specified levels. In a well designed directional coupler the spurious mode power will always be low compared to the power that is transferred from one wanted mode to the other. It therefore suffices to consider only one of the possibly many spurious modes at a time, and in particular pay attention to that particular spurious mode that of all the spurious modes is likely to be most strongly excited. An analysis that takes into account the interaction of the two wanted modes and just one spurious mode will reveal, what characteristics render a spurious mode the most critical one. This interaction is described by the following system of three coupled wave equations:

$$\frac{d}{dz}\begin{bmatrix} A_1 \\ A_2 \\ A_3 \end{bmatrix} = \begin{bmatrix} -j(\beta_1 + j\kappa_{11}) & \kappa_{12} & \kappa_{13} \\ \kappa_{21} & -j(\beta_2 + j\kappa_{22}) & \kappa_{23} \\ \kappa_{31} & \kappa_{32} & -j(\beta_3 + j\kappa_{33}) \end{bmatrix} \begin{bmatrix} A_1 \\ A_2 \\ A_3 \end{bmatrix} \tag{8.1.8}$$

where A_3 represents the amplitude of the forward travelling spurious mode. Exact phase synchronism between the two wanted modes allows us to substitute for their amplitudes A_1 and A_2 the amplitudes a_1 and a_2 of the even and odd normal modes, respectively, of the coupling region according to eqn. 8.1.3. This substitution transforms eqn. 8.1.8 to

$$\frac{d}{dz}\begin{bmatrix} a_1 \\ a_2 \\ A_3 \end{bmatrix} = \begin{bmatrix} -j(\hat{\beta} + j\kappa_{12}) & 0 & (\kappa_{31} + \kappa_{32})/\sqrt{2} \\ 0 & -j(\hat{\beta} - j\kappa_{12}) & (\kappa_{31} - \kappa_{32})/\sqrt{2} \\ (\kappa_{31} + \kappa_{32})/\sqrt{2} & (\kappa_{31} - \kappa_{32})/\sqrt{2} & -j(\beta_3 - j\kappa_{33}) \end{bmatrix} \begin{bmatrix} a_1 \\ a_2 \\ A_3 \end{bmatrix} \tag{8.1.9a}$$

with no direct interaction between the even and odd modes.

If at $z = 0$ only the wanted mode 1 is incident and if, with $A_1 = 1$, it has unit power there we have the following initial conditions for the mode amplitudes in eqn. 8.1.9*a*:

$$a_1 = 1/\sqrt{2}, \quad a_2 = 1/\sqrt{2}, \quad A_3 = 0 \tag{8.1.9b}$$

With only little power conversion to the spurious mode, and $|A_3| \ll 1$ as a consequence, the spurious mode amplitude along the coupler may be obtained to a good approximation by substituting for the even and odd mode amplitudes the solution in eqn. 8.1.4*b* and by then integrating the last of eqn. 8.1.9*a* to obtain A_3 at the output of the coupler

$$\hat{A}_3 = A_3(z = L) = \exp\left(-j \int_0^L (\beta_3 - j\kappa_{33})dz\right)$$

$$\cdot \left[\int_0^L \frac{\kappa_{31} + \kappa_{32}}{2} \exp\left(-j \int_0^z (\hat{\beta} + j\kappa_{12} - \beta_3 + j\kappa_{33})dz'\right)\right.$$

$$\left. + \int_0^L \frac{\kappa_{31} - \kappa_{32}}{2} \exp\left(-j \int_0^z (\hat{\beta} - j\kappa_{12} - \beta_3 + j\kappa_{33})dz'\right)\right] \quad (8.1.9c)$$

This approximation takes account only of the conversion from the wanted modes to the spurious mode, but neglects any reconversion. The expression within the brackets of eqn. 8.1.9*c* determines the absolute value of $\hat{A}_3$. To estimate its magnitude we replace the integrands in the exponent integrals of both its terms by their appropriate average values over the length of the coupler

$$\hat{\beta} + j\kappa_{12} - \beta_3 + j\kappa_{33} \approx \hat{\beta}_m + j\kappa_{12m} - \beta_{3m} + j\kappa_{33m} = \Delta\hat{\beta}^{(+)}$$

$$\hat{\beta} - j\kappa_{12} - \beta_3 + j\kappa_{33} \approx \hat{\beta}_m - j\kappa_{12m} - \beta_{3m} + j\kappa_{33m} = \Delta\hat{\beta}^{(-)} \quad (8.1.10a)$$

In view of the different distributions of the $\hat{\kappa}$- and β-values as a function of z, this represents only a very crude approximation, which we are forced to make, however, because otherwise any evaluation of $\hat{A}_3$ would be quite cumbersome. Introducing the constant average values of eqn. 8.1.10*a* into eqn. 8.1.9*c* we obtain for the output amplitude of the spurious mode

$$|\hat{A}_3| = \left|\int_0^L \frac{\kappa_{31} + \kappa_{32}}{2} \exp(-j\Delta\hat{\beta}^{(+)}z)dz\right.$$

$$\left. + \int_0^L \frac{\kappa_{31} - \kappa_{32}}{2} \exp(-j\Delta\hat{\beta}^{(-)}z)dz\right| \quad (8.1.10b)$$

When we assume the coupler to have no loss and all modes to propagate throughout the coupling region the quantities $\Delta\hat{\beta}^{(+)}$ and $\Delta\hat{\beta}^{(-)}$ will be real. Furthermore the coupling coefficients $\kappa_{31}(z)$ and $\kappa_{32}(z)$ between the wanted modes and the spurious mode are always approximately proportional to each other.

To keep spurious mode excitation low we should select a distribution for

$$\kappa_{31}(z) \sim \kappa_{32}(z)$$

as a function of z that renders both integrals in eqn. 8.1.10*b* as small as possible for all possible values of $\Delta\hat{\beta}^{(+)}$ and $\Delta\hat{\beta}^{(-)}$ within the band of operating frequencies. These integrals have a form which is quite similar to that of the coupling integral in eqn. 5.2.1*a*. In Section 5.2 coupling distributions were found that keep the coupling integral as small as possible. The same principles apply here and may be utilised to find suitable distributions of $\kappa(z)$ for minimum spurious mode excitation. We should keep in mind, however, that because of eqn. 8.1.10*a* expr. 8.1.10*b* is possibly only a very crude approximation and that hence an elaborate optimisation of $\kappa(z)$ which minimises $|\hat{A}_3|$ from the approximation 8.1.10*b* will not keep the spurious mode level as low as eqn. 8.1.10*b* might predict it.

If several spurious modes couple to the wanted modes with coupling coefficients of the same order of magnitude, that particular spurious mode will in general be excited most strongly which is closest in phase constant to the wanted modes. For this spurious mode $\Delta\hat{\beta}^{(+)}$ and $\Delta\hat{\beta}^{(-)}$ are smallest, and eqn. 8.1.10*b* leads to the largest output amplitude of all spurious modes. This main spurious mode determines the length L of the coupler which eqn. 8.1.10*b* requires to keep the spurious mode amplitude below a specified level.

For the example in Fig. 8.1 of a rectangular-to-round waveguide coupler for power transfer to the circular electric H_{01}-mode, the H_{31}-mode is closest in phase constant to the H_{01}-mode. The H_{31}-mode is hence the most critical spurious mode and determines the length of the coupler for a given mode selectivity. The other spurious modes of the round waveguide have a larger difference of phase constants to the H_{01}-phase constant and on the average stay at a lower level.

One notable exception from this rule occurs in metallic waveguide couplers when a spurious mode is at, or near, cutoff. In a lossless waveguide the propagation constant γ_i of a mode i vanishes at cutoff. Coupling coefficients for small-hole coupling between metallic waveguides follow from eqn. 2.7.11*e*. They are proportional to $1/\sqrt{\gamma_i}$ and hence go to infinity for a mode at cutoff. For such a spurious mode the condition that all spurious modes have finite coupling coefficients is no longer satisfied.

The coupling between forward and backward travelling components of the same mode is proportional to $1/\gamma_i$ and, at cutoff of this mode, has a singularity of even higher order. Let us therefore first investigate the interaction between forward and backward components of a spurious mode near cutoff, and then see how it might influence the performance of the coupler. The coupled wave equations for interaction between forward and backward components of a mode in metallic waveguides may be written as

$$\frac{d}{dz}\begin{bmatrix} A_i \\ A_{(-i)} \end{bmatrix} = \begin{bmatrix} -\gamma_i + c/\gamma_i & c/\gamma_i \\ -c/\gamma_i & \gamma_i - c/\gamma_i \end{bmatrix} \begin{bmatrix} A_i \\ A_{(-i)} \end{bmatrix} \qquad (8.1.11a)$$

According to eqn. 2.7.11*e* the factor c of the coupling coefficients is a slowly-varying positive real function of frequency. As a function of z this factor has nearly the same distribution as the coupling coefficient $\kappa_{12}(z)$ between the two wanted modes

$$c = c(z) \sim |\kappa_{12}(z)| \qquad (8.1.11b)$$

In matrix notation eqn. 8.1.11*a* appears as

$$\frac{d}{dz}A = \gamma A$$

We diagonalise this equation system by transforming the amplitude vector A to the amplitude vector a of normal modes of the coupling region

$$A = \mathbf{w}a$$

with $\mathbf{w}$ as the eignvector matrix of γ. Substituting the normal mode amplitudes a for A in eqn. 8.1.11*a* leads to

$$\frac{d}{dz}a = (\mathbf{w}^{-1}\gamma\,\mathbf{w} - \mathbf{w}^{-1}\frac{d}{dz}\mathbf{w})a \qquad (8.1.11c)$$

where

$$\mathbf{w}^{-1}\gamma\,\mathbf{w} = \begin{bmatrix} -\lambda & 0 \\ 0 & \lambda \end{bmatrix} \qquad (8.1.11d)$$

and $\pm\lambda$ represent the eigenvalues of the coupling matrix γ. These eigenvalues are given by

$$\lambda = \{\gamma_i^2 - 2c\}^{1/2} = \{-k^2 + (x_i/a)^2 - 2c\}^{1/2} \qquad (8.1.11e)$$

with k as the wave number of the material inside the metallic waveguide and $k_{ci} = x_i/a$ as the wave number at cutoff of the particular mode in the circular guide. Along any uniform section of the coupling region with $d(\mathbf{w})/dz = 0$ the eigenvalues $\pm\lambda$ are propagation constants of the normal modes of this coupling section. Owing to the term $2c$ in the radicand of eqn. 8.1.11*e* the propagation constant of normal modes in the coupling region differs from the propagation constant γ_i without any coupling. This term shifts the cutoff of the mode to a smaller wave number than the cutoff wave number k_{ci} without any coupling. If, without any coupling, a mode is already at, or even below, cutoff, it might therefore still be propagating within the coupling region. Under these conditions it is trapped within the coupling region and resonates at frequencies for which an integer number of half its wavelength fits into the coupling region.

At such trapped mode resonances the respective spurious mode interacts strongly also with both the wanted modes and degrades the mode selectivity and directivity of the coupler. Degrading effects of trapped mode resonances are actually observed only in coupling structures with very low losses, and then occur only in very narrow frequency bands at the resonance frequency. Any losses due to finite-wall conductivity or dielectric absorption will render either k_{ci} or k complex with small imaginary components. The propagation constant λ of normal modes will then likewise become complex. The normal modes suffer attenuation and their trapped mode resonances are damped enough that they no longer interact so strongly with the wanted modes.

To outline the design procedure of mode-selective directional couplers we take as a typical example the rectangular-to-round waveguide coupler in Fig. 8.1 for complete power transfer from the H_{10}-mode of the rectangular guide to the H_{01}-mode of the round guide. We specify the nominal frequency band of the rectangular waveguide from 29 to 38 GHz as the band of operating frequencies for the coupler and choose the diameter of the round waveguide at $2a = 17$ mm for its H_{01}-mode to be sufficiently above cutoff at 29 GHz.

Designating the H_{10}-mode of the rectangular guide as mode 1 and the H_{01}-mode of the round guide as mode 2 the respective phase constants are given by

$$\beta_1 = \{k^2 - (\pi/2\hat{a})^2\}^{1/2}, \ \beta_2 = \{k^2 - (x'_{01}/a)^2\}^{1/2}$$

$$\text{with} \quad x'_{01} = 3{\cdot}832$$

For phase synchronism we choose $\hat{a} = 3{\cdot}5$ mm.

The H_{31}-mode of the round waveguide is closest in its phase constant

$$\beta_3 = \{k^2 - (x'_{31}/a)^2\}^{1/2} \quad \text{with} \quad x'_{31} = 4{\cdot}20$$

to the phase constant of the H_{01}-mode. It is the most critical spurious mode, and will hence be designated as mode 3.

The H_{10}-mode has only a longitudinal magnetic field at the side wall adjacent to the round waveguide. Only this magnetic-field component couples the H_{10}-mode to the modes in the round waveguide. The reduced quantities h_1, h_2 and h_3 for the longitudinal magnetic fields of mode 1, 2 and 3 are from eqn. 2.7.11 related by

$$h_2/h_1 \approx 0{\cdot}24 \quad h_3/h_1 \approx 0{\cdot}52$$

For a thin partition wall between both waveguides κ_{32} is so much smaller than κ_{31} that it may be neglected. Since β_3 differs only little from β_1 eqn. 2.7.11*e* relates κ_{31} to κ_{12} by

$$|\kappa_{31}| = |\kappa_{12}|(h_3/h_2)$$

For efficient mode selectivity we choose the following coupling distribution:

$$|\kappa_{12}(z)| = \frac{\pi}{2L}\left[1 - \frac{5}{6}\cos\left(\frac{2\pi}{L}z\right)\right] \tag{8.1.12}$$

If the coupler is made long enough for

$$\frac{L}{2\pi}|\Delta\hat{\beta}^{(\pm)}|_{min} = \frac{L}{2\pi}(\beta_2 - \beta_3 - |\kappa_{12m}| - j\kappa_{22m} + j\kappa_{33m}) > 2{\cdot}0 \tag{8.1.13}$$

to be satisfied, it follows from eqn. 8.1.10*b* that the H_{31}-excitation is smaller than $\hat{A}_3 < 0{\cdot}02$.

The phase constant β_3 differs least from β_2 at the upper limit of the band of operating frequencies near 38 GHz. At this frequency the real value of $j\kappa_{33m}$ is slightly larger than $|\kappa_{12m}|$, while $j\kappa_{22m}$ is negligible. Thus eqn. 8.1.13 requires a coupler length of $L = 450$ mm.

The coupling holes are all to be equidistantly spaced. We avoid any excitation of backward travelling waves by spacing them with a sufficient margin below half the shortest guide wavelength in the operating band

$$s < \pi/|\beta_2|_{max}.$$

By selecting $s = 3{\cdot}5$ mm we ensure such a margin.

To facilitate fabrication we arrange for the partition wall between both guides to be sufficiently thick. The evanescent field attenuation in the coupling holes due to the finite wall thickness, changes with frequency but it balances, at least partially, the change with frequency of the hole coupling which is due to the frequency dependence of h_1 and h_2. This equalisation turns out to be near-optimum if we choose the wall thickness at $t = 0{\cdot}5$ mm.

The varying hole sizes along the coupler must now be determined to conform to the coupling distribution as specified by eqn. 8.1.12. Since the coupling strength varies with frequency we choose a suitable design frequency within the operating band such that the power conversion deviates as little as possible from the desired complete power transfer. The hole sizes for a given coupling strength follow from formulas in Section 2.7.

We choose 35 GHz as the design frequency and obtain $2r = 3{\cdot}2$ mm for the diameter of the largest hole in the centre of the coupling section. With this choice of design frequency the integrated coupling strength $\int_0^L \kappa_{12} dz$ deviates from $\pi/2$ by -5% at 38 GHz and by 20% at 30 GHz.

A rectangular-to-round waveguide coupler for complete power transfer between H_{10}- and H_{01}--mode was fabricated according to these specifications.[2] Adjustments were provided in this model for the width $2\hat{a}$ of the rectangular waveguide to tune it to phase synchronism between the H_{10}- and the H_{01}-mode. The coupling loss between these modes measured at 1 dB on the average, and over the frequency band from 29 to 39 GHz deviated from this value by only $\pm 0{\cdot}25$ dB. Inspection of the measured scattering parameters showed that heat dissipation accounts for nearly 0·5 dB of this coupling loss. The H_{31}-mode was measured to be in its maximum only -26 dB down from the input level and is hence more strongly excited than eqn. 8.1.10*b* predicts it, also in the lower part of the frequency band. This degradation appears to be partly due to the variations in phase constant which the adjustments of $\hat{a}(z)$ introduce. Trapped mode resonances of the H_{71}-mode occur slightly below the cutoff of this mode at 35 GHz. But they cause only small peaks of H_{10}-reflection at the input port of the rectangular waveguide.

8.2 Adiabatic coupling

The first Section of this Chapter presented a coupler principle for complete or nearly complete power transfer between two coupled modes that requires these two modes to propagate with equal phase constants all along the coupler, and that imposes relatively strict tolerances on the coupling integral, that is on the total amount of coupling along the coupler. When these requirements are met this coupler principle achieves total power transfer with a minimum amount of total coupling between waves. Whenever there are reasons to keep this total coupling at a minimum, for example because of considerations of coupler size, one should adopt this design principle, provided one can also meet the requirement for phase synchronism, and the tolerance on the total amount of coupling.

Here we shall discuss a coupler principle that achieves total power transfer between two coupled waves, but neither requires phase synchronism of these waves nor imposes any tolerances on the total amount of coupling between them. Since this coupler principle lends itself particularly well for complete power transfer between optical waveguides we will concentrate on this application and consider as a simple but illustrative example the two parallel dielectric slabs of Fig. 8.2*a*, which gradually change their thickness along the coupling section. One of the slabs decreases in thickness when going from left to right, while the other slab increases its thickness in the same direction. Except for this change in thickness the arrangement corresponds to the two parallel dielectric slabs in Fig. 2.17, which in Section 2.8 served as an example for coupled dielectric waveguides.

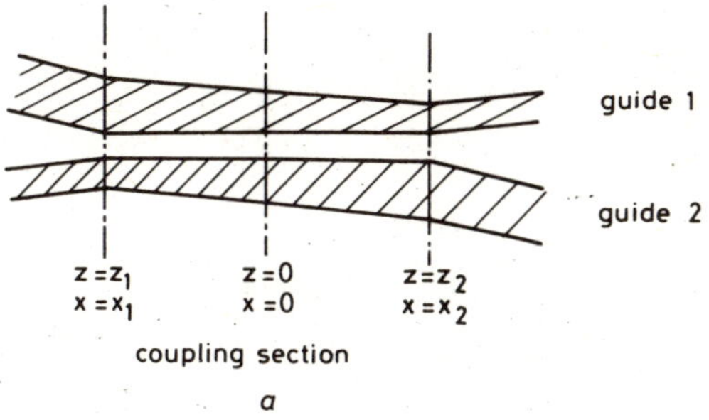

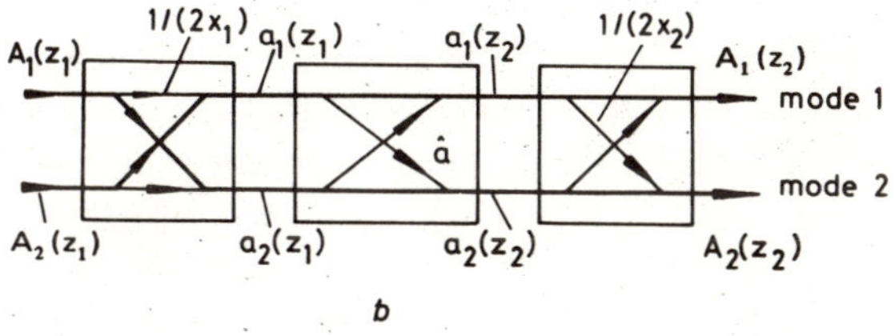

Fig. 8.2 *a* Adiabatic coupler composed of two dielectric slabs
b Flow diagram

The two dielectric slabs of Fig. 8.2*a* are assumed to be spaced closely enough in the region between $z = z_1$ and z_2 that their guided modes couple with each other. The change in slab thickness is, however, taken to be so slight that the coupling coefficients between z_1 and z_2 do not change to any significant extent. The slabs need to change their thickness only so slightly between z_1 and z_2, and in the opposite sense of one with respect to the other, that for the two modes, between which we wish to exchange all power, the difference in phase constant goes monotonically through zero about half way between z_1 and z_2.

Let us assume at first that only one mode in each slab with phase constant β_1 in slab 1 and β_2 in slab 2, respectively, interact with each other between z_1 and z_2. β_1 and β_2 designate the effective phase constants of each mode of the respective slab as modified by the presence of the other slab. The coupled wave equations for these two modes assume a more suitable form for our present analysis when, in defining the amplitudes of the two forward travelling modes, we extract a common propagation factor of the form

$$\exp\left[-j(1/2)\int_0^z(\beta_1+\beta_2)dz\right]$$

This leaves us with amplitudes A_1 and A_2 for the two forward travelling modes that change much less in phase along the coupling section. For them we obtain the following coupled wave equations:

$$\frac{d}{dz}\begin{bmatrix}A_1\\A_2\end{bmatrix}=\begin{bmatrix}-j\Delta\beta/2 & jC\\ jC^* & j\Delta\beta/2\end{bmatrix}\begin{bmatrix}A_1\\A_2\end{bmatrix} \tag{8.2.1}$$

In matrix notation they abbreviate to

$$\frac{d}{dz}A = \mathbf{X}A$$

By assuming one of the two coupling coefficients C and C^* to be the complex conjugate of the other, we restrict the analysis to lossless structures in which also the coupling mechanism entails no losses. For simplicity we assume C to be a real quantity. Under these conditions, the coupling factor C and the phase difference $\Delta\beta = \beta_1 - \beta_2$ are real functions of z. $\Delta\beta$ in particular changes monotonically with z. We place the origin of z at the crossover point of both phase constants, so that $\Delta\beta = 0$ at $z = 0$.

The mode amplitudes are assumed to be so normalised that the total power flow P in the z-direction follows from

$$P = A_1A_1^* + A_2A_2^* = A^TA^* \tag{8.2.2}$$

A^* denotes the complex conjugate of A, and the row vector A^T is the transpose of the column vector A.

The eqn. system 8.2.1 of coupled wave equations, with the phase difference $\Delta\beta$ diminishing to zero at $z = 0$, is reminiscent of the coupling between degenerate modes in Chapter 7. As opposed to the interaction between the degenerate modes in Chapter 7, however, we have here the possibility to choose the functions $C(z)$ and $\Delta\beta(z)$ independently of each other. For complete power transfer, moreover, we need sufficient mode interaction and hence a large coupling coefficient C.

For strong interaction with a large coupling coefficient the form (eqn. 8.2.1) of the coupled wave equations is awkward to deal with. We arrive at a more suitable form, when we transform from the wave amplitudes A_1 and A_2 to a new system of normal modes for the coupling region, for which the coupled wave equations exhibit much less interaction and are very nearly diagonal. Designating these normal mode amplitudes by a_1 and a_2 we let

$$A = \mathbf{w} \cdot a \quad \text{with} \quad a = \begin{bmatrix} a_1 \\ a_2 \end{bmatrix} \tag{8.2.3}$$

and $\mathbf{w}$ as the transformation matrix. If this matrix is normalised as shown

$$\mathbf{w}^T \cdot \mathbf{w}^* = \mathbf{1} \tag{8.2.4}$$

the total power in the z-direction follows as

$$P = a^T \mathbf{w}^T \mathbf{w}^* a^* = a^T a^* \tag{8.2.4a}$$

in terms of a_1 and a_2, just as it does in eqn. 8.2.2, in terms of A_1 and A_2.

To make the proper choice for the elements of the transformation matrix we substitute for A from eqn. 8.2.3 into eqn. 8.2.1

$$\frac{d}{dz} a = \left[\mathbf{w}^{-1} \mathbf{X} \mathbf{w} - \mathbf{w}^{-1} \frac{d}{dz} \mathbf{w} \right] a \tag{8.2.5}$$

We reduce the interaction between a_1 and a_2 in these equations to a minimum by requiring the first matrix product on the right-hand side of eqn. 8.2.5 to be diagonal

$$\mathbf{w}^{-1} \mathbf{X} \mathbf{w} = \quad = \begin{bmatrix} -j\lambda_1 & 0 \\ 0 & j\lambda_2 \end{bmatrix} \tag{8.2.6}$$

Under this condition only the off-diagonal terms of the second matrix product on the right-hand side of eqn. 8.2.5 cause interaction between a_1 and a_2. But these terms can be expected to be small, when the waveguides taper gently along the coupling region.

The eigenvalue problem which eqn. 8.2.6 represents, has as its solution the eigenvalues

$$\lambda_1 = \lambda_2 = C(1 + x^2)^{1/2} \tag{8.2.7}$$

where with x we abbreviate

$$x = \Delta\beta/(2C) \tag{8.2.8}$$

The eigenvector matrix **w**, when normalised according to eqn. 8.2.4*a*, follows from eqn. 8.2.6 as

$$\mathbf{w} = \frac{1}{\{1 + (\sqrt{1+x^2} - x)^2\}^{1/2}} \begin{bmatrix} 1 & (1+x^2)^{1/2} - x \\ -[(1+x^2)^{1/2} - x] & 1 \end{bmatrix} \tag{8.2.9}$$

When with these elements for **w** we evaluate $\mathbf{w}^{-1}$ and $d\mathbf{w}/dz$, we obtain with $x' = dx/dz$ the following equations from eqn. 8.2.5:

$$\frac{d}{dz}\begin{bmatrix} a_1 \\ a_2 \end{bmatrix} = \begin{bmatrix} -jC(1+x^2)^{1/2} & \dfrac{x'/2}{1+x^2} \\ -\dfrac{x'/2}{1+x^2} & jC(1+x^2)^{1/2} \end{bmatrix} \begin{bmatrix} a_1 \\ a_2 \end{bmatrix} \tag{8.2.10}$$

For an approximate solution of these equations we transform from z to

$$x = x(z) = \int_0^z x' dz$$

as the new independent variable, with values $x_1 = x(z_1)$ and $x_2 = x(z_2)$ at the input and output of the coupler, respectively. Eqn. 8.2.8 gives a one-to-one correspondence between x and z. In terms of x and with

$$\nu = 2C/x' \tag{8.2.11}$$

eqns. 8.2.10 transform to

$$\frac{d}{dx}\begin{bmatrix} a_1 \\ a_2 \end{bmatrix} = \begin{bmatrix} -j\nu(1+x^2)^{1/2}/2 & \dfrac{1/2}{1+x^2} \\ -\dfrac{1/2}{1+x^2} & j\nu(1+x^2)^{1/2}/2 \end{bmatrix} \begin{bmatrix} a_1 \\ a_2 \end{bmatrix} \tag{8.2.12}$$

When ν in these differential equations has large absolute values, the normal mode amplitudes a_1 and a_2, as functions of x, will rather rapidly turn their phase with respect to each other. Any coupling between them, as it is due to the off-diagonal terms of the matrix of eqn. 8.2.12, will then experience destructive interference and remain altogether quite small. The interaction between these normal modes is therefore only weak. To approximately evaluate the over-all coupling between them we refer to the previous coupling integral in eqn. 6.1.4.

If at $z = z_1$, corresponding to $x = x_1$, the normal modes are excited with an amplitude vector

$$a = \begin{bmatrix} 1 \\ 0 \end{bmatrix} \tag{8.2.12a}$$

then the normal mode amplitude $a_2 = \hat{a}$ at $z = z_2$, corresponding to $x = x_2$, follows from the approximate integration of eqns. 8.2.12 by way of the coupling integral eqn. 6.1.4 as:

$$\hat{a} = \int_{x_1}^{x_2} \frac{1/2}{1+x^2} \exp\left[j \int_0^x \nu(1+x^2)^{1/2} dx \right] dx \tag{8.2.13}$$

This integral representation for $\hat{a}$ disregards a phase factor which is of no significance here. It may readily be evaluated under the assumption that

$$\nu(1+x^2)^{1/2} = \nu_0 = \text{constant} \tag{8.2.14a}$$

does not change along the coupling region. Under these circumstances we may write for eqn. 8.2.13

$$\hat{a} = \left[\int_{-\infty}^{\infty} - \int_{-\infty}^{x_1} - \int_{x_2}^{\infty} \right] \frac{1/2}{1+x^2} \exp(j\nu_0 x) dx$$

in which the first integral has the solution $(\pi/2)\exp(-\nu_0)$. For the remaining two integrals we note that ordinarily, for adiabatic couplers, we have $-x_1 \gg 1$ and $x_2 \gg 1$. For these limits of integration and with partial integration of the last two integrals we find

$$\hat{a} \approx \frac{\pi}{2} \exp(-\nu_0) + \frac{1}{\nu_0 x_1^3} \exp(j\phi_1) + \frac{1}{\nu_0 x_2^3} \exp(j\phi_2) \tag{8.2.14b}$$

where ϕ_1 and ϕ_2 are phase angles, which we have no need to determine here.

The assumption (eqn. 8.2.14*a*) is somewhat arbitrary and does not reflect the situation in adiabatic couplers. For such couplers the coupling coefficient C changes only slightly, while for a constant rate of change of guide cross-section, the phase difference $\Delta\beta$ between the coupled modes changes also at an approximately constant rate. Therefore, for $C =$ constant as well as $d(\Delta\beta)/dz =$ constant we have also

$$\nu = \text{constant} \tag{8.2.15a}$$

along the coupling region for $x_1 < x < x_2$. Numerical integration of eqn. 8.2.13 under this condition leads to a solution which corresponds closely to eqn. 8.2.14*b* and which, for $-x_1 \gg 1$ and $x_2 \gg 1$, may be written in the following form:

$$\hat{a} \approx \exp(-\pi\nu/2) + \frac{1}{\nu x_1^3}\exp(j\phi_1') + \frac{1}{\nu x_2^3}\exp(j\phi_2') \tag{8.2.15b}$$

The phase angles ϕ_1' and ϕ_2' depend on x_1 and x_2 but their values are again of little significance here.

The solutions eqns. 8.2.14*b* and 8.2.15*b* have both been obtained from the coupling integral in eqn. 8.2.13. This coupling integral in turn approximates the actual solution of eqns. 8.2.12 under the initial condition (expr. 8.2.12*a*) with sufficient accuracy only, as long as $|\hat{a}| \ll 1$. The system eqn. 8.2.1 was also solved by Smith[3] under more general assumptions. His solution requires more elaborate mathematics, but from it we learn that our present approximation holds also for small values of the coupling parameter ν. It may hence not only serve to analyse adiabatic coupling with nearly complete power transfer, but also such tapered couplers which, due to limited values of ν, convert only part of the power from one mode to another at a crossover of their phase constant.

Waveguide couplers which operate on the principle of adiabatic coupling are usually required to convert power as completely as possible from one waveguide to the other. Let us therefore examine the conditions for such a complete power transfer from a particular mode of one waveguide to a mode of the other waveguide within the coupling region of Fig. 8.2 between z_1 and z_2. For this purpose we first transform by way of the eigenvector matrix $\mathbf{w}$ from the two modes of the individual waveguides to the normal modes of the tapered coupler. At $z = z_1$, corresponding to $x = x_1$, we assume again that $x_1 \ll -1$ and obtain from eqns. 8.2.3, 8.2.4 and 8.2.9 the following transformation:

$$a(z_1) = [\mathbf{w}(z_1)]^{-1} A(z_1) = \{[\mathbf{w}(z_1)]^T\}^* A(z_1)$$

$$\begin{bmatrix} a_1(z_1) \\ a_2(z_1) \end{bmatrix} \approx \begin{bmatrix} -1/(2x_1) & -\{1-1/(2x_1)^2\}^{1/2} \\ \{1-1/(2x_1)^2\}^{1/2} & -1/(2x_1) \end{bmatrix} \begin{bmatrix} A_1(z_1) \\ A_2(z_1) \end{bmatrix} \tag{8.2.16a}$$

The propagation of these normal modes from z_1 to z_2 is described by a transmission matrix $\mathbf{v}$ such that

$$a(z_2) = \mathbf{v}a(z_1) \tag{8.2.17a}$$

Since, from eqn. 8.2.15*b*, a_2 at $z = z_2$ has approximately the absolute value of $\hat{a}$,

$$|a_2(z_2)| \approx |\hat{a}|,$$

the off-diagonal elements of $\mathbf{v}$ have to the same approximation also this absolute value,

$$|v_{12}| = |v_{21}| = |\hat{a}|. \tag{8.2.17b}$$

Conservation of power in the loss-less coupling structure requires the diagonal elements of $\mathbf{v}$ to have as absolute values

$$|v_{11}| = |v_{22}| = (1 - |\hat{a}|^2)^{1/2} \tag{8.2.17c}$$

The eigenvector matrix $\mathbf{w}$ at $z = z_2$ transforms the normal mode amplitudes of the coupler at its output back to the amplitudes of the individual waveguide modes. Assuming again that $x_2 \gg 1$ eqns. 8.2.3 and eqn. 8.2.9 lead to

$$A(z_2) = \mathbf{w}(z_2)a(z_2)$$

$$\begin{bmatrix} A_1(z_2) \\ A_2(z_2) \end{bmatrix} = \begin{bmatrix} \{1 - 1/(2x_2)^2\}^{1/2} & 1/(2x_2) \\ -1/(2x_2) & \{1 - 1/(2x_2)^2\}^{1/2} \end{bmatrix} \begin{bmatrix} a_1(z_2) \\ a_2(z_2) \end{bmatrix} \tag{8.2.16b}$$

Altogether the mode amplitude vector $A(z_2)$ at the output of the coupler is related to the mode amplitude vector $A(z_1)$ at the coupler input by way of a transmission matrix $\mathbf{T}$ which represents the product of the two transformation matrices $\mathbf{w}^{-1}(z_1)$ and $\mathbf{w}(z_2)$ and the transmission matrix $\mathbf{v}$ for the normal modes of the tapered couplers. We have

$$A(z_2) = \mathbf{T}A(z_1) \tag{8.2.18}$$

where

$$\mathbf{T} = \mathbf{w}(z_2)\mathbf{v}\mathbf{w}^{-1}(z_1)$$

Fig. 8.2*b* shows a graphical representation of the transmission matrix $\mathbf{T}$ in the form of a flow diagram. It consists of three blocks, one for each matrix, of the triple product for $\mathbf{T}$. If we evaluate this product with the help of the above approximations for the elements of $\mathbf{w}$ and $\mathbf{v}$, we obtain the following expressions for the absolute values of the elements of $\mathbf{T}$:

$$|T_{11}| = |T_{22}| \approx \frac{\exp(j\varphi_1)}{2x_1} + \frac{\exp(j\varphi_2)}{2x_2} + \exp(-\pi\nu/2)\exp(j\varphi_3)$$

$$|T_{12}| = |T_{21}| = (1 - |T_{22}|^2)^{1/2} \qquad (8.2.18a)$$

The phase angles φ_1, φ_2 and φ_3 of the three terms in eqn. 8.2.18*a* that make up T_{11} and T_{22} have not been determined here. As long as x_1 and x_2 are large in absolute value compared to unity and as long as ν has substantial values, both T_{11} and T_{22} will also be small in absolute value compared to unity, and T_{12} and T_{21} will differ only little from unity. These are then the conditions for adiabatic coupling with nearly complete power transfer. They have been derived here under the assumption that the coupling parameter

$$\nu = \frac{4C^2}{\mathrm{d}(\Delta\beta)/dz}$$

changes only slightly over the coupling region from $z = z_1$ to $z = z_2$, and that no coupling takes place outside this region.

If this latter condition of no coupling for $z < z_1$ and $z > z_2$ is not entirely satisfied we may infer also from eqn. 8.2.18*a* what effect on the power transfer this interaction still has outside the actual coupling region near $z = 0$ and $x = 0$. Let us displace for example the limit z_2 of the coupling region and assume that ν is constant over the extent of this displacement. The corresponding shift of x_2 will first of all change the phase angle φ_2 in eqn. 8.2.18*a*. This change of phase accounts for the propagation of the normal modes and follows from the diagonal terms of the matrix in eqn. 8.2.12 as

$$\frac{d\varphi_2}{dx_2} = -\nu(1 + x_2^2)^{1/2}$$

Because of this phase change the absolute values of T_{11} and T_{22} (eqn. 8.2.18*a*) will fluctuate with a ripple of amplitude $1/x_2$. At $z = z_2$, and depending on the value of the coupling parameter ν there, the normal modes of the two separate waveguides still interact with each other. The amplitude ripple $1/x_2$ is a consequence of this residual interaction.

To generalise these results we note that eqn. 8.2.15*b* represents a good approximation for $\hat{a}$, also when ν is not quite constant, but when it changes gradually with x. As one such case, for which ν varies quite substantially with x, we considered the assumption of eqn. 8.2.14*a* and obtained for it expr. 8.2.14*b* for $\hat{a}$. This expression is not too much different from eqn. 8.2.15*b*. The controlling factor for $\hat{a}$, according to eqn. 8.2.15*b*, is always the magnitude of ν near $x = 0$. As long as ν is an even function of x the result for $\hat{a}$ will not differ greatly from eqn. 8.2.15*b* with ν evaluated at $x = 0$. Substantially larger values for $|\hat{a}|$ than that from eqn. 8.2.15*b* will only result if ν as a function of x has strong odd components near $x = 0$.

So far, in the adiabatic coupler, we have considered only two coupled modes, one in each of the two parallel waveguides. Quite often, and in particular for optical waveguide couplers, each of the waveguides is able to guide more than one mode. Typical for some micro-optical circuits is even the case of adiabatic coupling between multimode optical waveguides in which the individual guiding structures can guide a large number of normal modes. We will therefore investigate now how the adiabatic power exchange between two coupled normal modes is degraded by the interaction with other guided modes, and how this degradation can best be avoided.

We assume here, that all guided modes of one of the two waveguides, which take part in the interaction, are never degenerate with respect to each other, and consider as an example one particular unwanted or spurious mode, which the secondary waveguide 2 guides. This unwanted mode we designate by the mode index 3. We let the adiabatic coupler in Fig. 8.2 be so designed that the normal mode 1 of waveguide 1, when incident from the left-hand side, converts nearly all of its power to the normal mode 2 of waveguide 2. This power appears at the output port of waveguide 2 on the right-hand side of Fig. 8.2.

Fig. 8.3 shows how the phase constants β_1, β_2 and β_3 of these three modes 1, 2 and 3 vary along the coupler. The coupling factors C_{12}, C_{13} and C_{23} between these modes we assume to vary so little along the coupler that they may be taken as constant quantities. Under these conditions the transformed independent variables

$$x = \frac{\beta_2 - \beta_1}{2|C_{12}|}, \quad \hat{x} = \frac{\beta_3 - \beta_1}{2|C_{13}|}, \quad \hat{\hat{x}} = \frac{\beta_3 - \beta_2}{2|C_{23}|} \qquad (8.2.19a)$$

vary along the coupling region as shown in Fig. 8.3*b*.

For the power exchange between modes 1 and 2 to be nearly complete the coupling parameter

$$\nu = \frac{4|C_{12}|}{|d(\beta_2 - \beta_1)/dz|} \qquad (8.2.20a)$$

must be large enough so that the term $\exp(-\nu\pi/2)$ in eqn. 8.2.18*a* is sufficiently small. Let us specify this lower limit for ν as

$$\nu > \nu_0 \qquad (8.2.20b)$$

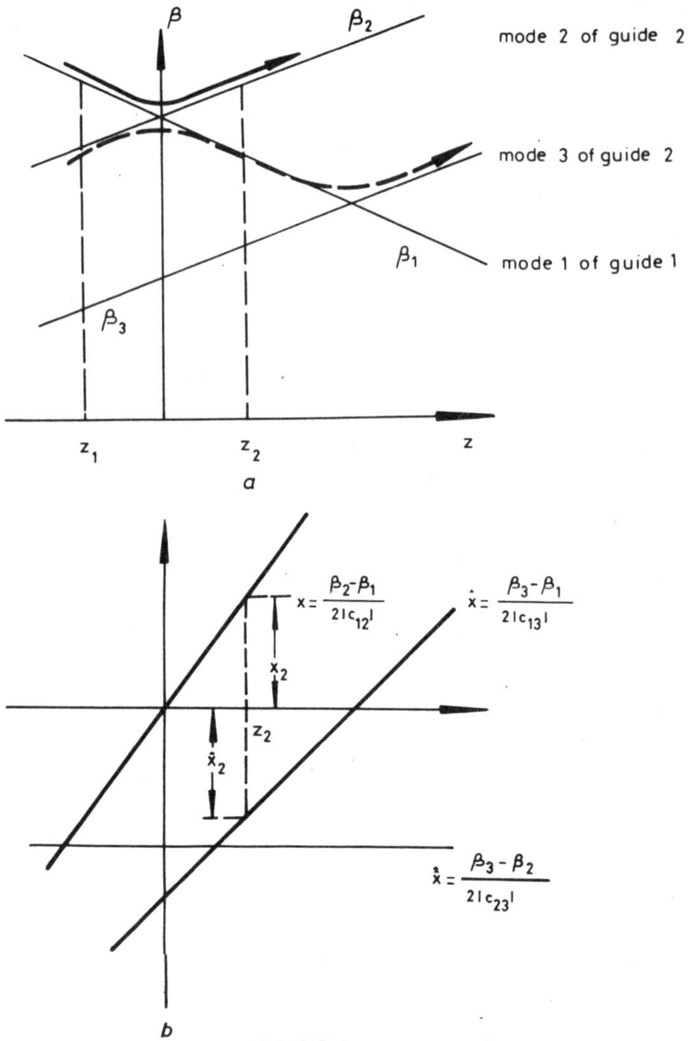

Fig. 8.3 *Adiabatic coupling with a spurious mode contained in guide 2*
a Variation of phase constants along the coupler
b Variation of values of x along the coupler

The degree to which both waveguides are tapered, one against the other, determines the rate $d|\beta_1 - \beta_2|/dz$ at which the phase difference changes in eqn. 8.2.20*a*.

For the coupler to convert only little power to the unwanted mode 3, the variable $\hat{x}$ must remain large in absolute value wherever the incident mode 1 has not yet converted its power to mode 2. The quantity $\hat{\hat{x}}$ on the other hand. must remain large in absolute value wherever mode 2 has picked up any power from the incident mode 1. At $z = z_2$ in Fig. 8.2, where $x = x_2$ equals $-\hat{x} = -\hat{x}_2$, we should therefore require

$$x_2 = -\hat{x}_2 > x_0 \tag{8.2.19b}$$

where x_0 is large enough, so that the term in eqn. 8.2.18*a* which is proportional to $1/(2x_2)$ remains insignificant for $x_2 = x_0$. The phase difference $\beta_2 - \beta_3$ varies only slightly along the coupler, and the quantity $\hat{\hat{x}}$ remains therefore nearly constant. This quantity should also be larger in absolute value than x_0,

$$|\hat{\hat{x}}| > x_0 \tag{8.2.19c}$$

to avoid too much power conversion from mode 2 to mode 3.

Each of the parameters x, $\hat{x}$ and $\hat{\hat{x}}$ is inversely proportional to the respective coupling factor C_{12}, C_{13} or C_{23}. If we keep the tapers of the coupled waveguides at a constant rate, but change only their spacing, then we change only the scale for $x(z)$ and $\hat{x}(z)$ but maintain their zero crossings as well as their slopes relative to each other. By adjusting the spacing between the tapered waveguides we can therefore achieve the desired magnitude for $x_2 = -\hat{x}_2$ as eqn. 8.2.19*b* requires it.

Should this spacing for the particular waveguide tapers and their rate of change $d|\beta_2 - \beta_1|/dz$ result in too small a value for the coupling parameter ν at $z = 0$, then we should choose gentler waveguide tapers with a lower rate of change $d|\beta_2 - \beta_1|/dz$, and adjust their spacing for a sufficient magnitude of $x_2 = -\hat{x}_2$. The results of Section 2.8 provide us with formulas to calculate the coupling factors C_{nm} of optical waveguides. These formulas, although derived for cylindrical structures may also be applied to tapered structures if the tapers are gentle enough.

In the adiabatic coupler, that meets these conditions for complete power transfer between modes 1 and 2 and for low degradation by other modes, we obtain under normal operating conditions the power flow which the solid arrow in Fig. 8.3*a* indicates. The power is incident in mode 1, and near where the phase constants β_1 and β_2 cross over transfers adiabatically to mode 2. At the crossover of the phase constants β_1 and β_3 no power is left in mode 1 to transfer to mode 3. If the power is incident in mode 2 , instead of in mode 1, then it follows the broken-line arrow in Fig. 8.3*a*. At the crossover of β_1 and β_2 the power transfers from mode 2 to mode 1, at the subsequent crossover of β_1 and β_3 it transfers from mode 1 to mode 3. This latter power transfer, however, is only complete when the respective coupling parameter

$$\nu = \frac{4|C_{13}|}{|d(\beta_1 - \beta_3)/dz|}$$

is sufficiently large. The power transfers under these conditions back to the same waveguide in which it first was launched, but into another mode. Should we want the power to transfer only from mode 2 to mode 1 in this particular direction then the waveguide tapers should only extend to $z = z_2$, and connect to cylindrical output waveguides for $z \geqslant z_2$.

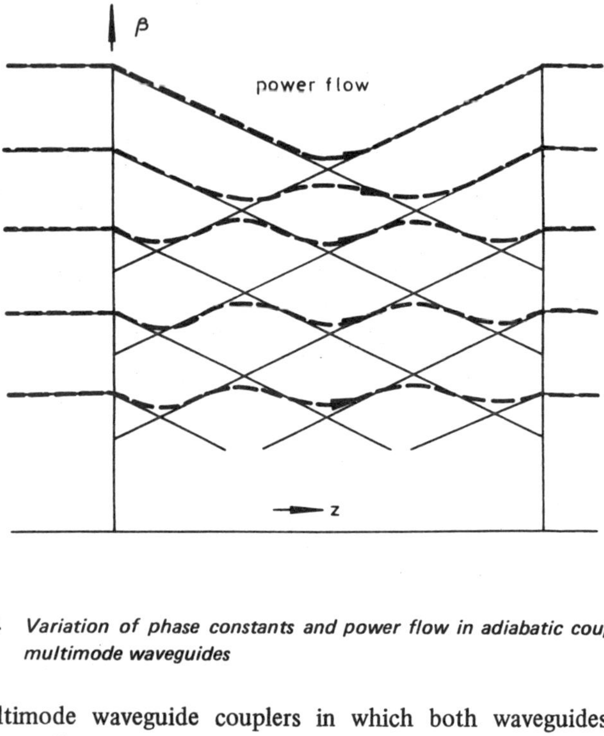

Fig. 8.4 *Variation of phase constants and power flow in adiabatic coupling of multimode waveguides*

Multimode waveguide couplers in which both waveguides taper symmetrically, one against the other, have also a mode spectrum in their two tapered waveguides of corresponding symmetry. The phase constant of each of the modes in one of the waveguides when plotted as a function of z, and the phase constant of the same mode in the other waveguide also as a function z, form curves which are symmetrical, with the enter plane $z = 0$ as the plane of symmetry. Fig. 8.4 shows a number of such phase curves for a symmetrical adiabatic coupler. If all the modes n of one waveguide couple strongly enough to all the modes m of the other waveguide, and if the waveguides taper gradually enough for all coupling parameters

$$\nu = \frac{4|C_{nm}|}{|d(\beta_n - \beta_m)/dz|}$$

to be sufficiently large, then complete power transfer occurs at each crossover of phase constants in Fig. 8.4. The broken-line arrows in Fig. 8.4 indicate the power flow in such a coupler. Power that enters the coupler in one particular mode of one waveguide leaves the coupler in the corresponding mode of the other waveguide.

References

1 MILLER, S.E.: 'Coupled wave theory and waveguide applications', *Bell. Syst. Tech. J.*, 1954, **33**, pp. 661–719

2 SPORLEDER, F.: 'Broadband modeselective directional couplers'. IEE Conf. Publ. 71, 1971, pp. 218–221

3 SMITH, R.B.: 'Coupling efficiency of tapered coupler', *Electron. Lett.* 1975, **11**, pp. 204–206

Chapter 9

Optical waveguide tapers

Optical waveguides in the form of glass fibres serve as the transmission medium in optical communications. Planar optical waveguides such as films and strips and strip-derived structures are needed in optical communications and optical data processing to form distributed components and to connect components and subsystems.

Quite often in these applications an optical waveguide with one particular cross-section must be joined to a waveguide of another cross-section. The optical waveguides to be connected differ sometimes not only in size but also in their cross-sectional form. Usually these waveguide connectors are required to launch as much as possible of the power that is incident in one waveguide into the other waveguide. If both waveguides are monomode, meaning that both of them guide only the lowest-order fundamental mode with its lowest order, or even zero cutoff, then all of the fundamental mode power incident in one waveguide should be launched into the fundamental mode of the other waveguide.

In such monomode waveguide transitions power may be lost to reflection and radiation. The transition should be designed to keep radiation and reflection loss at a minimum.

If one of the waveguides is monomode but the other multimode and if we require the transition to transmit with low power loss in both directions, then it should again be designed to launch as much as possible of the fundamental mode power which is incident from the monomode waveguide into the fundamental mode of the other waveguide. Should the transition split this incident power into several modes of the multimode waveguide, then, by the law of reciprocity, just this particular combination of mode amplitudes and phases must be incident from the multimode waveguide to launch all its power into the fundamental mode of the monomode waveguide.

If, however, the monomode-to-multimode waveguide transition is required to transmit optical power only in the direction from the monomode to the multimode waveguide, then it need only be designed to lose no power by reflection or radiation.

A fourth case is the transition between two different multimode waveguides. Usually such multimode waveguide transitions are required to transmit again as much as possible of the power incident from one waveguide into the other waveguide. For the transition to lose no power to reflection or radiation each guided mode on one side of the transition must launch all its power into guided modes on the other side of the transition. Two multimode waveguides that differ neither too much in form nor in size will also have nearly the same normal mode spectra. Each normal mode in one of the waveguides has then a corresponding normal mode in the other waveguide. The safest way to accomplish total power transfer between both is to have each of the modes in one waveguide convert all its power to the corresponding mode in the other waveguide.

If the two waveguides are similar in cross-sectional form but different in size, such as two round fibres of different core diameters, then each guided mode of the smaller waveguide has a corresponding mode in the larger waveguide; but a number of higher-order modes of the larger waveguide do not exist in the smaller size. Through a transition for complete power transfer from the smaller to the larger waveguide these higher-order modes would never be excited. By the law of reciprocity no power from these higher-order modes would convert to any mode of the smaller size waveguide. They will rather radiate their power near those regions of the taper where they approach cutoff.

The most obvious and, for all practical purposes, simplest form for such waveguide transitions between monomode and multimode waveguides, as well as from monomode-to-multimode waveguides, are gradual tapers. These waveguide tapers should have a change in cross-sectional form and size so gentle, that in the direction in which the mode volume of the local cross-section increases, all local normal modes propagate approximately independently of each other and transform adiabatically from the normal mode on one side of the taper to the corresponding mode on the other side. What residual interaction there remains in such tapered dielectric waveguides may be analysed with the coupled wave equations of Section 2.9.

Transitions between different dielectric waveguides in the form of gradual waveguide tapers are particularly well suited for optical waveguides because of their microscopic dimension. For local normal modes to interact only little in such a waveguide transition the taper must be longer than the beat wavelength

$$\Lambda = 2\pi/(\beta_m - \beta_n) \tag{9.1}$$

between two interacting modes with phase constants β_m and β_n. It is even desirable to have the taper to be much longer than any of these beat wavelengths. For waveguide tapers which extend over at least one beat wavelength or even over several beat wavelengths the conversions to wanted modes along the taper interfere destructively and the respective unwanted modes hardly collect any significant amount of power.

Optical waveguide tapers may, in most practical cases, extend over such distances without requiring too much space in optical circuits and systems.

Even though we have this taper length available, such tapers should nevertheless be given a form with which they perform best for a given length of the taper. Of the transitions between different optical waveguides, which we have mentioned above, those are best built in the form of tapers, that differ not too drastically in cross-sectional form, but only in size. Typical examples are fibre-to-fibre transitions, rectangular strip guide-to-fibre transitions, or transitions between fibre guides of different thickness. For such transitions in the form of waveguide tapers we distinguish broadly between three different kinds:

(*a*) monomode-to-monomode taper
(*b*) monomode-to-multimode taper
(*c*) multimode-to-multimode taper.

The design of these tapers requires special care if, first, radiation is to be avoided; secondly, if the fundamental mode of the monomode waveguide is to launch all its power only into the fundamental mode of the multimode waveguide, and thirdly, if the excitation and interaction of modes must be so limited that the multimode taper transmits in both directions with little loss of power.

9.1 Monomode waveguide tapers

Transitions between optical waveguides that differ only in cross-sectional size or also in form and size of their cross-section, but in both cross-sections guide only the fundamental mode, occur quite often in optical components and systems. A typical example is the junction of the rectangular strip guide of planar and integrated optics circuits and the round core of a monomode glass fibre.

Usually there is enough space available for these transitions to taper them gradually over some length and thereby avoid excessive radiation loss. This is particularly true for transitions between waveguides that do not differ too drastically in their cross-sectional form. Under these conditions they also cannot differ greatly in cross-sectional size, otherwise the wider cross-section would guide higher-order modes, or the more narrow cross-section would not confine the fundamental mode fields well enough. We have hence only a moderate change in cross-sectional dimensions in such monomode-to-monomode waveguide tapers, and even a very gradual transition accommodates this change in a relatively short taper.

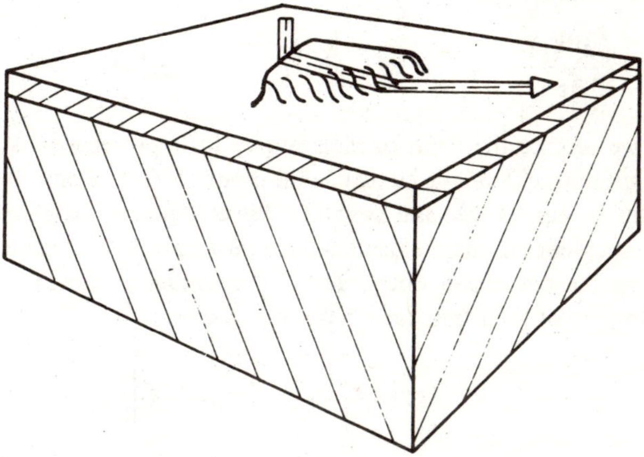

Fig. 9.1 *Film-guide prism with film tapers at the edges of the prism*

Very short tapers may, however, be required for monomode waveguide transitions, when they form integral parts of optical components. A typical example is the filmguide prism in Fig. 9.1 for the deflection of film waves. At the edges of the film prism the filmguide increases in thickness, with a larger phase constant β of the film modes as a consequence. The prism edges refract an incident filmwave so that altogether the prism deflects the filmwave from its direction of incidence. For the prism to perform satisfactorily the edges should be tapered smoothly enough to avoid reflection and radiation; but the tapers should not be any longer than is necessary to meet these requirements or they add too much to the prism size.

To analyse wave propagation and radiation in monomode-to-monomode optical waveguide tapers we refer to the coupled wave equations for dielectric waveguide transitions in Section 2.9. In gradual transitions, which guide only the fundamental mode, this mode interacts to any significant extent only with radiation modes. Reflections into the fundamental mode may be neglected as well as interaction between radiation modes. Under these conditions the rate of change of the fundamental mode amplitude $a_0^{(+)}$ follows from eqn. 2.9.80 as

$$\frac{da_0^{(+)}}{dz} = \kappa_{00}^{(++)} a_0^{(+)} + \int_0^{\sqrt{\epsilon} k} \kappa_0^{(++)}(\tau) a^{(+)}(\tau) \exp\left[j \int_0^z (\beta_0 - \beta_\tau) dz \right] d\tau \tag{9.1.1}$$

while from eqn. 2.9.81 the amplitude $a^{(+)}(\tau)$ of radiation modes changes along the taper as given by

$$\frac{\partial a^{(+)}(\tau)}{\partial z} = \kappa_0^{(++)}(\tau) a_0^{(+)} \exp\left[j \int_0^z (\beta_\tau - \beta_0) dz \right] \tag{9.1.2}$$

Since we require the taper to radiate only little power we may regard the amplitude $a_0^{(+)}$ of the fundamental mode to remain approximately constant in eqn. 9.1.2, and integrate this equation directly. With unit power incident in the fundamental mode at $z = 0$, we have $a_0^{(+)} \simeq 1$ all along the taper, and obtain for the amplitude of radiation modes as a function of their transverse phase constant τ:

$$a(\tau) = \int_0^l \kappa_0^{(++)}(z) \exp\left[j \int_0^z (\beta_\tau - \beta_0) dz' \right] dz \tag{9.1.3}$$

For the example of radiation modes on dielectric slabs in Section 2.9, eqns. 2.9.31 and 2.9.35 normalise radiation modes with respect to power. In such a normalisation for $a(\tau)$ the total power which radiates from the taper follows by integrating the product $a(\tau)a^*(\tau)$ over all relevant values of τ. $a^*(\tau)$ denotes the complex conjugate of $a(\tau)$. Since a gradual taper scatters fundamental mode power mainly in the forward direction the radiated power is given by

$$P_r = \int_0^{\sqrt{\epsilon} k} a(\tau) a^*(\tau) d\tau \tag{9.1.4}$$

At the lower limit of integration ($\tau = 0$) the radiation mode scatters power in the forward direction parallel to the waveguide; at the upper limit, where τ equals the wavenumber $\sqrt{\epsilon} k$ of the material embedding the optical waveguide, it scatters power in the transverse direction.

To evaluate the radiation of a monomode waveguide taper, we need to specify its cross-section as a function of z along the taper and determine also, as functions of z, the longitudinal phase constants β_0 and β_τ of the fundamental mode and of the radiation modes with which it interacts. We, furthermore, must determine the coupling coefficient $\kappa_0(\tau)$ of this interaction as it depends on z and τ. We then must evaluate the integrals of eqns. 9.1.3 and 9.1.4.

Instead of numerical evaluations which are feasible only for very specific examples, we learn more from an approximate solution of these equations, which reveals how, in general, radiation depends on the taper contour and its length.

For this approximation we assume the size of the waveguide cross-section to change not too much along the taper. Since the cross-sectional dimensions of a particular waveguide are rather limited at a specific light wavelength to a range where the waveguide still confines the fundamental mode fields and yet guides no higher-order modes, we are justified in making such an assumption. If the waveguide dimensions change very little from one waveguide to the other all quantities that make up the coupling coefficients according to eqn. 2.9.73 stay approximately constant along the taper, except for the derivative $\partial b/\partial z$. Even this derivative is constant, when the transition tapers at a constant rate. Whatever the dependence of $\partial b/\partial z$ on z, we may write, in this case, the coupling coefficient between the fundamental mode and the radiation field in the form

$$\kappa_0(z) = K_0 \frac{db}{dz} \tag{9.1.5}$$

where the factor K_0 is independent of z. By extracting $\partial b/\partial z$ from the contour integral of eqn. 2.9.73 and replacing it by db/dz in eqn. 9.1.5, we have restricted attention to such waveguide transitions, that change only one cross-sectional dimension along the taper. For fibre tapers, for example, the dimension b denotes the radius of the fibre core; For film waveguides with coupling coefficient given by eqns. 2.9.76–2.9.78 the dimension $b = d$ denotes half the thickness of the film.

Furthermore, for the taper in which the waveguide dimensions change very little from one waveguide to the other, the difference $\beta_\tau - \beta_0$ in axial phase constants of the fundamental mode and any coupled radiation mode of a specific τ-value stays approximately constant along the taper. Under these conditions (eqn. 9.1.3) for the amplitude $a(\tau)$ of a radiation mode reduces to

$$a(\tau) = K_0 \int_0^l (db/dz) \exp\left[j(\beta - \beta_0)z\right] dz \tag{9.1.6}$$

The radiation-mode amplitude according to this approximate formula can, for specific values of the phase difference $\beta_0 - \beta_\tau$, the taper length l, and the change in waveguide size $b(l) - b(0)$, be reduced to a minimum, by choosing a suitable taper contour $b(z)$, according to the principles that were outlined in Chapter 5. Such a design, however, considers of necessity only one particular value τ of the continuum of radiation modes with τ-values in the range $0 < \tau < \sqrt{\epsilon}k$. It will therefore hardly result in a taper of good overall performance. We will hence not pursue this design method any further, and instead consider two rather simple taper contours, and formulate for them general design rules.

As the first example we assume the taper to be linear, with db/dz constant between $0 < z < l$ and equal to

$$db/dz = (b_2 - b_1)/l \tag{9.1.7}$$

where $b_1 = b(0)$ and $b_2 = b(l)$. In this case eqn. 9.1.6 integrates to

$$a(\tau) = K_0(b_2 - b_1)\frac{1 - \exp\,[j(\beta_\tau - \beta_0)l]}{j(\beta_0 - \beta_\tau)l} \tag{9.1.8}$$

Any overall phase difference $(\beta_0 - \beta_\tau)l$ between the coupled modes, that is an integral multiple of 2π reduces the radiation mode amplitude to zero by destructive interference. The maxima which $|a(\tau)|$ exhibits as a function of $(\beta_0 - \beta_\tau)l$ decrease inversely proportional to this overall phase difference.

The coupling factor K_0 as a function of τ has, under normal conditions, its largest absolute value at $\tau = 0$. Here, the phase difference $\beta_0 - \beta_\tau$ is smallest and, since it appears in the denominator of the coupling coefficient of eqn. 2.9.66, renders K_0 relatively large. If therefore we choose

$$l(\beta_0 - \beta_\tau)|_{\tau=0} = 2\pi \tag{9.1.9}$$

then destructive interference suppresses these most strongly coupled modes of the radiation field, and the radiated power remains quite small. From eqn. 9.1.9 follows the smallest length of a linearly tapered transition with low radiation loss given by

$$l_{min} = 2\pi/(\beta_0 - \beta_\tau)|_{\tau=0} \tag{9.1.10}$$

A linear taper of this length between two monomode optical waveguides of not too different sizes reduces the forward scattered radiation field to a minimum by destructive interference.

If a linear taper of this length has still too large a radiation loss, then it should be made twice this length or an even larger integral multiple of it. Any integer multiple of the smallest taper length given by eqn. 9.1.10 reduces the forward scattered radiation field again to a minimum by destructive interference. At the same time, owing to the factor $(\beta_0 - \beta_\tau)l$ in the denominator of eqn. 9.1.8, the overall radiated power decreases, on the average, inversely proportional to the square of the taper length.

The linear taper exhibits discontinuities of its slope at both ends, where it connects to uniform waveguides with zero slope. A smooth transition from the uniform waveguides to the finite slope of the taper can be expected to perform much better, in particular when it is gradual enough to transform the normal modes of the uniform waveguides, with their plane phase fronts, to horn waves of the tapered region with curved phase fronts.

We will therefore discuss as a second example for a monomode waveguide transition between $z = 0$ and $z = l$ the cosine taper with

$$b(z) = \frac{b_1 + b_2}{2} - \frac{b_2 - b_1}{2} \cos\left(\frac{\pi}{l} z\right) \tag{9.1.11}$$

This taper has the slope

$$\frac{db}{dz} = \frac{\pi}{2} \frac{b_2 - b_1}{l} \sin\left(\frac{\pi}{l} z\right) \tag{9.1.12}$$

We assume again that $b_2 - b_1$ is small enough compared to $b_2 + b_1$, so that eqn. 9.1.6 approximates the amplitude of radiation modes at the end of the taper, when the fundamental mode enters it with unit amplitude and power. When we substitute for db/dz from eqn. 9.1.12 into eqn. 9.1.6 and solve the integral, we obtain at the taper end radiation modes of amplitude

$$a(\tau) = \frac{\pi^2 K_0(b_2 - b_1)}{2[\pi^2 - (\beta_0 - \beta_\tau)^2 l^2]} \{\exp [j(\beta_\tau - \beta_0) l] + 1\} \tag{9.1.13}$$

The radiation mode amplitude at a particular value of the transverse phase constant τ depends similarly on taper length l as in the linear taper. For certain values of the accumulated phase difference $(\beta_0 - \beta_\tau)l$ it vanishes due to destructive interference. These phase-difference values for zero radiation mode amplitude differ, however, from those for the linear taper. For the linear taper they are given by eqn. 9.1.9 and integer multiples of it. For the cosine taper they follow from

$$(\beta_0 - \beta_\tau)l = (2n + 1)\pi \quad \text{with} \quad n = 1, 2, 3 \ldots \tag{9.1.14}$$

as odd multiples of π. The smallest over-all phase difference $(\beta_0 - \beta_T)l$, at which the radiation mode amplitude cancels, is 3π.

If we follow the same design rule, as in case of the linear taper, but now for the cosine taper, we obtain

$$l = 3\pi/(\beta_0 - \beta_T)|_{T=0} \tag{9.1.15}$$

for the minimum length at which the forward scattered radiation field cancels by destructive interference.

The cosine taper must hence be 50% longer than the linear taper, when it is adjusted to operate in the first radiation loss minimum.

Compared to the linear taper the radiation mode amplitude at the end of the cosine taper shows another significant difference as a function of taper length. Besides the exponential phase factors in both eqns. 9.1.8 and 9.1.13 depending on taper length, the radiation mode amplitude of the linear taper is inversely proportional to the taper length, whereas for the cosine taper of sufficient length it is inversely proportional to the square of length. The overall radiation loss in long cosine tapers decreases on the average inversely proportional to the fourth power of taper length, whereas in the linear taper it decreases only inversely proportional to the taper length squared. This $1/l^4$-characteristic of average radiation loss occurs not only for the cosine taper, but whenever the taper slope db/dz has no discontinuities. To gain significant advantage from it the taper must be sufficiently long.

Before we conclude this rather general discussion of monomode optical waveguide transitions let us estimate how many light-wavelengths a linear or cosine taper must at least be long, to cancel the forward scattered radiation by destructive interference. The radiation modes which propagate nearly parallel to the waveguide have as their longitudinal phase constant the wave number $n_2 k = 2\pi n_2/\lambda$ of the surrounding medium with refractive index n_2. In the case of a film waveguide n_2 is the substrate index in case of a fibre-guide it is the refractive index of the cladding that surrounds the core.

The effective index $N = \beta/k$ of any guided mode of an optical waveguide varies between $N = n_2$ at cutoff and $N = n_1$ very far above cutoff, where n_1 designates the maximum refractive index of the guiding region. The fundamental mode of an optical waveguide near the upper limit of its monomode range has an effective index $N_0 = \beta_0/k$ nearly half way between n_2 and n_1

$$N_0 \simeq (n_1 + n_2)/2 \tag{9.1.16}$$

The longitudinal phase constants of the fundamental mode and the forward scattered radiation mode differ under these circumstances by

$$\beta_0 - \beta_T = (n_1 - n_2)k/2 \tag{9.1.17}$$

To satisfy eqn. 9.1.9 for the linear taper its length l relative to the light wavelength λ should be

$$l/\lambda = 2/(n_1 - n_2) \tag{9.1.18}$$

For the cosine taper to satisfy eqn. 9.1.14 with $n = 1$ its relative length should be

$$l/\lambda = 3/(n_1 - n_2) \tag{9.1.19}$$

Optical waveguides differ in the refractive index of their guiding region and the surrounding medium usually by only a fraction of a percent up to a few percent. Such tapers must extend over hundreds of wavelengths to effectively reduce the forward scattered radiation by destructive interference.

9.2 Monomode-to-multimode taper

Transitions between optical waveguides that taper the size of their cross-section from monomode to multimode dimensions are of considerable practical interest. A typical example is a monomode fibre that is to be matched with its minute core cross-section to the much larger spotsize of a laser. If the laser oscillates in its lowest transverse order the transverse field and intensity distribution of its output beam correspond quite closely to those of the fundamental fibre mode, only its spotsize is sometimes much larger than the fibre core and the spotsize of the fundamental fibre mode.

One therefore joins a multimode fibre of much larger core size to the laser and subsequently tapers the fibre down to a monomode core size. The laser excites with its lowest transverse order output beam mainly the fundamental fibre mode. The taper must transform this mode of relatively large spotsize to the small spotsize in the monomode core, without losing to much power by radiation. Along the taper the fundamental mode interacts with higher-order modes, notably the next higher-order one, and with the radiation field. For weak interaction and low radiation loss the taper must be long and gradual. To design such a taper we need to consider first of all the next higher-order mode. Fundamental mode power converts mainly to this next higher-order mode and, when this mode reaches cutoff near the monomode end of the taper, this power radiates.

A taper design that considers only the next higher-order mode could follow the procedure that was outlined for overmoded waveguide transitions in Section 6.1. Such a design requires, however, that the transition contour be produced within very tight tolerances to achieve the utmost in spurious mode suppression by destructive interference. For optical waveguides with their microscopic cross-section it is not really feasible to produce transition contours so accurately.

We will therefore resort here to a simplified design procedure that relies on destructive interference only over fractions of the taper length and yet keeps the overall mode conversion low. To this end we consider at first only a section of a long and gradual taper which is short enough so that in the coupled wave equations the coefficient κ for coupling between the fundamental and next higher-order mode as well as their difference in phase constants $(\beta_0 - \beta_1)$ remain nearly constant along this section. If only the fundamental mode is incident onto this section, then the fraction of its power that maximally converts to the higher-order mode is given by eqn. 8.1.6*b*. For weak coupling with $|\kappa| \ll |\beta_0 - \beta_1|$ this power fraction is simply given by

$$p = 4\kappa^2/(\beta_0 - \beta_1)^2 \qquad (9.2.1)$$

In any such section along the taper, if only the fundamental mode is incident on it, the power conversion does not exceed this maximum. To accommodate the total amount of coupling between the fundamental and next higher-order mode, that a transition between two given waveguide sizes requires, it therefore seems appropriate to choose a taper contour that keeps the coupling ratio $\kappa/(\beta_0 - \beta_1)$ constant along the taper.

We examine the interaction between the fundamental and next higher-order mode under this condition of constant coupling ratio by referring to the coupled wave equations 2.9.80 and 2.9.81 for dielectric waveguide transitions. If we disregard radiation modes and all the other higher-order modes, only the following two coupled equations for the amplitude a_0 of the fundamental mode and a_1 of the next higher-order mode remain to be considered

$$\frac{da_0}{dz} = \kappa a_1 \exp\left(j2\int \Delta\beta dz\right)$$

$$\frac{da_1}{dz} = -\kappa a_0 \exp\left(-j2\int \Delta\beta dz\right) \qquad (9.2.2)$$

with $\kappa = \kappa_{01}^{(++)}$ as the coupling coefficient between these two modes, when both travel in the positive z-direction, and $2\Delta\beta = \beta_0 - \beta_1$ as their difference in phase constants. For half the integrated phase difference

$$\varphi = \int \Delta\beta dz \tag{9.2.3}$$

as the new independent variable, and

$$\begin{aligned} A_0 &= a_0 \exp(-j\varphi/2) \\ A_1 &= a_1 \exp(j\varphi/2) \end{aligned} \tag{9.2.4}$$

as new wave amplitudes the coupled wave equations 9.2.2 transform to

$$\frac{d}{d\varphi}\begin{bmatrix} A_0 \\ \\ A_1 \end{bmatrix} = \begin{bmatrix} -j & \dfrac{\kappa}{\Delta\beta} \\ -\dfrac{\kappa}{\Delta\beta} & j \end{bmatrix} \begin{bmatrix} A_0 \\ \\ A_1 \end{bmatrix} \tag{9.2.5}$$

These equations correspond quite closely to eqns. 8.1.5*c* for uniform coupling between two waves which differ by $2\Delta\beta$ in their phase constants. Their solution, for $\kappa/\Delta\beta =$ constant along the transition corresponds likewise to the solution 8.1.6*a* of eqns. 8.1.5*c*. For the initial conditions $A_0 = 1$ and $A_1 = 0$ of only the fundamental mode incident at $z = \varphi = 0$ with unit power, the power that converts to the next higher-order mode is given by

$$|A_1|^2 = \frac{\kappa^2}{\Delta\beta^2 + \kappa^2} \sin^2 \{\varphi(1 + \kappa^2/\Delta\beta^2)^{1/2}\} \tag{9.2.6}$$

It fluctuates according to the sine function between zero and the maximum value

$$|A_1|^2_{max} = \kappa^2/(\Delta\beta^2 + \kappa^2) \tag{9.2.7}$$

An optical waveguide taper, in which the fundamental mode interacts significantly only with the next higher-order mode converts the power fraction according to eqns. 9.2.6 and 9.2.7 to this mode, only if the ratio $\kappa/\Delta\beta$ of coupling coefficient to difference in phase constants $2\Delta\beta$ is kept constant along the taper. If we allow only the small fraction p of the fundamental power to convert maximally to the next higher mode then the taper should be designed to obey

$$\kappa/\Delta\beta = \sqrt{p} \tag{9.2.8}$$

The coupling coefficient κ as given by eqn. 2.9.73 or 2.9.76 depends on the rate of change of waveguide dimensions as well as on the waveguide dimensions themselves. $\Delta\beta$ depends also on waveguide dimensions. Eqn. 9.2.8 therefore constitutes a differential equation for the taper contour. Under general circumstances this differential equation can only be solved numerically. In monomode-to-multimode tapers, however, the fundamental as well as the next higher-order mode are usually very far above cutoff at the multimode side of the taper, and simple approximations apply on this side for κ and $\Delta\beta$. In many such tapers of practical interest this far-above-cutoff condition prevails over most of the taper length, even quite close to its monomode end. We are therefore well justified to base a solution of eqn. 9.2.8 on far-above-cutoff approximations for κ and $\Delta\beta$. The taper region in which κ tends to be large and $\Delta\beta$ small and hence the interaction more pronounced, is anyway towards the multimode end of the taper, where far-above-cutoff approximations hold particularly well.

As a representative example we consider the symmetrical film or dielectric slab of relative permittivity $\epsilon_s = n_s^2$ and thickness $2d$, embedded in a medium of relative permittivity $\epsilon = n^2$. The even TE_m-modes of this slab have the characteristic equation 2.9.9 which, far-above-cutoff, for $\alpha \to \infty$, reduces to

$$\tau_s d = (m+1)\pi/2 \tag{9.2.9}$$

Their phase constants are approximately given by

$$\beta_m = \sqrt{\epsilon_s k^2 - \tau_s^2} \simeq \sqrt{\epsilon_s} k - \tau_s^2/(2\sqrt{\epsilon_s} k) \tag{9.2.10}$$

so that differences between them follow from

$$\beta_m - \beta_n = \frac{\pi^2}{8\sqrt{\epsilon_s} k d^2} [(n+1)^2 - (m+1)^2] \tag{9.2.11}$$

When the symmetrical slab tapers symmetrically between both sizes the even TE-modes couple only to even TE-modes. The coupling coefficients (eqn. 2.9.76) reduce in the far-above-cutoff approximation to

$$\kappa_{mn} = \frac{(m+1)(n+1)}{(n+1)^2 - (m+1)^2} \frac{2}{d} \frac{d(d)}{dz} \tag{9.2.12}$$

The fundamental mode has the mode index $m = 0$ and the next higher-order mode that couples to it in the symmetrical transition has $n = 2$. They couple mutually with

$$\kappa_{02} = \frac{3}{4d} \frac{d(d)}{dz} \tag{9.2.13}$$

and differ in phase constants by

$$\beta_0 - \beta_2 = \pi^2/(\sqrt{\epsilon_s} k d^2) \tag{9.2.14}$$

The condition (eqn. 9.2.8) for maximum power loss p results under these circumstances in the following differential equation:

$$\frac{3\sqrt{\epsilon_s}}{2\pi^2}kd\,\frac{d(d)}{dz} = \sqrt{p} \qquad (9.2.15)$$

Its solution by separation of variables for the initial slab thickness d_1 leads to a taper contour which, from

$$d^2 - d_1^2 = \frac{2\pi\sqrt{p}}{3n_s}\lambda z \qquad (9.2.16)$$

is parabolical. A parabolical taper contour was also obtained by Burns *et al.*[1] who consider the propagation angle θ in Fig. 2.18 of the rays that superimpose to form the fundamental mode of the slab. They require this propagation angle to remain along the taper a constant fraction of the taper angle ϕ in Fig. 2.20.

On its narrow side where, according to eqn. 9.2.14, the TE_2-mode differs much more in phase constant from the fundamental TE_0-mode than on its wider side, the taper starts with a relatively steep slope. Towards its wider multimode end both phase constants differ much less and, to keep the coupling ratio $\kappa/\Delta\beta$ constant, the transition tapers with a gentler slope.

For illustration we consider as a numerical example the transition from a slab of $2d_1 = 4\,\mu$m thickness to $2d_2 = 40\,\mu$m and with $n_s = 1{\cdot}5$ as its refractive index; at $\lambda = 1\,\mu$m wavelength we require it to convert not more than $p = 10\%$ of the fundamental mode power to the TE_2-mode. Such a transition, when tapered parabolically according to eqn. 9.2.16, must be 0·9 mm long.

9.3 Multimode-to-multimode taper

Tapered transitions between multimode optical waveguides can transmit power without radiation loss only between those modes that each waveguide on both sides of the transition guides. If a higher-order guided mode on the wide side of the transition does not have a corresponding guided mode in the waveguide at the narrow end it passes cutoff within the tapered transition and converts nearly all its power to radiation. To transmit power through the transition from the wide side into the narrower waveguide it must all be launched into those lower-order modes that the narrower waveguide still guides. In addition the transition must be gradual enough, and tapered in such a fashion, that these lower-order modes which are guided on both sides of the transition interact only weakly with higher-order modes and convert not too much power to them. Any power that is converted to such high-order modes is liable to be radiated, when these higher-order modes approach cutoff.

For the design of such low loss transitions between multimode mode waveguides we consider transmission of optical power from the smaller waveguide with its lower number of guided modes to the larger waveguide. We assume that the incident power distributes evenly over all guided modes of the smaller waveguide. As the waveguide tapers to a wider cross-section the number of guided modes increases, and these newly guided higher-order modes couple to those that were initially excited. Their interaction should, however, be weak enough not to lose too much power to them. To keep the maximum power conversion from any of the initially excited modes to any of the higher-order modes at a low and constant level throughout the transition we proceed again as for the monomode-to-multimode waveguide taper.

Just as by eqn. 9.2.8 we require the ratio $\kappa_{mn}/(\beta_m - \beta_n)$, of coupling coefficient between these modes to their difference in phase constants, to remain unchanged all along the taper.

Since this coupling ratio depends differently on waveguide dimensions for different mode pairs we can meet the condition for $\kappa_{mn}/(\beta_m - \beta_n)$ to be constant in general only for one particular mode pair.

In the far-from-cutoff approximation, however, all the $\kappa_{mn}/(\beta_m - \beta_n)$ depend likewise on waveguide size and its rate of change. By a suitable choice of taper contour the maximum power conversion between any of these modes may then be kept at a constant level all along the transition. This maximum power conversion level remains, however, still different from pair to pair of coupled modes.

The symmetrical taper between symmetric dielectric slabs serves us again as a representative example. Any film guide taper, even if it is not symmetrical, couples only TE-modes with TE-modes and TM-modes with TM-modes. No interaction occurs between TE- and TM-modes. A symmetrical slab taper, moreover, couples only even modes with even modes and odd modes with odd modes. Eqn. 9.2.12 represents the far-from-cutoff approximation of the coupling coefficient for interaction between the even TE_m-mode and the even TE_n-mode. The corresponding approximation for the difference in phase constants between these modes appears in eqn. 9.2.11. From them we obtain the ratio of one to the other under far-from-cutoff conditions as

$$\frac{\kappa_{mn}}{\beta_m - \beta_n} = \frac{16 n_s k (m+1)(n+1)}{\pi^2 [(n+1)^2 - (m+1)^2]^2} d \frac{d(d)}{dz} \tag{9.3.1}$$

This quantity has the same dependence on the thickness $2d$ of the slab and its rate of change $2d(d)/dz$ for all the interacting modes. We keep the maximum power conversion between any coupled pair at a constant level all along the taper by requiring

$$d \frac{d(d)}{dz} = C \tag{9.3.2}$$

to be a constant. Separating the variables d and z of this equation and integrating it from d_1 at $z = 0$ to d_2 at $z = l$ leads again to a parabolic taper contour

$$d^2 - d_1^2 = 2Cz \tag{9.3.3}$$

with

$$C = (d_2^2 - d_1^2)/(2l) \tag{9.3.4}$$

For this parabolic taper we obtain for any pair of coupled modes the coupling ratio

$$\frac{\kappa_{mn}}{\beta_m - \beta_n} = \frac{16 n_s (m+1)(n+1)(d_2^2 - d_1^2)}{\pi [(n+1)^2 - (m+1)^2]^2 \lambda l} \tag{9.3.5}$$

If the modes of such a coupled pair would not interact with other modes then the maximum power conversion from one to the other would, according to eqn. 9.2.6, be given by

$$p_{mn} = \{1 + (\beta_m - \beta_n)^2/(2\kappa_{mn})^2\}^{-1} \tag{9.3.6}$$

To control the power conversion from low-order modes to higher-order unwanted modes, we examine first for which pair of these modes the coupling ratio according to eqn. 9.3.5 is largest. The highest-order m of a mode which a slab of thickness $2d$ still guides, follows from eqn. 2.9.5. For even TE-modes it is the largest even integer smaller than

$$m_c = 4(\epsilon_s - \epsilon)^{1/2} d/\lambda \tag{9.3.7}$$

We denote this largest even number for $d = d_1$ by m_1. The even TE-mode of next higher order $n = m_1 + 2$ is the first one to become a guided mode near the narrow end of the taper. For these two modes the coupling ratio (eqn. 9.3.5) has the value

$$\frac{\kappa_{m_1, m_1+2}}{\beta_{m_1} - \beta_{m_1+2}} = \frac{n_s (m_1^2 + 4m_1 + 3)(d_2^2 - d_1^2)}{\pi (m_1^2 + 4m_1 + 4) \lambda l} \tag{9.3.8}$$

All other mode pairs $m \leqslant m_1$ and $n \geqslant m_1 + 2$ have smaller such coupling ratios. The mode pair with $m = m_1$ and $n = m_1 + 4$ comes closest in its coupling ratio to eqn. 9.3.8. Its value is

$$\frac{\kappa_{m_1,m_1+4}}{\beta_{m_1} - \beta_{m_1+4}} = \frac{n_s(m_1^2 + 6m_1 + 5)(d_2^2 - d_1^2)}{4\pi(m_1^2 + 6m_1 + 9)\lambda l} \qquad (9.3.9)$$

For $m_1 \gg 1$ it amounts to only a quarter of the largest coupling ratio according to eqn. 9.3.8. For such high mode orders, moreover, this largest coupling ratio reduces approximately to

$$\left(\frac{\kappa}{2\Delta\beta}\right)_{max} = \frac{n_s(d_2^2 - d_1^2)}{\pi\lambda l} \qquad (9.3.10)$$

With the coupling ratio of eqn. 9.3.9 amounting to only a quarter of this largest ratio, the maximum power conversion in the parabolic taper from the TE_{m_1}-mode to the TE_{m_1+4}-mode will be only the 16th part of the maximum power conversion between the TE_{m_1}-mode and TE_{m_1+2}-mode. The latter power conversion is hence so much larger than the power conversion between any other pair of wanted and unwanted modes, that with it we control the performance of the taper.

The power which is incident from the narrow side of the multimode taper is assumed to distribute evenly among all the guided modes of the narrow waveguide. For $m_1 \gg 1$ this slab guides approximately as many as m_1 even and odd TE-modes, and in addition m_1 even and odd TM-modes. Each of these modes transmits the $(2m_1)$th part of the total power. We assume the TE_{m_1}-mode and the TM_{m_1}-mode to convert maximally the same fraction p_1 of their power to the respective next higher-order mode.

A similar situation prevails for the odd TE- and TM-modes. For equal power distribution among all guided modes at the narrow end of the taper each of the odd modes transmits the same fraction of the total power as each of the even modes. We assume the highest order odd TE- and TM-modes which are still guided at the narrow end of the waveguide to each of them convert maximally the same fraction p_1 of their power to the respective next higher-order mode.

Of the $2m_1$ guided modes at the narrow end of the taper there are thus 4 modes which lose maximally the fraction p_1 of their power. All these four fractions together amount to a fraction of the total incident power as given by

$$p = 2p_1/m_1$$

If we specify this total power loss p, we can tolerate p_1 to be as large as

$$p_1 = m_1 p/2 \tag{9.3.11}$$

because in a gradual taper the even TE_{m_1}- and TM_{m_1}-modes and the corresponding odd modes lose much more power to their next higher-order neighbours than to modes of still higher order; modes of lower order than m_1 lose also much less power to these higher-order modes.

If we substitute p_1 (eqn. 9.3.11) for p_{mn} into eqn. 9.3.6 and also introduce the maximum coupling ratio from eqn. 9.3.10 into this equation we obtain, by solving it for l,

$$l = \frac{2n_s(d_2^2 - d_1^2)}{\pi\lambda}\left\{\frac{2}{m_1 p} - 1\right\}^{1/2} \tag{9.3.12}$$

A parabolic taper needs to have this length to keep, under present assumptions, below p the fraction of the total power that it maximally loses by mode conversion or to radiation. This equation applies of course only as long as $m_1 p/2 < 1$. When either the number $2m_1$ of guided modes for $d = d_1$, or the specified power loss p is so large that $m_1 p/2$ approaches unity, power conversion between other modes must also be considered.

As a representative example for illustration, we let $n_s = 1{\cdot}5$, $\lambda = 1\,\mu m$, $2d_1 = 20\,\mu m$, and $2d_2 = 100\,\mu m$. We further more specify $m_1 p/2 = 0{\cdot}5$, which gives us a safe margin for our simplified theory still to apply. From eqn. 9.3.12 the parabolic taper must be $l \approx 2{\cdot}3$ mm long to fill these specifications.

The total power fraction p which this taper maximally loses by mode conversion or to radiation still depends on the number $2m_1$ of modes over which the incident power distributes evenly. The highest-order even and odd TE- and TM-modes among them lose maximally $m_1 p/2 \triangleq 50\%$ each. With respect to the total incident power this represents only the fraction $p = 1/m_1$.

Reference

1 BURNS, W.K., MILTON, A.F., and LEE, A.B.: 'Optical waveguide parabolic coupling horns', *Appl. Phys. Lett.*, 1977, **30**, pp. 28–30

Index

Abrupt
rectangular waveguide junction, 125
TEM junction, 154
waveguide-to-horn junction, 133, 218, 227
Antenna feed systems, 221
Array of coupling elements, 53, 256

Butt joint, 187

Characteristic
equation of slab modes, 69, 73
impedance of a matching section, 162
Cornu spiral, 243
Coupled
wave equations, 25, 53, 60, 83, 104, 210, 232, 257
written in matrix notation, 28
Coupling
coefficient, 25, 35, 40, 54, 60, 69, 82, 88, 103, 120, 143, 192, 233
distribution function, 163, 186, 198, 267
factors, 43, 235
Crosspower of radiation fields, 76
Current
coefficient of horn mode, 119
densities, 10, 14
mode, 40
Curved section, 34
Cutoff
cross-section, 212
frequency, 230
wave number, 146, 178, 182
Cylindrical guides, 10, 12

Dielectric taper
linear, 288
parabolic, 295, 299
Discrimination of the coupling distribution function, 70, 163
Dolph-Chebȳshev distribution, 168, 171

E, H-wave, 38
Eigenfields, 13
Eigenfunctions, eigenvalues, eigenvectors of cylindrical waveguide modes, 36, 37
Even mode, 258

Field
discontinuity, 11
matching of an abrupt waveguide transition, 128
matching of an abrupt waveguide-to-horn, transition, 136, 140, 215
vector, 10
Film guide prism, 285
Flare angle of a horn, 99, 110

Higher-order coupling terms, 138, 194
Hole coupling, 50
Horn
junction, 101

modes, 93
nonuniform, 93
round, 93
sector, 93
uniform, 93

Input reflection coefficient of an abrupt waveguide-to-horn transition, 135
Integration of the coupled wave equations, 179
Interference
constructive, 258
destructive, 185, 284, 292

Kirchhoff's law, 154

Locus of spurious mode amplitude, 241

Main spurious mode
of a coupler, 264
of a transition, 184
Matching section
continuous, 159, 175
λ/4, 156, 175
two-line, 158, 175
Maxwell's equations, 10, 58, 99
longitudinal part, 79
transverse part, 79
Metrical coefficients of natural co-ordinates, 96, 111
Mode
degeneracies, 230
evanescent, 19
excitation by currents, 16
normal, 13
propagating, 19
Modes of a dielectric guide
guided, 74, 286
radiation, 75, 286
Multisection transmission-line transformer, 158

Natural co-ordinates of nonuniform horns, 94, 111
nonorthogonality of, 96
Neumann's number, 146
Normalisation of eigenfields
of horn modes, 106

Normal mode of
dielectric slabs, 67, 71
hollow tube waveguide, 38
horns, 97, 117
optical guides with opaque walls, 58

Odd mode, 259
Orthogonality
of eigenfields, 14, 18
of radiation modes, 76
Orthonormality relations of eigenfields, 38, 57, 129

Periodic structures, 13
Permeability, permittivity, 56
tensor of, 10
Phase synchronism of coupled mode, 256
Planar optical waveguide, 282
Power, complex,
carried by any mode, 19
carried by horn waves, 105
Power relations for lossless uniform waveguides, 19
Propagation, phase constant
of cylindrical waveguide mode, 13, 38
of horn mode, 99

Radiation mode of dielectric waveguide, 75
Reaction, 12
Reciprocity
law of, 10, 12, 14
theorem, 10, 12, 14
Reflection symmetry of a guide, 18

Scattering
differential equation for the, matrix, 212
matrix of a nonuniform section, 126, 131
of a small obstacle, 16
Sector transducer
linear, 242
Slab guide, 67, 71
Slot
coupling, 48
wave, 48

Source
fictitious, 11, 20, 99
impressed, 10
Spotsize of a laser, 291
Spurious mode
generation, 22
near cutoff, 149, 265
Striplines, 13

TE-, TM-modes, 38, 71
TEM-mode, 40
Total reflection of waves at
slab guide, 72
Transmission-line equations, 40
generalised, 43, 124
of TEM-line, 152
Transmission matrix
differential equations for the, 210
of a nonuniform waveguide, 17, 20, 209
Trapped mode resonances, 266, 268

Voltage
mode, 40, 119, 129

Wanted mode of a transition, 179
Wave admittance of horn mode, 98, 107, 117, 121, 219
Wave equation, 36, 68
Wave impedance of cylindrical waveguide mode, 38
Waveguide
circular, coaxial, rectangular, 37
dielectric, 56, 70
dielectric lined circular, 13
rectangular, with dielectric slab, 144
sector, 230
slab, 67, 71
Waveguide-to-horn transition, 221
abrupt, 133, 218, 227